ALSO BY SALLY SHAYWITZ, M.D.

Overcoming Dyslexia

OVERCOMING DYSLEXIA

2ND EDITION

OVERCOMING DYSLEXIA

2ND EDITION

———

A major update and revision of the essential
program for reading problems at any level, incorporating
the latest breakthroughs in science, educational methods,
technology, and legal accommodations

Sally Shaywitz, M.D.
and
Jonathan Shaywitz, M.D.

ALFRED A. KNOPF

NEW YORK

2020

Library of Congress Cataloging-in-Publication Data

Names: Shaywitz, Sally E. | Alfred A Knopf, Inc.
Title: Overcoming dyslexia : completely revised and updated /
Sally Shaywitz, M.D. and Jonathan Shaywitz, M.D.
Description: Second Edition | New York : Alfred A. Knopf, 2020. |
First edition published in 2003. |
Includes bibliographical references and index.
Identifiers: LCCN 2019028000 (print) | ISBN 9780385350327
(Hardcover) | ISBN 9780307558893 (ebook)
Subjects: LCSH: Reading disability. | Dyslexia. | Reading—
Remedial teaching. | Dyslexic children—Education.
Classification: LCC LB1050.5 .S42 2020 (print) |
LCC LB1050.5 (ebook) | DDC 371.91/44—dc23
LC record available at https://lccn.loc.gov/2019028000
LC ebook record available at https://lccn.loc.gov/2019028001

Jacket design by Carol Devine Carson

Manufactured in the United States of America

Published April 15, 2003
Second Edition, March 25, 2020

For my children, David, Diana, Jon, Adam, Marina,
and their children, who give so much and mean so much to me,

and

for Bennett, who is my life and who makes me whole

CONTENTS

PART IV
TURNING STRUGGLING READERS
INTO PROFICIENT READERS

PART V
CHOOSING A SCHOOL

PART VI
SUCCESS AFTER HIGH SCHOOL AND BEYOND

PART VII
MAKING IT WORK FOR THE DYSLEXIC CHILD AND ADULT

PART I

THE NATURE OF READING AND DYSLEXIA

1

THE POWER OF KNOWING

This book is about reading—an extraordinary ability, peculiarly human and yet distinctly unnatural. It is acquired in childhood, forms an intrinsic part of our existence as civilized beings, and is taken for granted by most of us. The unspoken belief is that if, as a child, you are sufficiently motivated and come from a home in which reading is valued, you will learn to read with ease. But as with so many other assumptions that appear to make intuitive sense, the assumption that reading comes naturally and easily to all children is simply not true. A substantial number of well-intentioned boys and girls—including very bright ones—experience significant difficulty in learning to read, through no fault of their own. This frustrating and persistent problem in learning to read is called *dyslexia*.

Most children look forward to learning to read and, in fact, do so quickly. For dyslexic children, however, the experience is very different. For them, reading, which seems to come effortlessly for everyone else, appears to be beyond their grasp. These children, who understand the spoken word and may love to listen to stories, cannot decipher the same words when they are written on a page. They grow frustrated and disappointed. Teachers wonder what they or the child is doing wrong, often misdiagnosing the problem or giving bad advice. Parents question themselves, feeling alternately guilty and angry.

It is for these parents, teachers, and children that I am writing this book. Exhilarated by new scientific discoveries about reading and dyslexia and frustrated by the relative lack of dissemination and practical application of these remarkable advances, I want to share with you all that I know about the science of reading. I want to make it very clear that it is now possible with a high degree of accuracy to screen for and then to identify children with dyslexia early on and to treat and remediate their difficulties, helping them learn to read. We can also do more than ever before for teenagers, young adults, and older adults.

As virulent as any virus that courses through tissues and organs, dyslexia can infiltrate every aspect of a person's life. It is often described as a *hidden* disability, but dyslexia is hidden only to those who do not have to live with it and suffer its effects. If you have a broken arm, an X-ray provides visible evidence; if you are diabetic, a blood glucose measurement confirms it. Heretofore, reading difficulties could be explained away in any number of ways. Now, however, men and women with dyslexia can point to an image of their brain's internal workings made possible by new brain imaging technology and say, *Here, look at this, this is the root cause of my problem.* We know now exactly where (and how) dyslexia manifests itself in the brain.

The harsh realities of the day-in, day-out experience of living with dyslexia can often clash dramatically with the perceptions of those teachers, administrators, acquaintances, and an array of self-appointed opinion-makers who question the very existence of the disorder that holds so many captive. Some still claim that dyslexia doesn't exist. They ascribe children's reading problems entirely to sociological or educational factors and totally deny the biology. Those who question the validity of dyslexia declare that there is no scientific evidence supporting either a biological or a cognitive basis to the disorder and contend that students with dyslexia reap the benefits of special treatment associated with their misdiagnosis.

These benefits seem to have been lost on Matthew,* a fifty-four-year-old marketing manager, who said to me, "I feel I have been sentenced for a crime I didn't even know I had committed." George, a student at the University of Colorado, described dyslexia as "the beast," an unknown predator that silently stalks him, continually disrupting his life. Not knowing why causes him great pain, including grades on examinations that reflect neither the

---

* Throughout this book I have changed the names and some identifying characteristics of many individuals in order to maintain their confidentiality. All those who are well known to the public are identified by their real names.

hours he spends studying nor his considerable fund of knowledge, nor his high level of intelligence. George wants to see "the face of the beast," to understand why this is happening to him. In addition to this deep desire to comprehend the nature of the mysterious problem enveloping him, he harbors a pervasive fear of being discovered, and of being severely embarrassed, as a dyslexic. Not knowing what to expect from his dyslexia or when the difficulty will manifest itself fills George with a constant sense of apprehension: "It's like it's lying there waiting for me to make a wrong move, and then all of a sudden it's there, taunting me again."

This all-too-common experience is now unnecessary. We know why dyslexics, no matter how bright and motivated, experience reading difficulties. Dyslexia is a complex problem with its roots in the very basic brain systems that allow humans to understand and to express language. By discovering how a disruption in these fundamental neural circuits for coding language gives rise to a reading impairment, we have been able to understand how the tentacles of the disorder reach out from deep within the brain and affect not only how you read but, surprisingly, a range of other important functions as well, including your ability to spell words, to retrieve words, to articulate words, and to remember certain facts. For the first time since dyslexia was initially described more than a century ago, scientists can see the "face of the beast," and we are now well on our way to taming and taking command of it.

I remember spending hours trying to convince Charlotte, a first-year law student who is dyslexic, to seek accommodations for her upcoming final exams. Charlotte was brilliant but was an incredibly slow reader who needed and deserved more than the allotted time. Her professors respected her, and she was sure to make *Law Review*—unless, she thought, it became known that she was dyslexic. With all the stereotyped views of dyslexia, she reasoned, her professors would have second thoughts about her abilities. Charlotte anguished over the decision: "If I ask for or take extra time, they'll all think I don't deserve my grade and that I'm really not so smart. If I don't take the extra time, I'll never finish." For Charlotte and others like her, so-called "special treatment" is a cruel irony.

I often lecture about dyslexia, and each time I am asked, *Where can I read about what you have just said? Where can I get this information? Have you a book to recommend?*

I originally wrote *Overcoming Dyslexia* in response to these questions and all the other questions I've never been able to answer in person. I am extremely pleased now to share with you in this second edition the phenomenal outpouring of cutting-edge scientific data increasing our understanding of dyslexia still further, and with it our ability to improve the lives of all children and now adults, too, who are dyslexic. I want to lift the barrier of ignorance surrounding dyslexia and replace it with the wonderful comfort of knowledge, and to empower each and every parent to know, first, what is best for your child, and second, what you can do to ensure that he or she becomes a reader.

In this second edition, my aim has been to bridge the enormous chasm that exists between what we are learning in the laboratory and what is being applied in the classroom. The field of neuroscience is exploding. Recent advances in our understanding of brain mechanisms underlying reading are nothing short of revolutionary. Alas, much of the time it feels like this new information is a well-kept secret. In an era when we can image the brain as an individual reads and literally see the brain at work, it is unacceptable to have children (and adults) struggle to read without the benefit of what this modern neuroscience has taught us.

As a working scientist and physician specializing in dyslexia and attention disorders, I have cared for dyslexic children for more than four decades. It is these boys and girls and their parents who serve as the inspiration for all my work. In science there are those whose interest is keenest in the theoretical questions surrounding a disorder rather than the clinical entity itself. Similarly, there are fine physicians who understand the clinical disorder and its effect on patients but who are far less familiar with the latest scientific developments. From my experience, I know that to help children and adults with dyslexia most effectively, we need the contributions of both of these often disparate worlds of knowledge. So another of my goals in writing this book is to bring to the reader a new level of scientific understanding of dyslexia. Once you understand dyslexia, its symptoms and treatment will make sense to you. There will be no mystery, and you will be in charge. You will be liberated to reason and to determine on the basis of this new knowledge what is best for you, your child, or your student. It has been an enormous satisfaction to me that I have been able to help so many parents to understand, in very clear and logical terms, what dyslexia is, how to identify it, what causes it, and, most critically, what can be done to help.

We now know that dyslexia is a very common problem, affecting one out of every five children—10 million in America alone. In every neighborhood and in every classroom worldwide, there are children struggling to read. For many affected children dyslexia has extinguished the joys of childhood.

Caitlyn was almost one of these children. I received a call from her grandfather, Adam, a college friend of mine, now a pediatrician in northern California, asking if I could see his grandchild. Caitlyn, seven years old and just then completing first grade, couldn't seem to catch on to reading. As Adam told me, "I always thought she was as smart as a whip. It just doesn't make any sense to me. Her mother, Esther, is beside herself. She would do anything for Caitlyn, but neither she nor anyone else seems to know what to do."

Once I saw Caitlyn, it was apparent why her mother was so upset. Caitlyn, after two full years of formal schooling, had not yet mastered beginning reading. She had memorized a few words that she could read by rote, but if shown a new word, Caitlyn could barely begin to sound it out. Instead she would call out words she knew, which generally had no relation to the word on the page. Sometimes she knew the first letter; for example, when shown the word *boy,* Caitlyn blurted out *bat.* Overall, out of twenty-four words that a first-grader should have mastered, Caitlyn knew four. Perhaps most frustrating to Esther was the attitude of the school. The principal acted as if Esther had some unspecified emotional problems; the school guidance counselor suggested that she was an overly anxious mother. But no one at the school seemed to be doing anything to address Caitlyn's lack of progress in reading. All her school reports spoke of her good behavior, noting that her progress was "good for someone at her level." "Now," Esther asked, "what does that mean?" Her requests for meetings with school personnel were ignored or slow to be acted upon. The few conferences that were held focused more on Esther's "emotional needs" than on Caitlyn's academic needs. Esther came to question herself, and watched her once cheerful daughter grow increasingly withdrawn.

The breaking point came during Caitlyn's seventh birthday party. Esther had worked hard to make it special. All through the party, Caitlyn kept asking, "When do I get to blow out my candles?" Suddenly the room was darkened and Esther came out with a cake with seven glittering candles sparkling brightly. Caitlyn ran over to the table, climbed up on a chair, and bent over the cake. She closed her eyes tightly, concentrated very hard, and then blew out all the candles. Then she ran up the stairs and closed her bedroom door. After some time had passed, Esther found Caitlyn on her bed, her favorite storybook, *Goodnight Moon,* open on her lap and tears streaming down her face. "You said I would get any wish I wanted and you were wrong—my wish didn't come true." Caitlyn still could not read any of the words on the page.

Caitlyn, of course, is dyslexic. She had difficulty pronouncing long names and trouble finding the right word to say. She had long pauses and many *um*'s when she was speaking. Her inability to identify some letters of the alphabet persisted. All these symptoms were consistent with the results of a series of state-of-the-art tests specifically designed to determine if a young child is dyslexic. I sat down with Esther and talked to her at length about Caitlyn and dyslexia. From experience, I knew how important it was for Esther to understand, at the most basic level possible, what a diagnosis of dyslexia meant for her daughter. I knew that once Esther grasped the information, she would become a powerful and highly effective advocate for her child.

Esther and Caitlyn went back to California with not only a diagnosis but, most important, a plan of action, a detailed program designed to overcome Caitlyn's reading difficulty. Esther now fully understood why Caitlyn could not read and exactly what needed to be done to fix that. Equally important, based on our new understanding of dyslexia, Esther was now also aware of Caitlyn's significant strengths and how these, often overlooked, could be called into play to help her read.

Esther followed through on our plan. She made sure that her daughter received the specific reading instruction that Caitlyn required. A year later, Esther was excited to share with me the news that Caitlyn had made dramatic progress. No longer a little girl for whom reading was an unfathomable mystery, Caitlyn was now a self-assured girl who understood how printed letters represent the sounds of spoken words. She could read, if not perfectly.

Caitlyn and her parents were flying to Connecticut for a scheduled return visit to my office. I couldn't wait to see this transformation. Rather than a sad, slouching Caitlyn, a spirited, smiling young girl literally leaped into my office, wasting no time to show off her new skills. She came prepared, quickly reaching into her backpack and holding up two books along with a pad of paper and a pencil. Caitlyn was justly proud of her gains. She had overcome her old nemesis—reading and pronouncing words she had never before encountered. She confidently pronounced one word after another, even difficult words. "*Sk, sk-oo-oo-l, school,*" she sounded out. "Oh," she said, "I can spell it, too. Wanna see?" Then, reaching for her pad and pencil, very carefully and with great determination, she wrote *s-c-h-o-o-l* in large, bold letters. Finally she reached into her backpack, pulled out the second book, and proceeded to read with great pride and concentration:

In the great green room
there was a telephone
and a red balloon
and a picture of
the cow jumping over the moon

I was impressed by Caitlyn's progress, but I was also struck by the dramatic turnaround her mother had made. I strongly believe that behind the success of every disabled child is a passionately committed, intensely engaged, and totally empowered parent. Esther was a changed person. She was smiling and self-assertive. She was at ease and communicated a sense of quiet but sizable confidence. As she told me:

> Now I am in charge of my daughter's destiny. I never again have to just stand there and wait for her school principal to determine my daughter's future. Now that I know, now that I understand, I never have to be at that man's mercy again. I understand her problem, I know what she needs, and I now have the power—irrespective of anyone—to act in my daughter's best interests. I feel so in control. I am a different person. To be freed of this absolute dependence on others for your child's entire future is the most exhilarating feeling imaginable. Last year I just didn't know what she needed. Now I know what to ask for her. I am no longer in the dark.

Esther was an easygoing, soft-spoken mother of four who had become a tiger, a power to be reckoned with when it came to protecting and ensuring Caitlyn's future.

Reading is often the key to realizing a parent's dreams for his or her child. Early on, children are tracked, and their futures are often laid out within the school setting. In the classroom, reading is king, essential for academic success. Reading problems have consequences all across development, including into adult life. This is why it is so important to be able to identify dyslexia accurately and precisely early on and take the appropriate steps without delay to ensure that the child learns both to read and to enjoy reading.

Most children who might not have been able to learn to read or to read well just a few years ago can now become competent readers. There may be a very small group of children for whom reading will continue to be extremely difficult, but even these boys and girls can benefit from the application of the remarkable advances in our understanding of the reading process.

Reading, as I've said and as I will explain later, is not a natural or instinctive process. It is acquired; it must be taught. How reading is taught can dramatically affect the ease with which a child learns to transform what are essentially abstract squiggles on a page into meaningful letters and then sounds and then words and then entire sentences and paragraphs. Reading represents a code—specifically, an alphabetic code. About 70 to 80 percent of children are able to break the code after a year of instruction in school, almost irrespective of the particular method of instruction used to teach reading. However, for the rest, reading remains beyond their reach after one, two, or even more years of schooling. Now we have the key to unlocking the reading code for these children as well.

What is so exciting about our new level of understanding of dyslexia is that it explains reading and reading difficulties at all ages and at all levels of education. By identifying the primary or core cognitive weakness responsible for dyslexia, scientists now understand how children acquire the ability to read and why some do not. The model of dyslexia that has emerged can be applied to understanding (and treating) reading difficulties in children just entering school, in children enrolled in primary and middle grades, and in young adults attending high school, college, or even graduate or professional school. The model has relevance, too, for the legions of adults who go through life without the ability to enjoy reading. These frequently neglected men and women can also benefit from our new understanding of reading. No matter who the child or adult is, what his background is, what kind of home he comes from, or what his intelligence level is, all the possible influences that may affect his ability to read are routed through the same pathway deep within the brain. This pathway has been identified. In practical terms this means we know what *functional* system in the brain is involved. Furthermore, we have now pinpointed its exact *neuroanatomic location*. Even more, we now know that dyslexia impacts reading very early on. As a result of these discoveries, it is now possible: (1) to screen for and to identify with an extremely high degree of precision the children who are at highest risk for dyslexia—even in kindergarten, before they develop reading problems; (2) to diagnose dyslexia accurately in children, young adults, and adults; and (3) to manage the disorder with highly effective evidence-based treatment programs.

While some cases are recognized when the child can't learn to read, perhaps in the first or second grade, currently the majority of cases of dyslexia are not identified until at least third grade. In fact, it is not unusual for cases of dyslexia to go unrecognized until adolescence or adulthood. So I will address the following questions as well: How do you go about identifying a reading problem in high school or beyond? Equally important, what do you do about it? Should an older student be offered special accommodations in school and for examinations, and if so, why? What is the best school setting for a dyslexic child?

According to one father who contacted me, this kind of information "saved" his son. He wrote:

> You saw my son Michael when he was a sophomore in high school and doing very poorly. Michael has turned around academically and emotionally. He is now a confident and secure young man. He has become his own best advocate. He knows what he needs and why. This allows him to speak up for himself and run his own interference. It is Michael who speaks to his teachers and they respect him for it. This sense of autonomy has rescued my son.
>
> P.S. This summer we are going to visit colleges.

It is never too late. The new knowledge is so basic and fundamental it is applicable to people across the entire lifespan, from the smallest of children to their grandparents. Take Rachel, a highly successful, self-made businesswoman, owner of a flourishing kennel and pet grooming business, who was unable to read beyond a fourth-grade level.

> It's so embarrassing. Salespeople come around and leave brochures for me. How can I tell them I can't read? Going into a restaurant, I can't read the menu, so I'm always reduced to saying, "Hey, what's your special for the day?" During our Passover seders, I simply want to die when it's my turn to read. Now one of my sisters comes to my rescue and begins to read my passage. But in spite of her good intentions, there is always someone new and well-meaning at the table who says, "Oh, didn't we just skip Rachel? Shouldn't she have a turn, too?" By now I have memorized a lot of sight words, but show me a new word and it could just as well be Greek. Can you believe I was so desperate to read, I even ordered a reading

program advertised on television? Here was a program for six- and seven-year-olds and I bought it, but the sad part was that even this program didn't help me.

Now that I'm married and looking forward to having children, I want to be a reader. I want to do the normal everyday things everyone else does: read the newspaper, read a recipe, read the instructions on a bottle of medicine.

Rachel began an intensive new program for adults with dyslexia. When I recently saw her, she reported:

I am going to have a baby. Each day when I get up in the morning, I open my pregnancy book and read about how my baby is growing. And when Joy is born, I am going to be able to read to her, too. It feels so good. I can read!

Caitlyn, Michael, Rachel, and their families experienced the new sense of hope that I want to share with you. I want to take you into my laboratory and show you a revolutionary new kind of science—our new ability to watch the brain at work: thinking, speaking, reading, and remembering.

In the pages that follow I will examine both the science and the human side of dyslexia. In Part I, I want to make clear what dyslexia is and how it evolves over time. I want to explain the cognitive basis for dyslexia and then what the newest brain research is teaching us about the neurobiology of dyslexia and reading. I want to share with you the substantial progress in identifying the underlying neural mechanisms responsible for reading and for dyslexia. These studies address the most fundamental questions of all, the most abstract and yet most compelling challenge facing researchers: How does the mind work, and what is the relationship between brain and behavior, between thinking and reading, and between brain structure and function?

Parts II through VI take what we have learned in the laboratory and apply it not only to the classroom, across development from primary school through college, but also to the child's life and future. I will talk about the impact of science on how we approach, diagnose, and treat children and adults with dyslexia. To ensure that this knowledge is used wisely, I will also discuss practical issues of special concern to many of you, including early diagnosis as well as diagnosis in older children and adults; special considerations in the diagnosis of bright young adults; the most effective treatment for children, young adults, and adults; the relation between dyslexia and standardized tests; the relation between dyslexia, anxiety, and attention-deficit/hyperactivity disorder (ADHD); accommodations for older children; and what dyslexia does and does not mean in terms of choosing a school, a college, and a career. We will begin by examining how the disorder was discovered more than a century ago.

2

THE HISTORICAL ROOTS OF DYSLEXIA

In the late nineteenth century, physicians in rural Seaford, England, and in the heart of Scotland wrote in medical journals about children in their Victorian society who were bright and motivated, who came from concerned and educated families, and who had interested teachers, but who nevertheless could not learn to read. The very innocence of the descriptions and the deep perplexity expressed by the physicians offer a unique understanding of dyslexia, one that cannot be obtained by reading the current literature.

On November 7, 1896, Dr. W. Pringle Morgan of Seaford wrote in the *British Medical Journal* about Percy F., fourteen years old.

> He has always been a bright and intelligent boy, quick at games, and in no way inferior to others his age.
>
> His great difficulty has been—and is now—his inability to read.
>
> He has been at school or under tutors since he was 7 years old, and the greatest efforts have been made to teach him to read, but, in spite of this laborious and persistent training, he can only with difficulty spell out words of one syllable. . . .
>
> I next tried his ability to read figures, and found he could do so easily. He read off quickly the following: 785, 852, 017, 20, 969,

and worked out correctly: (a+x) (a−x)=a²−x². . . . He says he is fond of arithmetic, and finds no difficulty with it, but that printed or written words "have no meaning to him," and my examination of him quite convinces me that he is correct in that opinion . . . he is what [Adolf] Kussmaul [a German neurologist] has termed "word blind." . . .

I might add that the boy is bright and of average intelligence in conversation. His eyes are normal . . . and his eyesight is good. The schoolmaster who has taught him for some years says that he would be the smartest lad in the school if the instruction were entirely oral.

Morgan captures the basic elements underlying what we refer to today as developmental dyslexia. Percy seems to have all the intellectual and sensory equipment necessary for reading, and yet he cannot read. The deficit seems only to involve reading letters and words, and Percy does quite well in math. As Morgan goes on to say, the numeral 7 is easily discerned and read, but not the written word *seven*.

Morgan was the first to appreciate word-blindness as a developmental disorder occurring in otherwise healthy children. However, the intriguing observation that men and women with good eyesight and strong intelligence could still lack the ability to read had been made by physicians centuries before—but always in adults who had suffered some sort of brain insult. What is perhaps the earliest recorded case of word-blindness dates back to 1676, when a German physician, Dr. Johann Schmidt, published his observations of Nicholas Cambier, a sixty-five-year-old man who lost his ability to read following a stroke. As the concept evolved, cases appearing in the medical literature described men and women like Cambier who had once read normally but then had suffered a stroke, tumor, or traumatic injury resulting in the loss of the ability to read, a condition termed *acquired alexia*. As more cases were reported, there was an increasing interest in the nature of the reading difficulty and its associated symptoms. For example, in 1872 the distinguished British neurologist Sir William Broadbent reported a case of acquired alexia, noting that his patient also suffered from a profound difficulty in naming even the most familiar objects. As the patient was brought into a London hospital, he was able to say, "I can see [the words] but cannot understand them."

Although Broadbent's contribution was important in providing a descriptive account of acquired reading difficulties, it was not until 1871 that Adolf Kussmaul came to the seminal realization that "a complete text-blindness may exist, although the power of sight, the intellect and the powers of speech are intact." Kussmaul is credited with coining the term *word-blindness* (*wortblindheit*) for this perplexing condition. He narrowed the clinical entity of word-blindness to that of an isolated condition affecting the ability to recognize and read text but with both intelligence and expressive language intact. Kussmaul then went even further in tracing the cases to lesions in the back of the brain, around the left angular gyrus.

Another German physician, Professor Rudolf Berlin, in Stuttgart, further refined our perceptions of these acquired reading problems. In his monograph *Eine besondre Art der Wortblindheit* (A Particular Kind of Word-blindness), published in 1887, Berlin describes six cases that he had personally observed over a period of twenty years. Berlin uses the term *dyslexia* to refer to what he perceives as a special form of word-blindness found in adults who lose the ability to read secondary to a specific brain lesion. If the lesion was complete, there would follow an absolute inability to read, acquired alexia. However, if the disruption was only partial, "there may be very great difficulty in interpreting written or printed symbols (dyslexia)." He conceptualizes dyslexia as a member of the larger family of language disorders called aphasia, in which there is difficulty in either understanding or producing spoken language, or both. According to Berlin, on March 4, 1863, a Herr B. complained that he had to stop work because

> reading of printed and written characters had become so very difficult to him. . . . He had exactly the same difficulty with [Jaeger] types of all sizes [types of increasing sizes used to assess visual acuity]. . . . There was no pain or discomfort in the eyes . . . the letters did not become dim or confused—he could simply not read further. . . . Neither the eye nor its muscular apparatus showed any abnormality on the most careful examination.

Imagine what a startling observation this must have been in mid-nineteenth-century Germany. The very idea that you could have perfect vision and yet could not *see* the words on the page in order to read them—that you could see and read a numeral written in a type as small as Jaeger 1 and still not be able to read a simple word in a type as large as Jaeger 16—must have been chilling. So it is not surprising that cases of word-blindness were often referred for consultation to specialists in the eye and vision—ophthalmologists. And it was a report of acquired alexia by an ophthalmologist at the Glasgow Eye Infirmary, Dr. James Hinshelwood, that served as the direct catalyst for Morgan's subsequent paper describing congenital word-blindness.

In the December 21, 1895, issue of the medical journal *The Lancet*, we can read Hinshelwood's report of the case of a highly educated fifty-eight-year-old man, a teacher of French and German, who suddenly awoke one morning to discover that

> he could not read the French exercise which a pupil gave him to correct. On the previous day he had read and corrected the exercises as usual. Greatly puzzled . . . having summoned his wife, he asked her if she could read the exercise. She read it without the slightest difficulty. He then took up a printed book to see if he could read it, and found that he could not read a single word. He remained in this condition until I saw him. On examining his visual acuity with the test types I found that he was unable to read even the largest letters of the test types. He informed me that he could see all the letters plainly and distinctly, but could not say what they were. . . . I found on examining further with figures that he did not experience the slightest difficulty in reading any number of figures quite fluently and without making any mistakes whatever. He could read figures printed on the same scale as Jaeger No. 1, the smallest of the test types, and from other tests it was evident that there was no lowering of his visual acuity. His inability to read was thus manifestly not due to any failure of visual power. . . . No other mental defect could be ascertained on the most careful examination.

It is easy to appreciate how, after reading this report, Morgan became excited when he encountered virtually the same constellation of symptoms and findings in his young patient Percy F. Although Hinshelwood's report concerns an adult who had once been a good reader, the similarities of the reading difficulties experienced by Percy F., who had never learned to read, and Hinshelwood's patient are impressive. Both had symptoms of word-blindness and each was unable to read words, but as Morgan and Hinshelwood emphasize, each could read figures—in the smallest of types—and perform mental calculations without any hesitation. Hinshelwood's 1895 report is important for several reasons, not the least of which is the clarity of his clinical description of acquired difficulty in reading as opposed to those problems in reading that are secondary to specific ophthalmologic impairments affecting a person's sight or visual acuity. And, of course, it was Hinshelwood's paper that prompted Morgan in 1896 to submit his own report of word-blindness, but in this instance of congenital origin.

It is not at all surprising that, historically, cases of acquired word-blindness in adults were noted prior to cases of congenital word-blindness. Mainly this is because in acquired cases the occurrence of word-blindness is abrupt; there is a dramatic change—a sudden loss of the ability to read. Acquired word-blindness primarily affects adults and occurs secondary to a brain insult occurring during a person's lifetime. Such insults, whether from a stroke or a tumor, typically affect the left side of the brain, where they may damage multiple brain systems. In addition to problems reading, affected patients may experience muscle weakness on the right side of the body, difficulty pronouncing words, or problems naming objects. In contrast, congenital word-blindness occurs in children and reflects an inherent dysfunction, one present since birth. Here the clinical picture is more subtle; it evolves gradually as the child meets with continuing problems in reading as he progresses through school. As we shall see, the reading difficulty may be overlooked for long periods of time. The congenital form is much more circumscribed, affecting primarily reading, sometimes spoken language, but never muscle strength.

From a neurological perspective, the difference in the two forms of the disorder is in the timing of the disruption to the laying down of neural systems within the brain. In the congenital form there is a glitch in the wiring when it is first laid down during embryonic development, and this miswiring is confined to a specific neural system (for reading). In the acquired condition, a lesion blocks an already working neural system and may extend to impact other systems as well.

Morgan's seminal report marked a watershed in our appreciation of unexpected reading difficulties in children. There soon came a flurry of reports of similar cases encountered by other physicians, almost exclusively eye surgeons, mostly from Britain but also from Europe, South America, and eventually the United States. But no one embraced or understood the significance of the disorder as fully as Hinshelwood did, nor was anyone as committed to bringing the disorder to the attention of as many of his colleagues.

Although Hinshelwood initially reported cases of acquired word-blindness, he soon became absorbed with the congenital form of the condition. By 1912 he had reported in a series of papers and monographs at least a dozen cases of congenital word-blindness. These reports are noteworthy for the similarities shared by the children they describe. For example, in 1900, Hinshelwood details the cases of two children who simply could not learn to read, although everything about each of them suggested that they should not only be able to learn to read but should become capable readers as well.

The first case describes an eleven-year-old boy who had been at school for four years before he was dismissed because "he could not be taught to read." According to the boy's father, the child was at the school for a number of years before his problem was even noticed, because

> he had such an excellent memory that he learned his lessons by heart; in fact, his first little reading-book he knew by heart, so that whenever it came to his turn he could from memory repeat his lesson, although he could not read the words. His father also informed me that in every respect, unless in his inability to learn to read, the boy seemed quite as intelligent as any of his brothers and sisters.

To demonstrate the child's sharp mind, Hinshelwood describes stating the address of his eye clinic and also writing it on an envelope that the boy's father then misplaced. Fortunately, the young boy's quickness of mind allowed him and his father to keep their clinic appointment; Hinshelwood was quite impressed when the little boy was able immediately to repeat the address after "hearing me state it once."

A second case deals with a ten-year-old boy who did well in every academic subject with the exception of reading. As Hinshelwood comments, "He was apparently a bright, and in every respect an intelligent, boy. . . . In all departments of his studies where *the instruction was oral* he made good progress." The boy's father, a physician, thought the reading problem might be due to some visual defect and thus sought a consultation with Hinshelwood. Careful examination indicated that the difficulty in learning to read was "not owing to any defect of his intelligence or to any diminution of visual acuity. His father noted that the boy never reads for amusement. As his father expresses it, 'it seems to take a great deal out of him.'"

In 1902, Hinshelwood reported two additional cases of young, intelligent children who could not read at all well. One was a ten-year-old girl who worked very hard in school and yet, "after four years of laborious effort" still read even the most basic book "with the greatest of difficulty." He was impressed both by the extraordinary effort reading required for this child and by the incredible patience required of someone teaching her, so much so that "on several occasions [even] her mother abandoned the task in despair." He remarks on the isolated nature of the difficulty, pointing out her general intelligence, her good vision, and her ability to do arithmetic.

In the second case, that of a seven-year-old boy, Hinshelwood notes the stress on dyslexic children of trying to read and underscores the importance of patience and support:

> He does not even know all the letters of the alphabet, but if gently told when he was wrong and if given time, as a rule he can name the letters correctly at last. He can repeat the letters of the alphabet rapidly by heart.

Consistent with previous case reports, the boy experienced no difficulties if he was taught orally; in fact, the young man was so good at learning that he, too, "for a time . . . concealed the fact of his disability." The person who, perhaps, knows the child best sums up his ability:

> His mother says that he is a smart, intelligent boy, even smarter and quicker in many respects than her other children, his one defect, according to her, being that he cannot be taught to read.

THE CONTEXT OF CONGENITAL WORD-BLINDNESS

As clinicians, Hinshelwood and his colleagues were more than mere observers or cataloguers of their findings; they were concerned with the implications of the disorder: how long it lasts, how common it is, which groups of children are at highest risk, what treatment is best.

In addition to providing clear descriptions of his patients, Hinshelwood focuses on the central concept underlying developmental dyslexia: an *unexpected* difficulty in learning to read. From a practical perspective this means that the weakness in reading is isolated and circumscribed, reflecting, according to Hinshelwood, a "local" rather than a generalized cerebral dysfunction. *A child who is slow in all cognitive skills would not be eligible for consideration as dyslexic; a dyslexic child has to have some cognitive strengths, not only depressed reading functions.*

Having examined an ever-growing number of children with congenital word-blindness, Hinshelwood was convinced that this disorder was far more common than generally appreciated:

> I had little doubt that these cases were by no means so rare as the absence of recorded cases would lead us to infer. Their rarity, I thought, was accounted for by the fact that when they did occur, they were not recognized.

Accordingly, Hinshelwood worked hard to publicize his observations through lectures and published reports. Since the clinical picture is so characteristic, he knew that once a physician was aware of the disorder, diagnosis would be facilitated.

As I will discuss in detail later, the diagnosis of dyslexia is a clinical one, based on the synthesis of information gleaned mainly from observations of the patient and from his or her history. Hinshelwood and other physicians of his day were able to make this diagnosis comfortably based solely on clinical presentation. Both Hinshelwood's and Morgan's reports predated the publication of the earliest version of the first standardized test (for IQ), the Binet-Simon Intelligence Scale, which did not become available until 1905.

At about the same time, and after seeing a succession of children with congenital word-blindness, E. Treacher Collins, an eye surgeon at the Royal London Ophthalmic Hospital in Moorfields, England, concluded that the core symptoms of the disorder "were frequently overlooked and put down to mere stupidity, or some error of refraction, very much to the disadvantage of the individual, because the individual was often blamed, bullied, laughed at, for a defect which was not his fault but his misfortune."

Physicians treating patients with congenital word-blindness were also impressed with its impact on the entire family and across generations. Many commented on the important role played by a family member, usually the wife/mother, in assisting her spouse/child. For example:

> The father . . . who is very involved in writing and has a leading position in industry has to have all his manuscripts corrected in detail by his wife. . . . Out of 5 children, 4 require every written paper to be corrected by their mother.

Above all, Hinshelwood was a physician, a practitioner to whom patients in a state of distress came desperately seeking help. Congenital word-blindness was more than a curiosity to him; he wanted to make sense of the disorder so that he could help his patients. He realized that for congenital word-blindness, treatment meant extending his perspective to include the educational establishment itself. In particular, he recognized the urgent need for early identification of children with congenital word-blindness:

> It is a matter of the highest importance to recognise as early as possible the true nature of this defect, when it is met with in a child. It may prevent much waste of valuable time and may save the child from suffering and cruel treatment. When a child manifests great difficulty in learning to read and is unable to keep up in progress with its fellows, the cause is generally assigned to stupidity or laziness, and no systematised method is directed to the training of such a child. A little knowledge and careful analysis of the child's case would soon make it clear that the difficulty experienced was due to a defect in the visual memory of words and letters; the child would then be regarded in the proper light as one with a congenital defect in a particular area of the brain, a defect which, however, can often be remedied by persevering and persistent training. The sooner the true nature of the defect is recognised, the better are the chances of the child's improvement.

A model of the complete clinician, Hinshelwood was highly responsive to the needs of his patients. He urged schools to establish procedures for screening populations of children for signs of congenital word-blindness and to provide appropriate teaching to those children identified with the disorder. In 1904 he wrote:

> In these days of scholastic reform, it is evident that there should be a systematic examination of all [these] children by a medical expert who would be able to differentiate the various defects and report as to the best means of dealing with the different groups. In our educational scheme this is a subject which has met with but scant attention, and, in my opinion, special provision for dealing with the education of [these] children on a scientific basis is one of the most crying needs in our present educational methods.

Hinshelwood made specific recommendations for children with dyslexia, as in this case:

> On my advice no further attempts were made to teach him in the class, but he was advised to have special reading lessons by himself. The lessons were not to be too long, but were to be repeated frequently during the day at intervals so as to refresh and strengthen the visual impressions made in the first lesson. This plan was adopted, and succeeded in a degree which surpassed all our expectations.

In another case,

> I advised them to make no further attempt to make him read in class with other boys. His mistakes and difficulties aroused the ridicule of the others, and this excited him and made him worse. I advised frequent short lessons by himself, both at the school and on his return home.

Dr. E. Nettleship, another turn-of-the-century ophthalmologist who saw dyslexic patients, understood the challenges in serving all dyslexics equally:

> The detection of congenital word-blindness is easy in the children of well-educated parents, whose children receive much individual attention. It must be much more difficult, both to recognize and deal with, in the children who crowd our Infant Elementary Schools. That the condition had been differentiated, and is receiving attention from medical men, should lead presently to its being dealt with by tutors . . . and by the teachers in all Infant Schools. The education of [these] children, by more or less special methods, is already receiving more attention than formerly. If from amongst such children, those can be sifted whose only, or principal, difficulty is real inability to learn to read, the result cannot but be useful both to the individuals and the community.

Today, as in Nettleship's time, reading difficulties are often overlooked in children from disadvantaged circumstances. It is not that children from enriched backgrounds are "overidentified" as dyslexic but rather that far too few poor children with the same difficulties are ever noticed, much less treated, for their reading problems.

Dyslexia was increasingly noted and reported by physicians not only in Great Britain but also in Holland (1903), Germany (1903), and France (1906). Awareness of the disorder soon traveled across the Atlantic to South America (Buenos Aires, 1903) and then to the United States. Here the first report of childhood reading difficulties came in 1905 from a Cleveland ophthalmologist, Dr. W. E. Bruner, followed within a year by a second from a Denver physician, Edward Jackson, who described two cases of "Developmental Alexia (Congenital Word Blindness)." By 1909, E. Bosworth McCready, a Pittsburgh physician interested in congenital word-blindness, was able to locate forty-one reported cases of the disorder worldwide. As he noted, "While the majority of cases have been reported by ophthalmologists they have not in a single instance held the ocular [sic] conditions responsible for the word-blindness." McCready, too, takes note of the seeming paradoxical association of dyslexia with creativity and intellectual superiority. He describes one man who "could not read but little in spite of every advantage . . . though he is now a brilliant and prominent member of a profession in which much reading is a *sine qua non*," another dyslexic who is a judge, and still another who is a "writer of verse of unusual excellence." Later I will talk about the contemporary counterparts of these gifted dyslexics.

These cases of unexpected reading difficulties experienced by so many children (and adults) and observed through the eyes of an ever-growing cadre of concerned physicians can essentially be superimposed on one another; taken together, they create a composite image even stronger and sharper than the individual pictures. As such, these case reports represent a valuable legacy: They provide indisputable evidence of the unchanging and enduring nature of the characteristics of dyslexia in children.

When I first came across these reports, I was stunned by the convergence between the historical and contemporary accounts of dyslexia. The basic template provided by these early reports remains intact. Now I will tell you about extraordinary scientific advances that have added a depth and a precision to our knowledge that was unimaginable just a few years ago.

3

THE BIG PICTURE:
WHO IS AFFECTED AND
WHAT HAPPENS OVER TIME

Research over the last several decades has produced unparalleled results, particularly for our understanding of reading and dyslexia. Given that I am most intimately aware of the details of our own work at the Yale Center for Dyslexia & Creativity, and because our center's research strategy and findings reflect and provide valuable insights into the state of current research in the field, I will use our studies as a model for discussion.

At first we focused on the basic characteristics of the disorder, designing a research study aimed at answering these questions: Who is dyslexic? How many children are affected? Are boys and girls both affected? What happens to a child with dyslexia over time, and how long does the problem last? When does the problem go away, or is it with someone for his or her entire life? Many of the answers have emerged from the Connecticut Longitudinal Study, which has been ongoing for over three decades.

THE CONNECTICUT LONGITUDINAL STUDY

In 1978, when I was asked to care for patients with reading difficulties, it was unclear to me just how a physician takes care of children who are dyslexic. There were many basic questions for which I could find no satisfactory answers. For example, is dyslexia a common problem or a rare one? Most of the information about dyslexia was based on studies of children who had already been identified—either by their schools or by a clinic—as having a reading problem. I thought there might be large numbers of children sitting at their desks in classrooms all over the country who were unable to read but who had never been accounted for in studies of dyslexia. To understand the full scope of the problem we needed to count *all* children with reading problems and not only those who were already visible and receiving help.

In order to differentiate between children who were experiencing reading problems and those who were learning to read with ease, it was necessary to establish exactly what rate of reading development, what behaviors, and what characteristics are typical for children. That would require studying large numbers of children over an extended period—as it turned out, from the time they entered kindergarten well into mature adulthood and the workplace.

To begin the study we selected children attending kindergartens in twenty-four randomly chosen Connecticut public schools in the 1983–84 school year. We worked with expert statisticians to select a study sample that represented the geographic and demographic diversity of Connecticut and the nation as a whole at that time. Since we wanted to develop a more nuanced understanding of each of these children, we gathered information representing a range of qualities, including cognitive ability, academic achievement, behavior at home and in the classroom, and self-perception. We obtained information from parents, teachers, and the children themselves. For the most in-depth understanding of each child, parents shared personal information about themselves, including their behaviors, family history, schooling, and employment. The criteria for those to be enrolled in the study were purposely left very broad; we did not want to influence the results by excluding any particular group of children.

We enrolled 445 children in the study, representing 95 percent of those eligible. The group's gender, ethnic, and racial composition was representative of the population of children entering public kindergarten that year. The participants have been regularly monitored since then. The boys and girls who began this study are now mature men and women. Some graduated from high school, some dropped out, some hold GED degrees; some hold college, graduate, and professional school degrees; some are working in a wide range of fields, others are not employed; some have married and are now parents; some are in jail. Although they now live in thirty-four different states and have lived in at least seven foreign countries, the vast majority of the original participants (80 percent) continue to be committed to the Connecticut Longitudinal Study. Their experiences provide a panoramic view of the process of learning to read. Each of these young men and women is owed a huge debt of gratitude.

MODELS OF READING AND DYSLEXIA

One of the first questions the Connecticut study addressed was the relation between good and poor readers: Do they form a continuum, or are they two distinct groups? Our educational policy for the identification of reading-disabled children in the past has often been based on the belief that there exists a gap in nature that allows us to easily separate dyslexic from all other readers. As Scottish psychiatrist R. E. Kendell said, "Classification is the art of carving nature at the joints; it should indeed imply that there is a joint there, that one is not sawing through bone." Results of our research suggest that in dyslexia, like virtually every other biological condition occurring in nature (except for pregnancy, extra digits, life and death), there is no natural joint separating dyslexic and good readers.

Evidence from the Connecticut study, as well as from Britain, Ireland, and New Zealand, provides a picture of an unbroken continuum of reading ability and reading disability, a conceptualization referred to as a *dimensional* model by researchers. The contrasting *categorical* model features discontinuity—a natural break in the linkage between good and poor readers. A dimensional model recognizes that there are no natural joints in nature to separate one group of readers from another and that while such cutoff points may be imposed, they are accepted as arbitrary.

In truth, the need to refer to disorders, even those that occur along a continuum, by a specific diagnostic label often obscures the fact that many, if not most, disorders in nature occur in gradations and thus conform to a dimensional rather than a categorical model. Hypertension, obesity, and diabetes represent common dimensional disorders; visual and hearing deficits also occur along a continuum. When blood pressure reaches a certain level, a patient is considered to have hypertension, but individuals just on the other side of the cutoff point, although not labeled as hypertensive, will share many traits in common with those said to have hypertension. For hypertension, as for dyslexia, there is no natural gap separating affected people from others, and a predetermined decision is made about where to place the cutoff point.

The demonstration that reading difficulties occur along a continuum brings with it important practical educational implications, especially since most current policies for the provision of educational services for reading disability reflect a different view of dyslexia—the categorical view, a cutoff point.

Not recognizing shades of grey represented by struggling children who haven't yet failed enough to meet a particular criterion, schools may be under-identifying many children who will go on to experience significant reading problems. Data from the Connecticut Longitudinal Study indicate that this is more than a theoretical possibility. As I will discuss later, science has made great progress in delineating the nature of dyslexia, including the underlying difficulty, and with this knowledge, the predictable symptoms that charac-terize children (and adults) who are dyslexic. Thus, the most appropriate approach to diagnosing dyslexia is what is referred to as a clinical synthesis of all the relevant information about the child—obtaining developmental and academic history, listening to him reading aloud, carefully reviewing scores on a battery of tests, and determining if the child demonstrates signs of dyslexia.

→ Many children whom an arbitrary cutoff does not "qualify" as dyslexic might still require and benefit from help in reading.

THE PREVALENCE OF DYSLEXIA

Figures provided by schools that indicate the number of children receiving educational services for learning disability provide only a crude approxi-mation of dyslexia's prevalence. While dyslexia is by far the most common and best-validated learning disability, schools are not asked to indicate how many students are identified as *dyslexic*. Hopefully this will change in the near future.

According to the most recent U.S. Department of Education statistics, last updated on April 20, 2018, 34 percent of the 6.7 million children receiving special education services (14 percent of the total public school enrollment of children ages three through twenty-one) are receiving these services for specific learning disabilities. Since dyslexia is estimated to comprise at least 80 percent of all learning disabilities, we can infer that about 2 million children (about 4 percent of the school population) are receiving special educational services for dyslexia.

In contrast, large-scale surveys that directly measure reading proficiency indicate that dyslexia may be far more prevalent. For example, the National Assessment of Educational Progress (NAEP), carried out by an arm of the U.S. Department of Education, tests a representative sample of over 270,000 students in reading and mathematics biannually and provides at each grade an indication of how many of these children are reading at or below a level considered to be standard or proficient for that grade. The NAEP assessment measures reading by asking students to read grade-level materials and then answer questions based on what they have read. Data from the 2019 NAEP indicate that overall less than half of all students are proficient in fourth-grade reading. Among some groups of students, the numbers are far worse. Only about one in five African American, Hispanic, and Native American students are proficient in fourth-grade reading. In a 1998 NAEP survey, 55 percent of the children of college graduates performed below proficiency levels in reading achievement in the eighth grade.

Moreover, according to the 2019 NAEP data, as many as 34 percent of fourth-graders had not achieved even the most basic or rudimentary skills in reading. In some groups, particularly African American students, more than 50 percent of fourth-graders have not achieved those levels. Concern about children's reading has a long history and led the Committee on Preventing Reading Difficulties in Young Children of the National Research Council to conclude in 1998 that "the educational careers of 25 to 40 percent of American children are imperiled because they don't read well enough, quickly enough, or easily enough."

Data from our Connecticut Longitudinal Study indicate that dyslexia affects approximately one child in five. From a national perspective, this means that there is not a family in America that has not been touched by dyslexia in some way—a child, a spouse, a grandchild, a sister or a brother, a niece or a nephew, a friend or a neighbor.

There is an important difference in how children are identified as dyslexic by schools and by research studies. For example, in the Connecticut study, each child was administered a test of intelligence and a reading test individually. Using this methodology, we found that about 20 percent of children are dyslexic. Contrast this with the approximately 4 percent of children currently diagnosed as dyslexic and it is obvious that schools are failing to diagnose the great majority of dyslexic students, children who could be helped. This constitutes one of the great tragedies in American education in the twenty-first century.

We were curious to know how many of the children identified as dyslexic by our research study had been identified as such by their schools. Accordingly, for each child in the study, we asked school personnel to tell us if that child had been identified as having a reading disability or as dyslexic and if he had received special help for a reading problem. We found that *less than one-third of children who were dyslexic were receiving school services for their reading difficulty*, strongly suggesting undiagnosed problems.

The apparent large-scale under-identification of dyslexic children is particularly worrisome because even when school identification takes place, it occurs relatively late—often past the optimal age for intervention. Dyslexic children are generally in the third grade or above when they are first identified by their schools; reading disabilities diagnosed after third grade are often extremely difficult to remediate. This failure to identify early takes on particular significance in view of our published data indicating that a large achievement gap between typical and dyslexic readers is already present in grade one. Early identification is important because the brain is much more plastic in younger children and potentially more malleable for the rerouting of neural circuits. Equally important, once a pattern of reading failure sets in, many children become defeated, lose interest in reading, and develop what often evolves into a lifelong loss of their own sense of self-worth. Importantly, inspired by these data indicating the early presence of the achievement gap between typical and dyslexic readers, I worked toward, and have developed, an evidence-based screening instrument: The Shaywitz DyslexiaScreen™ is for children in kindergarten to grade three and is completed by the child's teacher on a tablet. The result is a determination of at-risk or not at-risk for dyslexia. (For more details, see pages 169–174.)

Dyslexia occurs in children and adults living in northern Europe, southern Europe, Scandinavia, England, Australia, the Middle East, and North and South America. At one point it was thought that dyslexia affected only those who used alphabetic writing systems, such as English and German, and that those who used written language systems that were primarily logographic, such as Chinese and Japanese, were not at-risk for dyslexia. This assumption has proven to be false. Researchers have found comparable prevalence rates for dyslexia among American, Japanese, and Chinese children: for example, almost 13 percent in Hong Kong.

In 1996 I published an article, "Dyslexia," in *Scientific American.* In response, I heard from every part of the globe—Africa, Italy, Sri Lanka, Sweden, Israel, Thailand, England, Argentina—about children and adults experiencing problems exactly like those described in the article. Diplomats, scientists (including Nobel laureates), and CEOs have all told me about their difficulties with reading. Furthermore, the first edition of this book has been translated into many languages, including not only alphabetic scripts but also logographic languages such as Japanese, Korean, and traditional Chinese. Figure 1 is a picture of two visitors to the Yale Center for Dyslexia & Creativity from Korean television, including the anchorwoman and a teacher who is dyslexic, who understands the cost of dyslexia and is working to increase public awareness of the condition in Korea. (My husband and colleague, Bennett, and I are also in the photo.) Clearly dyslexia knows no boundaries, neither geographic nor ethnic nor intellectual.

Figure 1. The Drs. Shaywitz at the Yale Center for Dyslexia & Creativity with visitors from Korean television

WE KNOW SOME GIRLS ARE DYSLEXIC

The diagnosis of dyslexia presents a unique set of circumstances. While biologically based, dyslexia is expressed within the context of the classroom, so that its identification often depends on the school system's understanding of dyslexia. Since most research studies of dyslexic children are based on children who have already been identified by their schools, we wondered if school identification processes might be biased and result in the identification of certain groups of children and the exclusion of others. For example, it had been generally assumed that reading disability was far more common in boys than in girls. Could this be the result of some systematic bias in school identification procedures?

The data from the longitudinal study were ideally suited to address our question. First, the sample allowed us to examine a representative sample of children rather than only a school-identified group. Second, since every boy and girl participating in the study received individual ability and achievement tests, we were able to apply the criteria for reading disability set forth in the official school guidelines. In principle, then, our identification and the school's identification of a reading disability were ostensibly based on the same criteria. Theoretically both groups should comprise the same children. The children were in the second and third grades at the time of testing.

As shown in Figure 2, we found that according to school identification procedures, the prevalence of reading disability is three to four times more common in boys than in girls. These findings are in agreement with older reports in which the ratio of boys to girls with reading disability has varied from 2:1 to 5:1. A common thread uniting these past studies is that they

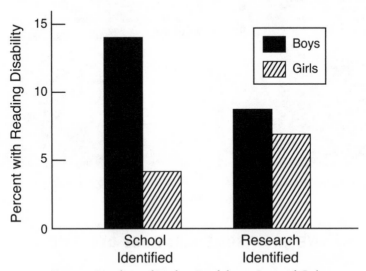

Figure 2. Prevalence of Reading Disability in Boys and Girls
Schools identify many more boys than girls; in contrast, when each child is tested (research identified), comparable numbers of boys and girls are identified as reading-disabled.

were all based on samples identified through either clinic or school identification procedures. In contrast, we found no significant difference in the prevalence of reading disability in boys and girls we identified using our research-based criteria. In general, when each child in a school (or school district) is individually tested, which is what we did, researchers report as many reading-disabled or dyslexic girls as boys. Consistent, too, are findings from still other studies that indicate girls with dyslexia are not as readily identified as boys and in fact are often more severely impaired in reading before they are identified for special education services.

Teachers were asked to complete the Multigrade Inventory for Teachers (MIT), which comprises a series of scales asking about, for example, Attention, Academic Achievement, Fine Motor, Language and Behavior. Analysis of teacher ratings of the children's behavior revealed why girls are less readily identified than boys. There are significant differences between how teachers rate *typical* boys and girls. Teachers have incorporated a norm for classroom behavior that reflects the calmer behavior of normal girls. Boys who are a bit rambunctious—although still within the normal range for the behavior of boys—may be perceived as having a behavior problem and referred for further evaluation. Meanwhile, the well-mannered little Jennifers and the shy Tanishas who sit quietly in their seats but who nevertheless are failing to learn to read are often overlooked and never, or only much later, identified as dyslexic by their school systems.

DYSLEXIA OVER TIME

For many years researchers and educators questioned whether dyslexia represented a developmental lag that children somehow outgrew or whether the disorder represented a more persistent deficit in reading. The question is important, for if dyslexia is simply a lag in reading development—a temporary snag—this difficulty will be outgrown and parents and teachers need not be so concerned about early reading difficulties. On the other hand, if dyslexia is not outgrown, there is real urgency in identifying children early on and, equally important, in ensuring that they receive help as soon as they are identified.

Here, too, we were able to use data provided by the Connecticut Longitudinal Study. Using two complementary strategies, we determined decisively that dyslexia is a persistent, chronic condition and that it does not represent a temporary lag in reading development. In one approach we compared individual growth rates in reading skills across grades one through twelve in two groups of readers: one a group of boys and girls who had never experienced any reading problems, the other a group of children who met criteria for dyslexia in the early grades. To no one's surprise, we observed that both groups increased their reading skills over time. However, and most important, as shown in Figure 3, the gap in reading ability between good and dyslexic readers remains. Dyslexic readers never catch up with their classmates who are good readers. If a child is dyslexic early on in school, that child will continue to experience reading problems.

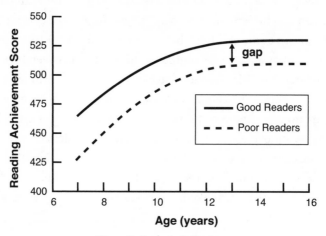

Figure 3. Dyslexia Is Persistent

Over time, reading performance improves in both good readers (upper curve) and poor readers (lower curve). However, the *gap* between the two groups remains.

Perhaps of even more interest and of particular relevance, as noted earlier, a more recent (2015) examination of the CLS data revealed that the differences in reading between typical and dyslexic readers not only appear as early as first grade but, importantly, persist through adolescence. These data indicate that *dyslexic readers do not catch up with typical readers primarily due to large differences observed as early as first grade.*

An important implication of these findings is that if this persistent gap between dyslexic and typical readers is to be narrowed, or even closed, effective reading interventions must be implemented very early, when children are still developing the basics for reading acquisition, such as phonological awareness, letter-sound understanding, and building a lexicon. Implementing effective reading programs early, even in kindergarten, offers the potential to reduce the observed persisting disparities between dyslexic and typical readers. The earliest possible intervention is vital.

One final point: About a third of the children in the longitudinal study were receiving special help, but this help was often erratic, occurring sporadically and consisting of what might best be described as a Band-Aid approach to a gushing wound. In general, we found that children received help for very limited periods of time, often from well-meaning but untrained teachers and with methods that did not reflect state-of-the-art, evidence-based instructional strategies.

Now that the pernicious nature of dyslexia has been uncovered, we must ensure that our approach to identification and intervention—in terms of age of screening, definition used in identification, intensity and frequency of instruction, and content of instruction—is consistent with the seriousness and emerging scientific knowledge of the disorder. I will now detail what we have learned about the basic nature of dyslexia. This knowledge provides the foundation for the most accurate diagnosis and effective interventions for reading difficulties.

4

WHY SOME SMART PEOPLE CAN'T READ

I want you to meet two of my patients, Alex and Gregory. Alex is ten years old; Gregory, a medical student, just celebrated his twenty-third birthday. Their experiences are typical of children and young adults with dyslexia. You will learn how Alex's and Gregory's seemingly diverse symptoms—trouble reading, absolute terror of reading aloud, problems spelling, difficulties finding the right word, mispronouncing words, rote-memory nightmares—all represent the expression of a single, isolated weakness. At the same time you will learn that other intellectual abilities—thinking, reasoning, understanding—are untouched by dyslexia. This contrasting pattern produces the paradox of dyslexia: profound and persistent difficulties experienced by some very bright people in learning to read. I am emphasizing the strengths of the dyslexic because there is often a tendency to underestimate his or her abilities. The reading problem is often glaringly apparent, while the strengths may be more subtle and overlooked. As I discuss later on, new insights into brain function in dyslexia tell us that it represents a very isolated weakness. The neural systems involved in thinking and reasoning are intact and perhaps even enhanced.

ALEX

In his first years of life, Alex was so quick to catch on to things that his parents were surprised when he struggled to learn his letters in kindergarten. When shown a letter he would stare, frown, and then randomly guess. He couldn't seem to learn the letter names. In first grade he was struggling to link letters with their sounds. By the third grade Alex continued to stammer and sputter as he tried to decipher what was on the page in front of him. Language had clearly become a struggle for him. He seemed to understand a great deal, yet he was not always articulate. He mispronounced many words, leaving off the beginning (*lephant* for *elephant*) or the ends of words, or inverting the order within a word (*emeny* for *enemy*). Alex had trouble finding the exact word he wanted to say, even though it seemed he could tell you all about it. One evening he was trying to explain about sharks living in the ocean: "The water, the water, lots of water, salty water with big fish, it's a lotion. No, no, that's not what I mean. Oh, you know, it's on all the maps, it's a lotion—ocean, that's what it is—a sea, no big sea, it's an ocean, an ocean!"

Looking at this handsome, very serious little boy who could spend hours putting together complex puzzles and assembling intricate model airplanes, his father could not believe that Alex had a problem. Alex, however, became increasingly aware of his difficulty reading, asking more and more frequently why all his friends were in a different reading group. He practiced, he tried, but it just never seemed to come out right.

His parents brought him to the Yale Center for Dyslexia & Creativity for evaluation. We learned that Alex was extremely smart, scoring in the gifted range in abstract reasoning and in logic. His vocabulary was also highly developed. Alex could learn, he could reason, and he could understand concepts at a very high level. Despite these strengths, his performance in reading words was dismal; for example, he was able to read only ten out of twenty-four words on a third-grade level. What gave Alex the most difficulty, however, was nonsense words (made-up words that can be pronounced; for example, *glim, rold*). He struggled to decipher these words. Sometimes he used the first letter to generate a response (such as *glim* for *gern, rold* for *ruck*); at other times it seemed as if he just gave up, making seemingly random guesses. In contrast, he was able to read a short passage silently and answer questions about it far better than he had been able to read and pronounce isolated single words. In reading silently he made good use of clues, such as pictures in the book and surrounding words that he could read, to help get to the meaning of sentences and passages that contained words he could not read. "I picture what it says," he explained. However, Alex sparkled when asked to *listen* to a story and then respond to a series of questions, scoring significantly above average. Reading aloud was particularly painful for Alex: He was reluctant to read in front of the class, and it was easy to understand why. His reading was labored; words were mispronounced, substituted, or often omitted entirely. Words that he correctly read in one sentence would be misread in a subsequent sentence. He read excruciatingly slowly and haltingly. Increasingly Alex would ask to go to the bathroom when it was nearing his turn to read. If called upon, he often acted silly, making the words into a joke or tumbling himself onto the floor and laughing so that he would have to be sent out of the room.

Poor spelling skills were compounded by his almost illegible handwriting. Letters were large, misshapen, and wobbly. In contrast, Alex's math skills, particularly problem-solving and reasoning abilities, were in the superior range. At the close of the testing, Alex diagnosed his own reading problem: "I don't know the sounds the letters make." Furthermore, he told the evaluator that it bothered him that his friends were in a different reading group. Sometimes, he said, this made him very sad. His one wish was to be a better reader, but he didn't know exactly how that would happen.

When I met with Alex's parents, they had many questions: Does he have a problem? If so, what is the nature of the problem? What could be done to help him? Above all, they asked, "Will he be all right?" I reassured them that not only would Alex survive, he would thrive.

GREGORY

In the course of my work, I have evaluated for reading disabilities not only hundreds of children but also scores of young adult men and women. Their histories provide a picture of what the future will be for a bright child like Alex who happens to be dyslexic. Gregory, a medical student in his early twenties, was a grown-up Alex. He came to see me after experiencing a series of difficulties in his first-year medical courses. He was quite discouraged.

Although he had been diagnosed as dyslexic in grade school, Gregory had also been placed in a program for gifted students. His native intelligence, together with extensive support and tutoring, had enabled him to graduate from high school with honors and to gain admission to an Ivy League college. In college Gregory had worked hard to compensate for his disability and eventually received offers from several top medical schools. Now, however, he was beginning to doubt his own ability. He had no trouble comprehending the intricate relationships among physiological systems or the complex mechanisms of disease; indeed, he excelled in those areas that required reasoning skills. More difficult for him was pronouncing long words or novel terms (such as labels used in anatomic descriptions); perhaps his least well developed skill was that of rote memorization, for example, of dates and names of places.

Both Gregory and his professors were perplexed by the inconsistencies in his performance. How could someone who understood difficult concepts so well have trouble with the smaller and simpler details? I explained that Gregory's dyslexia (he was still a slow reader) could account for his inability to name tissue types and body parts in the face of his excellent reasoning skills. His history fit the clinical picture of dyslexia as it has been traditionally defined: an unexpected difficulty learning to read despite intelligence, motivation, and education. Furthermore, I was able to reassure him that scientists now understand the underlying nature of dyslexia that forms the basis for highly effective strategies to help those with the disorder.

WHY ALEX AND GREGORY HAVE TROUBLE READING

Explanations of dyslexia put forth beginning in the 1920s and continuing until recently held that defects in the visual system were to blame for the reversals of letters and words thought to typify dyslexia. Eye training was often prescribed to overcome these alleged visual defects. Subsequent research has shown, however, that in contrast to a popular myth, children with dyslexia are not unusually prone to *seeing* letters or words backward, and that the deficit responsible for the disorder resides in phonology, a specific component of the language system. These poor readers, like Alex, do have significant difficulty, however, in *naming* the letters, often calling a *b* a *d* or reading *saw* as *was*. The problem is a linguistic one, not a visual one.

As noted earlier, dyslexia represents a specific difficulty with reading, not with thinking skills. Comprehending spoken language is often at a very high level, as it was for Alex, as are other higher-level reasoning skills. Dyslexia is a localized problem.

Understanding that dyslexia reflects a problem in a specific component of the language system and not a general weakness in intelligence or a primary visual impairment represented a major step forward. Further advances have clarified the nature of the language impairment. Dyslexia does not reflect an overall defect in language but rather a localized weakness within a specific component of the language system: the phonologic module. The word *phonologic* is derived from the Greek word *phone*, meaning *sound* (as in *phonograph* and *telephone*). The phonologic module is the language factory, the functional part of the brain where the sounds of language are put together to form words and where words are broken down back into these elemental sounds.

Over the past two decades a model of dyslexia has emerged that is based on phonological processing—processing the distinctive sounds of language. The phonological model is consistent both with how dyslexia manifests itself and with what neuroscientists know about brain organization and function. Over the past three decades, researchers at the Yale Center for Dyslexia & Creativity and elsewhere have had the opportunity to test and refine this model through reading and, more recently, brain imaging studies. We and other dyslexia researchers have found that the phonological model provides a cogent explanation of why some very smart people have trouble learning to read.

THE PHONOLOGICAL MODEL

To understand how the phonologic model works, you first have to understand how language is processed in the brain. Think of the language system as a graded series of modules or components, each devoted to a particular aspect of language. The operations within the system are rapid and automatic, and we are unaware of them. They are also mandatory. For example, if we are seated at a table in a dining room, we *must* hear what the person at the next table is saying if she is speaking loudly enough. It is nearly impossible to tune language out. That is why it is so difficult to study when others nearby are speaking.

Scientists have been able to pinpoint the precise location of the glitch within the language system (Figure 4). At the upper levels of the language hierarchy are components involved with, for example, semantics (vocabulary or word meaning), syntax (grammatical structure), and discourse (connected sentences). At the lowest level of the hierarchy is the phonological module, which is dedicated to processing the distinctive sound elements of language. Dyslexia involves a weakness within the language system, specifically at the level of the phonologic module.

The Language System:
Reading and Speaking

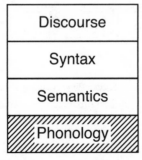

Figure 4. Pinpointing the Core Weakness in Dyslexia
Research pinpoints the weakness at the lowest
level of the language system.

The phoneme* is the fundamental element of the language system, the essential building block of all spoken and written words. Different combinations of just forty-four phonemes produce the tens of thousands of words in the English language. The word *cat*, for example, consists of three phonemes: *k*, *aah*, and *t*. Before words can be identified, understood, stored in memory, or retrieved from it, they must first be broken down into phonemes by the neural machinery of the brain. Just as proteins must first be broken down into their underlying amino acids before they can be digested, words must first be broken down into their underlying phonemes before they can be processed by the language system. Language is a code, and the only code that can be recognized by the language system and activate its machinery is the phonologic code.

This is critical for both speaking and reading. Let's first consider speaking (Figure 5). If I want to say the word *bat*, I will go into my internal dictionary or lexicon deep within my brain and first retrieve and then serially order the appropriate phonemes—*b*, *aah*, and *t*—and then I am ready to say the word *bat*.

In children with dyslexia, the phonemes are less well developed. Think of such a phoneme as a child's carved letter block whose face is so worn that the letter is no longer prominent. As a consequence, such children when speaking may have a hard time selecting the appropriate phoneme and may instead retrieve a phoneme that is similar in sound. Think of Alex's experi-

---

* The phoneme is defined as the smallest unit of speech that distinguishes one word from another.

SPEAKING

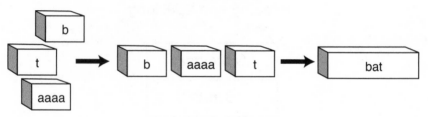

Figure 5. Speaking: Making Words
A speaker retrieves and then orders the phonemes to make a word.

ence in retrieving the word *lotion* when the word he was reaching for was *ocean*. Alex knew exactly what he wanted to say but could not retrieve the exact word, so instead he picked a close but incorrect phoneme. Alternately, a dyslexic might order the phonemes incorrectly, and the result might be, just as Alex said, *emeny* instead of *enemy*. Such sound-based confusions are quite common in the spoken language of dyslexics. For example, a child looking at his perspiring mother as she struggled in a traffic jam remarked, "You know, Mom, it's not the heat, it's the *humanity*." (The intended word, of course, was *humidity*.) On another occasion a politician greeting his supporters said, "Welcome to this lovely *recession*." Of course he meant to say *reception*. In each instance the confusion was a phonologic one (that is, based on the *sound* of the word) and did not reflect a lack of understanding of the meaning of the word in question. Unfortunately, such phonologic slips roll off the tongues of dyslexics fairly regularly and are often (incorrectly) attributed to a lack of understanding. In the next chapter I review what science has taught us about why some very smart people have trouble retrieving words as they speak.

Reading is the converse of speaking. In reading we begin with the intact printed word on the page—the blocks representing phonemes are all lined up correctly. The reader's job is to convert the letters to their sounds and to appreciate that the words are composed of smaller segments or phonemes (Figure 6).

Dyslexic children and adults have difficulty in developing an awareness that spoken and written words are composed of these phonemes or building blocks. Think of the little boy who got his first pair of glasses and then said, "I never knew that building was made of red bricks. I always thought its wall was just one big smudge of red paint." In the same way, while most of us can

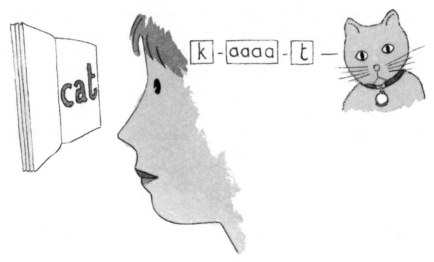

Figure 6. Reading: Turning Letters into Sounds
To read, a child converts the letters into sounds or phonemes.

detect the underlying sounds or phonemes in a word—for example *b, aah,* and *t* in *bat*—children who are dyslexic perceive a word as an amorphous blur, without an appreciation of its underlying segmental nature. They fail to appreciate the internal sound structure of words.

The phonologic model tells us the exact steps that must be taken if a child is to go from the puzzlement of seeing letters as abstract squiggly shapes to the satisfaction of recognizing and identifying these letter groups as words. Overall, the child must come to know that the letters he sees on the page represent, or map onto, the sounds he hears when the same word is spoken.

The process of acquiring this knowledge is orderly and follows a logical sequence. First, a child becomes aware that words he hears are not just whole envelopes of sound. Just as the little boy noted the bricks in the wall, the beginning reader starts to notice that words are made up of smaller segments—that words have parts. Next the child becomes aware of the nature of these segments, that they represent sounds. He realizes, for example, that in the word *cat* there are three segments of sound, *k, aah,* and *t.* Then the child begins to link letters he sees on paper to what he hears in spoken language. He begins to realize that the letters are related to sounds he hears in words and that the printed word has the same number and the same sequence of phonemes (sounds) as the spoken word. Finally he comes to understand that the printed word and the spoken word are related. He knows that the printed word has an underlying structure and that it is the same structure he hears in the spoken word. He understands that both spoken and written words can be pulled apart based on the same sounds, but in print letters represent these sounds. Once the child has made this linkage, he has mastered what is referred to as *the alphabetic principle*. He is ready to read.

Now I will consider the fundamental difficulty that is at the heart of dyslexia and why so many children have problems mastering the alphabetic principle. Later I will focus on how to teach these children most effectively to overcome their difficulties in reading.

5

EVERYONE SPEAKS, BUT NOT EVERYONE READS

Prospective readers must master the alphabetic principle in order to learn to read, yet as many as one in five children are unable to do so. Why should developing an awareness that spoken words are made of smaller segments—phonemes—present such a formidable challenge? The answer lies, surprisingly, in the very same mechanism that makes speaking so easy.

As linguists Noah Chomsky and Steven Pinker of the Massachusetts Institute of Technology have convincingly argued, spoken language is innate.* It is instinctive. Language does not have to be taught—all that is necessary is for humans to be exposed to their mother tongue. Through neural circuitry deep within our brains, a specialized phonological module *automatically* assembles the phonemes into words for the speaker and disassembles the spoken word back into its underlying phonemes for the listener. Thus spoken language, which takes place at a preconscious level, is effortless.

Some linguists, like Pinker, argue that language can be traced back as many as one million years. John DeFrancis, professor emeritus of Chinese at the University of Hawaii, suggests that speech emerged some fifty thousand

---

* Virtually all humans speak, save for those who have rare medical conditions such as profound deafness from birth.

years ago as the dominant mode of communication among *Homo sapiens.* Every human society has a spoken language, and man is the only species that communicates by speaking (although many other species communicate using a variety of signals—grunts, screeches, electric shocks, odors, whistles, and calls). If a baby is neurologically healthy, there is almost no way she can avoid learning to speak.

Language is open-ended. It is generative. Using phonemes we can create an indefinite number of words, and with these words an infinite number of ideas. We can tell jokes, we can muse, we can tell a story, we can imagine, we can describe. We can speak in the present or reflect wistfully about the past or project hopefully into the future.

In contrast, animal communication systems are closed and the signals fixed. They are holistic; they do not come apart and cannot be added to or rearranged to form a new message. With animals there are a limited number of signals, and each signal is yoked to a specific meaning. There is no possibility for novelty or for infinite variation.

THE PARTICULATE PRINCIPLE

In 1989, linguist William Abler offered a brilliant insight into the language system—what he called the particulate principle of self-diversifying systems. Although this sounds like jargon, the principle itself is elegant in its simplicity. Using biological inheritance as a model, Abler suggested that chemical compounds and human language adhere to the same principles as do combinations of DNA in forming a seemingly endless number of proteins. He reasoned that these natural processes share two basic features. Each has as its core element a "particle," and each is characterized by a hierarchical structure. The particles serve as the building blocks that give rise to open-ended hierarchical systems. For chemicals, it is atoms; for genetics, it is the nucleotides in DNA; and for language, it is the phonemes. At the next level of the hierarchy, the particles are now larger: Atoms combine to form molecules, nucleotides join to form proteins, and phonemes come together to form words.

By using a small number of phonemes, a speaker has the ability to create a seemingly infinite number of words and then sentences and paragraphs. To appreciate the vast number of combinations possible, consider that Shakespeare used 29,066 different words in his complete works (884,647 words in total).

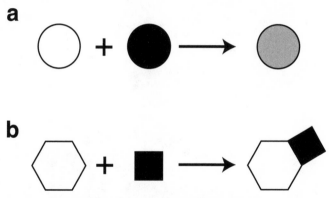

Figure 7. The Advantage of Particles Not Blending
When elements blend as they combine, as in the two circles shown,
the result is a mixture of the two, something in between. In contrast, when two phonemes (particles) combine, as shown here with
the two geometric shapes, they do not blend. Instead they combine
to form something entirely new. Endless combinations are possible,
which can produce an endless number of words.

What allows the particulate principle to operate so effectively is the
special nature of the particles. Whether they are atoms of sodium, strands
(nucleotides) of DNA, or phonemes, they don't change; they maintain their
original identity. This means that they are able to combine with each other
to form entirely new and larger units, be they compounds, proteins, or
words—and yet the particles themselves are not changed in the process.
(The advantage of *particles* over elements that do not retain their identity
is shown in Figure 7. If two elements combine and blend together, the
result is a mixture—somewhere in between the original two; however, when
two particles combine while still maintaining their original properties,
the new item formed is unique.) Forming words by combining—but not
blending—phonemes brings with it the extraordinary potential to form an
infinite number of words, limited only by the number of combinations. As
phonemes combine to form words and words join together to form phrases
and sentences, at each succeeding level of the hierarchy a larger and an
entirely novel structure is created.

A speaker can generate phonemes very quickly, with rates averaging ten
to fifteen phonemes per second. In fact the pace outstrips the capacity of
the listener's acoustic machinery, which cannot pick up or process a series of
incoming sounds that quickly. On the other hand, if the speaker uttered each
phoneme very slowly, spoken language would be interminable. A further

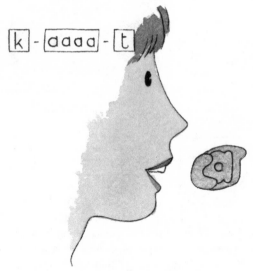

Figure 8. Coarticulation:
Overlapping Several Phonemes into One Burst of Sound
To say a word, a speaker first retrieves each phoneme
and then coarticulates or overlaps one over the other. As
shown in this illustration, the three phonemes forming
the word *cat* are overlapped into one packet of sound.

complication is that the listener has to receive the phonemes at a sufficiently fast pace so that several can be held in short-term memory at the same time and integrated to form the intended words and phrases. Phonemes can be held in the temporary memory storage bin for only one or two seconds, or about five to seven unrelated words, before each, like a bubble, vanishes.

Evolution has solved this problem through coarticulation: the ability to overlap several phonemes—while maintaining the integrity of each—into one bubble of sound (Figure 8). As a result of coarticulation, bursts of sounds come at a pace compatible with the auditory system's capacity to process them, and the phonemes arrive at a rapid enough pace to meet the constraints of the short-term memory system.

Critical to the entire process is the fundamental difference between sounds—the sounds of language and the sounds of noise—and the innate ability of the phonologic module to distinguish speech from nonspeech sounds. Phonemes are presented to the sense organ, the ear and its receptors, disguised in their outer packaging as an ordinary burst of sound. For example, if I were to say the word *cat*, this would appear on an oscilloscope as one pulse of sound—and that is what it is to the listener's ear. However, camouflaged within the packet of sound are three pieces of language: the phonemes, *k, aah,* and *t.* Once the sound is safely past the hearing machinery, the language system takes over, immediately recognizing the three phonemes as particles of language and processing them accordingly. Here the language system is different from other systems, like vision and hearing. The eye receives visual stimuli and the ear acoustic stimuli, but the ear is only a way station for language. Nature has provided a mechanism to allow sound, like a barge towing a valuable cargo, to guide the phonetic particles past the ear and into the safe haven of cerebral neural circuits specialized for receiving language. When the ear receives the parcel of sound, the specialized phonologic module in the brain immediately activates and recovers the phonemes contained within each pulse of sound, automatically translating the sound into particles of language. The listener receives the exact message sent by the speaker.

What makes human speech possible as a means of communication is the coarticulation referred to earlier. In producing a word, the human speech apparatus—the larynx, palate, tongue, and lips—automatically coarticulates, that is, compresses and merges the phonemes together. As a result of coarticulation, several phonemes are folded into a single pulse or bubble of sound, without any overt clue to the underlying segmental nature of speech. Consequently, spoken language appears to be *seamless.* It is as if coarticulation has lacquered over any fissures or gaps between the individual phonemes so that what is presented to the listener is a smooth, seamless stream of speech.

READING IS MORE DIFFICULT THAN SPEAKING

The effortless and seamless nature of spoken language has everything to do with why reading is so hard for dyslexic children. Although speaking and reading both rely on the same particle, the phoneme, there is a fundamental difference: *Speaking is natural and reading is not.* Herein lies the difficulty. Reading is an acquired act, an invention that must be learned at a conscious level. It is the very naturalness of speaking that makes reading so hard.

Just as our lungs breathe in and out for us and the chambers of our hearts contract rhythmically, highly refined neural circuitry within our brains allows us to speak and to listen without conscious thought or effort. For spoken language the phoneme comes all ready to go. This is the gift that evolution has delivered to humans. Not so for reading. While reading, too, relies on the phonologic code, the key to unraveling it is not so readily apparent and can be accessed only with effort on the part of the beginning reader.

Profound differences distinguish reading from speaking. Not only are reading and writing relatively recent human accomplishments (man has had a written language for only about five thousand years), reading is still relatively rare in the world. Reading is not built into our genes; there is no reading module wired into the human brain. In order to read, we have to take advantage of what nature has provided: a biological module for language. For the object of the reader's attention (print) to gain entry into the language module, a truly extraordinary transformation must occur. The reader must somehow convert the print on a page into a linguistic code—the phonetic code, the only code recognized and accepted by the language system. However, unlike the particles of spoken language, the letters of the alphabet have no inherent linguistic connotation. Unless the reader-to-be can convert the printed characters on the page into the phonetic code, these letters remain just a bunch of lines and circles totally devoid of linguistic meaning. This essential distinction between written and spoken language was best captured by linguist Leonard Bloomfield: "Writing is not language, but merely a way of recording language by visible marks." The written symbols have no meaning of their own but rather stand as surrogates for speech, or, to be more exact, for the sounds of speech.

Beginning readers throughout the world must learn how to decipher print—how to convert an array of meaningless symbols on paper so that they are accepted by a powerful language machinery that recognizes only the phonetic code. I remember observing a visiting student as she tried to use her euros to obtain a Coke from a vending machine at Yale. Although she literally had a pocketful of euros, her currency was essentially worthless in a machine that accepted only American coins. So it is with reading. The most eloquent of written prose is rendered meaningless if it cannot be transformed into the phonetic code recognized by that reader's language module.

BREAKING THE CODE

The very first discovery a child makes on his way to reading is the realization that spoken words have parts. Suddenly the child appreciates that the word he hears comes apart into smaller pieces of sound; he has developed *phonemic awareness*. It's a remarkable discovery. There is little reason for a child to notice this. Since spoken language is built into our genes and takes place automatically, its segmental nature is not part of our consciousness. Furthermore, as a result of coarticulation, spoken language is seamless, further obscuring its underlying segmental nature. But once a child becomes aware of the segmental nature of spoken language, he has the basic elements—the particles of spoken language, phonemes, and their sounds—to which he can now attach the appropriate written letters. Letters linked to phonemes are no longer meaningless marks on paper but have been transformed into something truly spectacular: language. Translated into the phonetic code, printed words are now accepted by the neural circuitry already in place for processing spoken language. Decoded into phonemes, words are processed automatically by the language system. The reading code is deciphered.

Seventy to 80 percent of American children learn how to transform printed symbols into a phonetic code without much difficulty. For the remainder, however, written symbols remain a mystery. These children are dyslexic. They, like Alex, cannot readily convert the alphabetic characters into a linguistic code.

Typically, a young child develops phonemic awareness if he is to become a reader. That is to say, he must understand that spoken words are made up of smaller units of speech sounds: phonemes. And it is these very same phonemes to which the letters of the alphabet must attach if the written word is to be brought into the language system. All readers—dyslexic readers included—must take the same steps. The difference is simply in the effort involved and the time it takes to master the alphabetic principle.

Children vary greatly in their ease of developing phonemic awareness. For some children the process is speedy and apparently effortless; exactly how these children develop phonemic awareness is currently not known. Exposure to a rich language environment, one in which children are given lots of opportunity to hear and to play with spoken words—for example, to hear rhymes and to practice rhyming songs—surely facilitates this awareness.

In dyslexic children a glitch within the language system, at the level of the phonological module, impairs the child's phonemic awareness and therefore his ability to segment the spoken word into its underlying sounds. As a result of this deficit, children have difficulty breaking the reading code.

The reading process consists of two major components: decoding, which results in word identification, and comprehension (Figure 9a). When, as in Figure 9b, we consider the language module and the components of reading side by side, it becomes clear how a bright child like Alex could have trou-

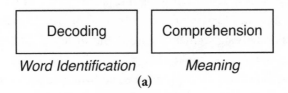

READING

| Decoding | Comprehension |

Word Identification   *Meaning*

(a)

DYSLEXIA

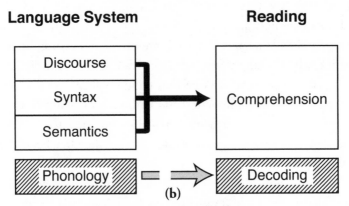

Language System   **Reading**

Discourse

Syntax → Comprehension

Semantics

Phonology → Decoding

(b)

Figure 9. The Paradox of Dyslexia I

a. Two major components of reading: decoding and comprehension.
b. A phonological weakness interferes with decoding; higher abilities necessary for comprehension are intact.

ble reading single words and yet comprehend what he is reading at a much higher level.

A phonologic weakness at the lowest level of the language system impairs decoding. At the same time, all the cognitive equipment, the higher-order intellectual abilities necessary for comprehension—vocabulary, syntax, discourse (understanding connected text), and reasoning—are intact. All the equipment Gregory or Alex or other dyslexics require for understanding, for forming concepts, for comprehending the written text, is there, unaffected by the phonologic deficit. The richness and depth of their intellectual abilities explain why identification is so often delayed in bright dyslexic children. As one of my patients' third-grade teacher remarked, "Madison is so smart; she knows the answers to the most difficult questions. She is the first one in our class to catch on to the most abstract concepts. I could never imagine that she had any problem at all." Luckily, we can now effectively treat pho-

DYSLEXIA
Going from Text to Meaning

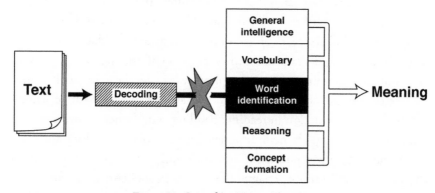

Figure 10. Going from Text to Meaning

A phonological weakness blocks decoding, which in turn interferes with word identification. This prevents a dyslexic reader from applying his higher-level skills to get at a word's meaning. But even if he can't identify the word specifically, he can apply these higher-level skills to the context around the unknown word to guess at its meaning.

nologic deficits, while, ironically, the complex reasoning and sophisticated thinking skills that dyslexic children often possess are almost impossible to teach. My husband, the Charles and Helen Schwab Professor in Dyslexia at Yale, theorizes that since the basic circuitry for linking letters to sounds is disrupted in dyslexic readers, they develop and come to rely on other neural systems, not only for reading but for problem-solving as well. Such individuals may see things in a different, perhaps more creative way and are able to think out of the box. (Later you will meet some of these people.)

As shown in Figure 10, although the language processes involved in comprehension and meaning are intact in someone like Alex, they cannot be called into play, because they generally can be accessed only after a word has been identified. Now we can understand why Alex, with a superior intelligence, excellent vocabulary, and boundless curiosity, cannot decipher even the simplest of words or read a passage aloud. When reading connected text silently, even if Alex can't quite decipher a specific word, he puts his ability to think and to reason to work and uses the context around the word to guess at its meaning. Recall that Alex did much better when asked to read a passage to himself and answer questions than when asked to read isolated words. Best for him is when he can *listen* to a story, because he can use all his higher-level thinking skills to follow the narrative and answer questions about it.

The impact of the phonological weakness is most obvious in reading, but it can also affect speech in predictable ways. Evidence began accumulating more than three decades ago that the core difficulty in dyslexia is getting to the sound structure of the spoken word. Researchers showed that children between four and six years of age develop an awareness that words come apart. By age six and a half to seven years most children (about 70 percent) can count the number of sounds (phonemes) they hear in small words. By this time many children have also had at least one full year of schooling, including instruction in reading. Reading and phonemic awareness are mutually reinforcing: Phonemic awareness is necessary for reading, and reading, in turn, improves phonemic awareness still further. The 30 percent of children who still cannot separate the sounds in spoken words after a year of reading instruction likely reflect the 20 to 30 percent of schoolchildren who go on to experience dyslexia.

In the 1980s, researchers began to address that connection explicitly. British researchers Lynette Bradley and Peter Bryant found that a preschooler's phonological aptitude correlates with his reading three years later. They (and other investigators) also found that training a young child to attend to the sounds in spoken words before he goes to school significantly improves his success in learning to read later on. In these studies, one group of preschool children received specific training in attending to the sounds of words. They learned to categorize words based on their first, middle, or last sounds. For example, using pictures, children were taught that words could share beginning sounds (*pig, pan*), middle sounds (*hen, pet*), and end sounds (*hen, pin*). Another group received general language training emphasizing the meaning of words. Without question, the group receiving the sound-based training showed the most improvement in reading and spelling. This study was also important in showing that the kinds of experiences a child has before he goes to school influence his ability to read years later.

In the 1990s we and other research groups demonstrated that phonological difficulties are the most significant and consistent markers of dyslexia in childhood. One type of test in particular seemed quite sensitive to dyslexia. This test asks a child to segment words into their phonemes and then delete specific phonemes from the words. For example, the child must say the word *clock* without the *k* sound (*lock*) or say the word *sour* without the *s* sound (*our*). A child's performance on this test was most related to his ability to decode single words and was independent of his intelligence, vocabulary, and reasoning ability. When we gave this and other tests of phonemic awareness to a group of fifteen-year-olds in our Connecticut Longitudinal Study, the results were the same: Even in high school students, phonemic awareness was the best predictor of the ability to read words accurately or quickly.

If dyslexia is the result of a phonologic weakness, then other consequences of impaired phonological functioning should also be apparent—and they are. When dyslexics misname objects, the incorrect responses tend to share phonological characteristics with the correct response; *misnaming is a result not of a lack of knowledge but of confusing the sounds of language.* For example, as in the series of drawings in Figure 11, a girl, Amy, shown a picture of a volcano, calls it a tornado. When given the opportunity to elaborate, Amy demonstrates that she knows what the pictured object is. She can describe the attributes and activities of a volcano in great detail and point to other pictures related to volcanoes. She simply cannot summon the word *volcano*.

Figure 11. The Paradox of Dyslexia II
Amy has difficulty reading the word *volcano*. When shown a picture of a volcano, she retrieves *tornado*, a word that sounds similar. Once Amy hears the word *volcano*, it is clear that she knows exactly what it means.

This finding converges with other evidence in suggesting that while the phonological component of the language system is impaired in dyslexia, the higher-level components remain intact. Phonologic abilities are not related to and in fact are quite independent of intelligence. Many children with superior intelligence develop dyslexia, while other children with much lower levels of intelligence catch on to reading with relative ease.

The phonologic model crystallizes exactly what we mean by dyslexia. As shown in Figure 12, a circumscribed, encapsulated weakness is often surrounded by a Sea of Strengths: reasoning, problem-solving, understanding concepts, critical thinking, empathy, and vocabulary. The phonologic weakness masks what are often excellent thinking and comprehension skills. Dyslexics like Gregory use the "big picture" of theories, models, and ideas as a framework to help them remember specific details. It is true that when details are not unified by associated ideas or a theoretical framework—when, for example, Gregory must commit to memory long lists of unfamiliar names—dyslexics can be at a real disadvantage. Even if Gregory succeeds in memorizing such lists, he has trouble producing the names on demand, as he often must when he is questioned on rounds by an attending physician.

Rote memorization and rapid word retrieval are particularly difficult for dyslexics. On the other hand, dyslexics appear to be disproportionately rep-

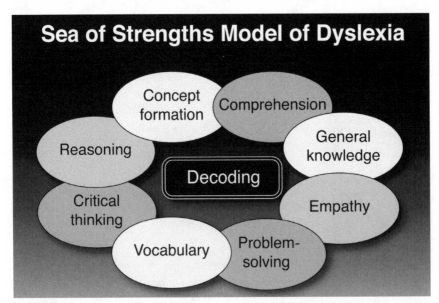

Figure 12. Sea of Strengths Model of Dyslexia
In dyslexia an encapsulated weakness in decoding is surrounded by many strengths.

resented in the upper echelons of creativity and in the people who, whether in business, finance, medicine, writing, law, or science, have broken through a boundary and have made a real difference to society. I believe that this is because a dyslexic cannot simply memorize or do things rotely; she must get far underneath the concept, understand it at a fundamental level. This need often leads to a deeper understanding and a different perspective than that achieved by many of those for whom things come easier because they just can memorize and repeat, without ever having to deeply and thoroughly understand.

Even when the dyslexic knows the information, needing to rapidly retrieve and orally present such information often results in retrieval of a related phoneme, such as in substituting *humanity* for *humidity*. As a result, the dyslexic may appear much less capable than he is; however, given time and when not pressured to provide instant oral responses, the dyslexic can deliver an excellent oral presentation. Similarly, dyslexics like Gregory lack reading fluency and frequently need to resort to the context to help identify specific words. This strategy slows them down further and helps explain why the provision of extra time as an accommodation is so necessary if dyslexics are to show their knowledge.

As you have just read, dyslexia is primarily a problem involving the individual sounds of speech, one that impacts not only written language but spoken language as well. We now understand at a fine-grained level how difficulty getting to the sounds of spoken language explains dyslexia in all its manifestations, accounting for and predicting the full range of difficulties experienced by a dyslexic child and adult.

For example, one of the most troubling, and until very recently least understood, primary symptoms of dyslexia is difficulty with word retrieval. Dyslexia impacts spoken language so that the brightest thinker is often a surprisingly struggling speaker. For instance, the person may be thinking and wanting to utter "An octopus has eight tentacles" and instead finds himself saying out loud "An octopus has eight testicles." Retrieving and uttering the wrong word impacts children and adults and brings with it great embarrassment and feelings of shame and unworthiness.

You can appreciate the sense of wanting to hide quickly after uttering one of the snafus featured on our Yale Center for Dyslexia & Creativity website: "The dinosaurs became distinct." "The monster is a pigment of my imagination." "I hope I win the constellation prize." "He had to use a fire distinguisher."

WHY SOME VERY SMART PEOPLE
UTTER SOME VERY SILLY THINGS

While this symptom of dyslexia has been known for decades, why and how this happens remained a puzzle until recently.

How could a person be so smart and yet not be able to express his thoughts accurately when speaking? Science has finally broken through and revealed the two-step neural mechanism responsible for the word-retrieval struggles of the dyslexic. At last scientific data explain how it can be that the brightest, most accomplished dyslexic—for example, a Pulitzer Prize–winning poet—can conjure up a wonderful poetic concept but when speaking to a group finds that he is tongue-tied and cannot utter the same poetic words he grasped so easily in his mind. The impact of word-retrieval difficulties cannot be emphasized enough. For this dyslexic poet, it has meant having to leave the stage, flushed with embarrassment, in the midst of a talk before an initially eager and enthusiastic and now disappointed audience. "I thought I knew exactly what I wanted to say. I can't understand why the words just wouldn't come out right. What's wrong with me?"

As is often the case with a person who is dyslexic, the difficulty is not in thinking or higher-level reasoning but in the lower-level mechanics of uttering the word, the actual physical act of forming the utterances leading to saying a word aloud (Figure 13).

The two-step mechanism accounts for how the dyslexic can use her intact higher-order thinking abilities to develop a concept and identify the abstract semantic representation of the word that represents this concept. No problem here. However, in order *to utter the word*, the speaker must go into her internal lexicon (dictionary) and access and retrieve each of the individual sounds that represent the word, for it is these smallest particles of speech that are necessary to program the articulatory muscles (lips, tongue, palate) that produce the spoken word. The result is misspeaking or a long series of *um*'s before the sought-after word is finally retrieved.

→ The first step of this mechanism—the cognitively demanding process of identifying the basic concept—is intact.

→ The second step—trivial for most, but seemingly insurmountable for the dyslexic—is where the trouble lies.

WORD RETRIEVAL IN TYPICAL AND DYSLEXIC READERS

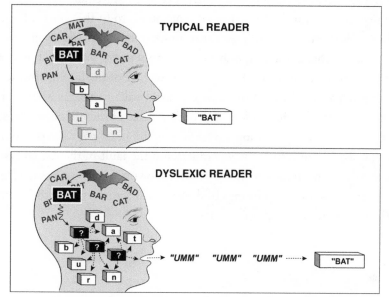

Figure 13. Word Retrieval in Typical and Dyslexic Readers

→ It is this fundamental difficulty in retrieving the tiny
individual sounds of spoken language that lies at the heart
of dyslexia and explains the dyslexic's major problem in
word retrieval.

This mechanism brings understanding together with profound relief
to men and women who are dyslexic and to dyslexic children and their
parents. The explanation proffered by this mechanism confirms that these
word-retrieval difficulties are not an indication of some higher-level cognitive deficit but rather reflect a lower-level mechanical error in being able to
program the muscles of articulation necessary to utter a word.

As illustrated in Figure 13, an utterance begins with an idea. For example,
speaking the word *bat* begins with conjuring up the concept, the idea of a
small flying nocturnal creature, and holding in mind the word that represents
it, *bat*. This is straightforward for both a typical reader and a person who is
dyslexic. In order to put voice to the concept, to actually say the word aloud,
the speaker must replace the concept with the actual sounds, the phonemes
/b/ /a/ /t/ that represent the spoken word. It is straightforward for the typical
reader—in a flash he goes from concept to speech. But for a child or adult

who is dyslexic, instead of the phonemes being crisp and clear, they appear fuzzy and indistinct. As a result, the dyslexic must search in his messy storehouse of sounds for the right sequence of sounds to fit the word, a difficult process with an unsure result. At times, in what may seem like an eternity to the speaker, what comes out of his mouth is a series of *um*'s which may end with a correct, albeit somewhat delayed, utterance of the intended word. At other times a close but not quite correct word will emerge, as when a child says, "I love to visit the *Oysters* Museum" in place of the intended *Cloisters*.

By identifying a phonologic weakness as the heart of dyslexia, we have taken a significant step. Now we understand the fault in the neural system for translating print into language. In the following discussion you will learn about the neural underpinnings of dyslexia and actually see, through brain activation patterns, what goes wrong.

6

READING THE BRAIN

Incredible as it may seem, today we can actually see someone's brain at work as she reads. Sophisticated imaging studies of the reading brain bring phonemes to life, allowing researchers to virtually track the printed word as it initially registers as a visual icon, then is transformed almost instantly into the sounds (phonemes) of language and simultaneously activates its meaning stored within the brain's internal dictionary. Within milliseconds—less than the blink of an eye—assorted lines and circles in the brain's circuitry make the vast journey to meaning, as a cacophony of abstract symbols is translated into a symphony of words. As recently as the beginning of the nineteenth century, scientists and philosophers weren't even sure that thinking, speaking, or reading had their roots in the brain. We've come a long way.

The earliest conceptions of the mind were grounded in religious and philosophical notions, often focused on the soul. By the second century, animal experiments were beginning to provide clues to the central importance of the brain. Most influential were the experiments of the Greek physician Galen which demonstrated that the brain was critical for movement, for sensation, and for thinking. This brilliant physician-scientist associated all varieties of higher cognitive activities—memory, imagination, and intelligence—with the brain, and he also related brain injury to disruptions of these processes. Galen viewed the brain as a holistic and undifferentiated organ, and this conceptualization reigned virtually unchallenged for hundreds of years.

LOCALIZING BRAIN FUNCTIONS

In the late eighteenth and early nineteenth centuries, physicians took a great interest in the origins of human personality traits as well as mental illness and were eager for a new conceptual framework, one that continues to influence neuroscientists to this day—the fundamental notion that cognitive functions could be localized within specific brain structures. Scientists and physicians had begun trying to localize brain functions using the only method available to them at the time: studying the brains of individuals who had suffered an accident of nature, such as a stroke or a traumatic injury to the brain. After a patient died, examination of the brain revealed the area of injury, which was in turn related to the patient's symptoms. It was just such a circumstance that allowed the first cognitive process, expressive language, to be localized in the brain.

In 1861 a fifty-one-year-old Frenchman named Leborgne was transferred to the clinical service of Paul Broca, a highly respected French physician. The long-suffering Leborgne had been hospitalized for at least two decades with a range of neurological ills, including epilepsy, loss of speech, and paralysis of his right side. Leborgne had been nicknamed Tan by the other patients because that was the only word he could say, and he repeated it over and over again. His ability to understand spoken language, however, was relatively unaffected. Leborgne was a patient of Broca's for less than a week when he passed away. Examining Leborgne's brain afterward, Broca was at once struck by a rather large irregular lesion on the surface of the left frontal region (the inferior frontal gyrus, to be precise; see Figure 14). Located just behind the left temple, this portion of the brain is now commonly referred to as Broca's area. Today neurologists classify a loss of language as an aphasia, and the specific kind of language difficulty observed in Leborgne—a loss of fluent speech while retaining the ability to understand language—is known as Broca's aphasia.

Broca's observation ushered in a new era in brain exploration of cognitive functions. There were to be no more quarrels over the brain's role in cognition; loss of a highly specific cognitive function—not just any speech prob-

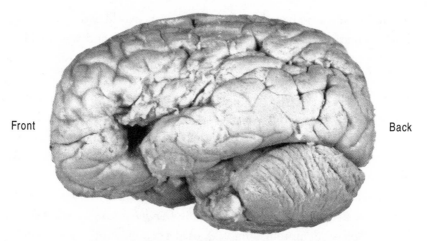

Figure 14. Tan's Brain Showing Defect in Broca's Area
A view of the left side of Broca's patient's brain indicates a large, irregular
defect in the inferior frontal gyrus, a region now referred to as Broca's area.

lem, but the loss of the ability to express words, for example—was linked to
the destruction of a particular brain region. Broca went on to define different
types of speech difficulties. For example, he noted that the expressive aphasia
he had initially described could be accompanied by retention of the ability
to swear. In describing the case of a beleaguered parish priest decades earlier
(1843), another physician, Jacques Lordat, had unknowingly provided one
of the most striking cases of this dissociation. Following a stroke, the priest
could say very few words, but he was able to utter "the most forceful oath of
the tongue, which begins with an 'f' and which our Dictionaries have never
dared to print."

Broca also clearly established that the roots of reading—language and
speech—originate in the cerebral cortex.* As self-evident as this may seem
today, the logical source of our ability to speak was long thought to be
obvious—the tongue. For example, in ancient Rome, the treatment for the
inability to speak was to tone and massage the tongue, like a weak or lazy
muscle, and to gargle vigorously. Although perceptive physicians seemed to
disentangle *loss of speech* (aphasia) from *loss of movement* (paralysis) of the
tongue, the view that the tongue was the primary organ for speech was firmly
entrenched and lingered through the seventeenth century at least. When

* Cortex refers to the *grey matter,* that is, the location of the nerve cells close to the surface
of the brain; *white matter* refers to the long tails or axons of the nerve cells that carry the
message from the nerve cells to other neurons in the brain or spinal cord.

William Harvey, who had electrified the scientific world with his discovery of the circulation of blood, suffered a stroke in 1651 and was unable to speak, his apothecary provided this renowned physician-scientist with what he considered to be the very best aphasia treatment available. The apothecary crisply clipped Harvey's frenulum, the thin sliver of tissue tying the tongue to the base of the mouth. The hope was that if his tongue were freed, Harvey's aphasia would improve. Although we can surmise that Harvey's tongue was now looser, we can also conclude with great assurance that his aphasia was not improved.

The precocious German neurologist Carl Wernicke, who by the age of twenty-six had completed the three-volume *Der aphasischer Symptomenkomplex* (The Aphasic Symptom Complex), used a combination of astute clinical observation and deductive reasoning to suggest that damage to an area along the upper part of the temporal lobe (the brain region behind the helix of the ear; see Figure 15) would produce still another type of aphasia. This speech disturbance, now referred to as Wernicke's aphasia, was almost the inverse of the expressive aphasia localized by Broca. Instead of the patient failing to get the words out but usually fully understanding language, in Wernicke's aphasia the patient speaks with ease but does not understand language and utters gibberish.

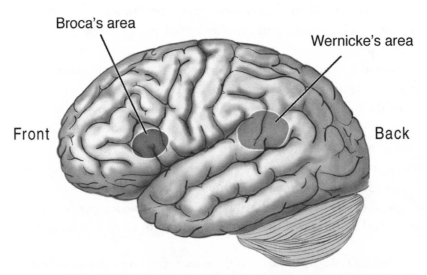

Figure 15. Brain Localization of Expressive and Receptive Language
The left side of the brain, with the two major areas associated with language high-lighted: expressive language (Broca's area) and receptive language (Wernicke's area).

HOW THE BRAIN READS

Broca's discovery opened the door to learning how the brain reads. Remember, in order to read we must enter the language system; at a neural level, this means that reading relies on the brain circuits already in place for language. Although not appreciated at the time, the identification of Broca's area as a site critical for language marked the first step in the quest to map out the neural circuitry for reading.

The strategy used by Broca and legions of neurologists afterward was straightforward. After a patient passed away, areas of damaged tissue were identified, localized according to anatomic landmarks, and correlated with the patient's symptoms. It is important to appreciate what this type of evidence can tell us as well as its limitations. To pinpoint a site that disrupts the circuitry for spoken language is not quite the same as mapping out the neural machinery responsible for language. Broca's contemporary, British neurologist John Hughlings Jackson, appreciated this distinction when he said, "To locate the damage which destroys speech and to locate speech are two different things." Let's say an adult suddenly loses his ability to read. His condition, acquired alexia, typically follows a stroke, in which destruction of brain tissue produces a break in the circuitry and, one could say, causes a power outage that interrupts the pathway for reading. The wiring, originally laid down correctly, is now disrupted. Locating the area of disruption tells us that this is an area that is necessary for reading, but it doesn't identify the complete neural circuitry responsible for reading.

In the condition of developmental dyslexia, where reading fails to develop normally, something has gone awry right from the beginning. Consequently, we would *not* necessarily expect to find a distinct lesion, a cut in the wiring; instead, the wiring may not have been laid down correctly in the first place, a glitch having taken place during fetal or early life, when the brain is hard-wired for language. As a result, the tens of thousands of neurons carrying the phonologic messages necessary for language do not appropriately coalesce to form the resonating networks that make skilled reading possible.

The enormous complexity of the brain in its initial development presents a myriad of opportunities for a misconnection or a false connection. In this case, the neural system necessary for phonologic analysis is somehow miswired and a child is left with a phonologic impairment that interferes with spoken and written language. Depending on the nature or severity of this fault in the wiring, we would expect to observe variations and varying degrees of reading difficulty.

But there are many possible scenarios for developmental dyslexia. What was needed to understand dyslexia is the ability to map out the full neural circuitry for reading. That was achieved in the 1980s. In the interim, physicians tried to make sense of dyslexia in the best way they could. Hypotheses offered by the American neurologist Norman Geschwind that dyslexia results from damage to or improper development of language regions in fetal life played an important role in focusing attention on the neurobiological substrate of dyslexia.

A REVOLUTION: READING THE MIND

In 1973, with the advent of computed tomography (CT, which is a computed series of X-rays that build up a three-dimensional image of the brain), scientists could now *see* the brain for the first time. For this groundbreaking achievement, the British scientists Godfrey N. Hounsfield and Allan M. Cormack were awarded the Nobel Prize for Physiology or Medicine in 1979.

Using CT and, later, magnetic resonance imaging (MRI), which is discussed below, neuroscientists could see the finest details of brain anatomy. These pictures, however, provide information about structure and not about function. It wasn't until functional brain imaging became possible in the early 1980s that scientists were able to see the brain at work in a healthy person. Brain *function* could be observed as a person read, spoke, thought, or imagined.

Positron emission tomography (PET) was the first technology developed to study the brain at work. It involves measuring blood flow to brain regions through the use of a radioactive compound injected into the bloodstream. Though much has been and continues to be learned about brain function from this technology, it is hampered by its invasive nature and the elaborate equipment required to make up the radioactive materials. So for the most part PET has been supplanted by a newer technology, functional magnetic resonance imaging (fMRI), which allows neuroscientists to visualize the inner workings of the human brain in a completely noninvasive way; there is no radiation, nor are there any injections. Currently fMRI is the most widely used method to study the brain at work.

In the seminal paper "On the Regulation of the Blood Supply of the Brain," published in 1890 in the *Journal of Physiology*, British scientists C. S. Roy and C. S. Sherrington argued for the principle of *autoregulation of cere-*

bral blood flow. Basically, they provided evidence that within the brain, local blood supply varies in response to functional activity in that region. In their own words: "These facts seem to indicate to us the existence of an automatic mechanism by which the blood supply of any part of the cerebral tissue is varied in accordance with the activity of the chemical changes which underlie the functional activity of that part."

In 1981, Louis Sokoloff, a scientist at the National Institutes of Health, showed that it is changes in energy metabolism that directly influence alterations in blood flow: "It is, therefore, clear . . . that energy metabolism and functional activity are closely coupled in the nervous system and that local blood flow is distributed and adjusted in the cerebral tissues according to local metabolic demand and thereby to local functional activity."

This makes it possible to envision the sequence of events at the associated brain region as a person reads: increase in functional activity of local neurons → increase in local metabolism → increase in local blood flow.

This concatenation of brain events makes sense. Performing a cognitive task like reading is work and consumes energy. As a child tries to determine whether two words rhyme, a chain of events is set in motion: The neural systems necessary for carrying out this task are turned on and consume energy, and to meet the increased energy requirement more blood flow is needed to bring additional fuel (oxygen) and nutrients to that location. This concept—the autoregulation of cerebral blood flow—is the underlying principle of functional brain imaging. This principle, together with the fact that increased blood flow produces alterations in the magnetic properties of blood, is what underlies the functioning of fMRI.

Functional MRI is based on the magnetic properties of a basic component of blood: oxygenated hemoglobin. In red blood cells, oxygen is bound to hemoglobin and transported throughout the body for delivery to working cells. The magnetic properties of the hemoglobin molecule change depending on the amount of oxygen bound to it. Blood with higher oxygen concentrations produces a stronger magnetic signal than blood with less oxygen. As a person carries out a specific cognitive task, responsible neurons in sites distributed throughout the brain become activated, blood flow to these brain regions increases, bringing with it rich, highly oxygenated blood, and the fMRI apparatus picks up its stronger magnetic signal.

Now you will learn how fMRI experiments are carried out and what fMRI is teaching us about how the brain reads.

7

THE WORKING BRAIN READS

Functional MRI is remarkably patient-friendly, for both adults and children. In our research we have imaged hundreds at all ages with uniformly positive feedback.* Virtually the same scanner used for MRIs to evaluate headaches or torn knee ligaments is used in functional imaging. All sorts of people have participated in our imaging studies of reading, including older adults, young adults, adolescents, and even children as young as six years old. For those readers who have not had an MRI, I want to explain briefly what it's like by telling you about the experience of Kacie, age eight.

When Kacie and her parents first arrived at the MRI center they were greeted by Marina, a research coordinator, who introduced the family to the center and helped Kacie feel at home. Marina then reviewed with Kacie the tasks she would be asked to perform as she was being imaged. Kacie quickly caught on and was soon ready for the imaging. In the scanning room she lay down on a sliding table with a large open tube at one end. At the start of the imaging, the table moved so that Kacie's head lay comfortably within the open tube (Figure 16).† As she lay in the scanner, Kacie responded to each stimulus by pressing a YES or a NO button on a device she was holding

* For this discussion I have chosen to focus primarily on our own imaging program at Yale, the studies I know best.
† The tube contains the electromagnet that produces the images.

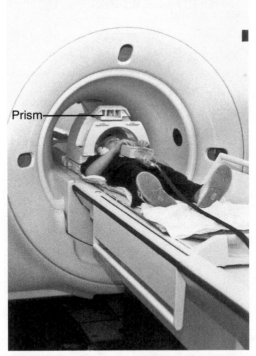

Figure 16. Kacie Getting Her Brain Imaged in the Scanner
Kacie is holding the response box and looking
through the prism just above her head.

in her hands. This device measured how accurately and how rapidly she
responded.

Once Kacie glided into the scanner, she looked up through a prism (peri-
scope) just above her head to see the stimuli, pairs of words, as they were
rapidly projected one after another on a screen in front of her. Nonsense
but pronounceable words such as *LETE* and *JEAT* were used as stimuli, and
Kacie was asked whether each pair rhymed. As shown in Figure 17, if the
words rhymed, she pressed the YES response button on the device she was
holding in her hands.

Since these words are made up, Kacie had never seen them before, so they
could not have been memorized and had to be sounded out, ensuring that
the neural pathways for phonologic analysis would be activated. As each
pair of words flashed up on the screen and Kacie sounded out the words to
herself, the MRI scanner recorded the underlying neural processes. If, for
example, the process activated neurons in Broca's area, fresh oxygenated
blood would flow into this area and into each region connected to the neural
network responsible for phonologic analysis. Within this circuit of actively

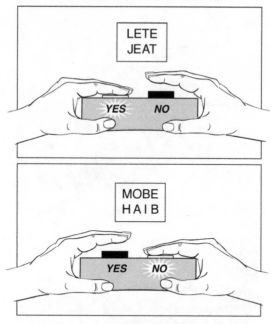

Figure 17. Reading and Responding While in the Scanner
Kacie reads the words and presses the YES button if the words rhyme and
the NO button if they do not.

firing neurons, newly arriving oxygenated blood was replacing oxygen-poor
blood, and the scanner was recording it all. After the imaging was completed,
Kacie was none the worse for wear (see Figure 18).

Immediately after imaging, we give each participant a copy of her *structural*
brain MRI—essentially a photograph of her brain anatomy, which is invariably normal in people who are dyslexic. While my colleagues and I anticipated that dyslexics would be reassured when shown their own brain images,
we didn't realize how meaningful and moving an experience this would prove
to be. Children and adults alike have embraced their brain images as if they
had just been reunited with a long-lost and deeply mourned relative. As I
point out the various normal anatomic structures, no matter how old our
subject is, he literally lights up with wide-eyed amazement. If an adult, the
remark is invariably "You may not believe this, but I always thought I was dumb
and that my brain would look abnormal, with some parts missing or
holes or something not right. I can't believe I have a normal brain. What
a relief."

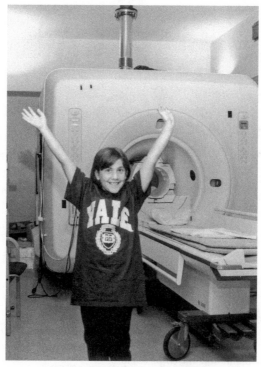

Figure 18. Mission Accomplished
Kacie after her imaging was completed.

BASIC LANDMARKS OF THE BRAIN

The brain is an extraordinary organ, and for all its complexity in functioning, its organizing structure is relatively straightforward. I want to digress here briefly and point out some of the basic landmarks that I will refer to as we move forward.

The brain is made up of two mirror-image sides, or hemispheres: right and left. As shown in Figure 19, the front of the brain, near the forehead, is referred to as *anterior,* and the back of the brain, behind the ears and further back, is called *posterior.* Each hemisphere of the brain is divided into four lobes or sections: *frontal, parietal, temporal,* and *occipital.* The frontal lobes are anterior, the occipital lobes posterior, and the parietal and temporal lobes somewhere in between. The parietal lobe sits above the temporal lobe. There are mirror-image lobes on each side of the brain—for example, the left temporal lobe and the right temporal lobe. Traditionally the left side of the brain has been associated with language.

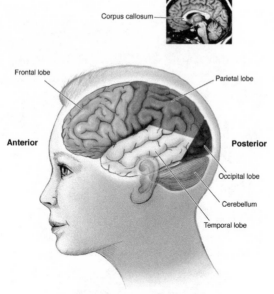

Figure 19. Brain Landmarks
The left side of the brain is shown. The inset is an
MRI image of the structures in the center of the
brain; note the corpus callosum.

Connecting the left and right hemispheres is a broad band of tissue made up of the tails, or axons, of nerve cells, which busily carry messages from one hemisphere to the other. This band of connecting, insulated fibers is called the *corpus callosum* and on typical anatomic MRI images appears as a white band. In contrast, the brain cells that are the command center and originate the message are not insulated and appear as grey matter. If the right and left lobes are the vertical poles of the letter *H,* the corpus callosum is the horizontal band connecting the two sides. Finally, tucked beneath the occipital lobes is the *cerebellum,* the part of the brain that controls movement and coordination.

WATCHING THE BRAIN READ

Our initial functional imaging studies were focused on mapping out the neural circuitry for reading. Naturally, in planning our research, our first goal was to identify the neural pathways for breaking words into their underlying sounds.

To begin we recruited a group of subjects, nineteen men and nineteen women, all of whom were good readers. To activate the circuitry for phonologic analysis, we asked our subjects to judge whether two nonsense words rhymed. After all the participants had been imaged, we examined the data and saw a surprising difference between the brain activation patterns in men and women (see Figure 20). As shown in the brain images, men activated the left inferior frontal gyrus, while women activated the right as well as the left; that is, while all nineteen men activated only the left side of the brain, most but not all women—eleven out of nineteen—showed activation on both sides of the brain (eight women activated only the left side). This represented the first demonstration of a visible sex difference in brain organization for language. Women performed the task as accurately and as quickly as men.

Our discovery of sex differences in brain organization was exciting, but what most excited us was the identification of specific neural sites for sounding out words. The very same brain region that demonstrated sex differences, the inferior frontal gyrus, was also involved in reading. Having established

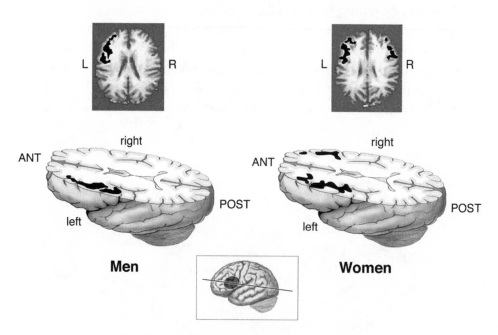

Figure 20. Sex Differences in Brain Organization for Language
The black areas indicate sites of activation in men and women as they rhymed nonsense words and show that men activate the left side of the brain and women activate both the left and right sides of the brain. The insets are the fMRI images showing the brain activations.

to our satisfaction the feasibility of using fMRI to study reading, we felt comfortable proceeding to the next step: a series of studies that would begin to teach us, at the level of brain organization and function, why some very smart people have trouble reading.

Progress in our understanding of the neurobiology of dyslexia over the past two decades has been breathtaking. First, the neural systems for reading in typical readers, both children and adults, have been mapped out. Next, and most important, careful examination of brain activation patterns reveals a glitch in this circuitry in dyslexic readers. Studies from around the world leave no doubt: Dyslexic readers use different brain pathways than good readers do.

Good readers activate highly interconnected neural systems for reading, encompassing regions in the back and the front of the left side of the brain (see Figure 21). Not surprisingly, the reading circuitry includes brain regions dedicated to processing the visual features, that is, the lines and curves making up letters, to transforming the letters of the alphabet into the sounds of language and getting to the meaning of words.

Most of the reading brain is in the back (posterior) of the brain. The posterior reading system, as noted above, is made up of two distinct pathways for reading words, one sitting somewhat higher in the brain than the other. The upper pathway is located primarily in the middle of the brain (technically, the parieto-temporal region, or PT), just above and slightly behind the ear. The lower or occipito-temporal (OT) path runs closer to the bottom of

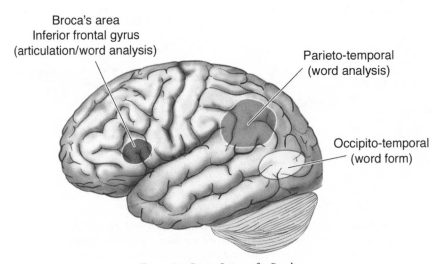

Figure 21. Brain Systems for Reading

the brain; it is the site where two lobes of the brain, the occipital and the temporal, converge. One can envision this hectic region as a hub, a place in the brain where incoming information from different neural systems comes together and all the relevant information about a word—how it looks, how it sounds, and what it means—is tightly bound together. The OT circuit is barely beneath and somewhat in back of the ear, near the area where children often get swollen glands associated with scalp or ear infections.

Scientists have long focused on these areas as important in reading. As early as 1891, the great French neurologist Jules Dejerine suggested that the PT region was critical for reading, and just one year later he was also the first to link the OT area to reading. These two subsystems have different roles in reading. Their functions make sense in light of the changing needs of the reader: Beginning readers must first analyze a word; skilled readers identify a word virtually instantaneously. The PT system works for the novice reader. Slow and analytic, its function seems to be in the early stages of learning to read, that is, in initially analyzing a word, pulling it apart, and linking

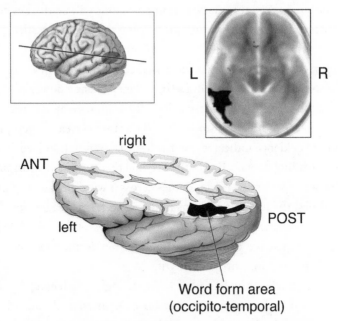

Figure 22. An Area for Skilled Reading: The Word Form Area
This is a correlation map. In the area shown in black there is a strong correlation between children's reading scores and activation in the word form area in the left side of the brain; the higher the score, the more activation is observed. The inset shows the correlation in an MRI image of the brain.

its letters to their sounds. In contrast to the step-by-step PT system, the OT region is the express pathway to reading and is the one used by skilled readers. The more skilled the reader, the more she activates this region (see Figure 22). It responds very rapidly—in less than 150 *milliseconds* (less than a heartbeat)—to seeing a word. Instead of analyzing a word, the OT region reacts almost instantly to the whole word pattern. One brief glance and the word is automatically identified, on sight. Not surprisingly, the OT region is referred to as the visual *word form area.*

A NEUROBIOLOGICAL EXPLANATION OF HOW WE COME TO READ WORDS AUTOMATICALLY— AND HOW WE TEACH READING

The ability to simply glance at a word and be able to identify it quickly is the hallmark of mature, pleasurable reading. Such automatic reading is currently out of the reach of a dyslexic reader. The hope is that a better neurobiological understanding of how automatic reading develops will be the key to developing newer and better techniques for helping the dyslexic reader grasp the elusive brass ring of reading: becoming an automatic reader.

Emerging brain imaging data have yielded an exciting new explanation of how automatic reading develops. These data present a powerful challenge to the prior theory that considered the left occipito-temporal region—the word form area—to function somewhat like a storage area for well-practiced words. Good evidence indicates that the left OT system is indeed critical for developing reading fluency. But in the original theory, investigators proposed that when an experienced reader comes across a word he has seen and read multiple times, the relevant neurons in the word form area become "tuned" to that word. As a result, the theory stipulates, the reader is then able to immediately match the printed word before him to the one stored in the word form area and instantly identify it.

This conceptualization has now been strongly challenged by a group of prominent British neuroscientists who have marshaled compelling evidence that the occipito-temporal region acts not as a storage area but rather as an "interactive" neural node where information from multiple inputs, including phonological (sound-based), orthographic (visual), and semantic (meaning-based), are integrated and synthesized. In this view, the neural mechanisms for the elements of reading (phonology, orthography, semantics) are best conceptualized as distributed patterns of neural activity representing each of these elements. In other words, multiple neural pathways must be activated and then joined together.

The controversy then is between whether the word is stored in toto, waiting to be recognized and identified (the old theory), or if this word form area functions more as a synthesizer or integrator, a neural node where all the critical parts or components making up the word are brought together (the new theory). That is, does rapid, fluent recognition of the word result from the word being read correctly over and over again, so that eventually the neurons in the word form area become tuned to the word and respond as if the word had been prefabricated and is immediately recognized, as the older theory proposes? Or is fluency for the word built up by experience, repeatedly bringing together and integrating the neural streams representing each of the three essential components of the reading process—orthography (how the word looks), phonology (how the word sounds), and semantics (what the word means)—as the interactive theory proposes, so that eventually this process takes place virtually instantaneously?

At this time, the evidence indicates that it is the latter, the interactive theory, where the integration of the visual perception together with the speech sounds it represents and its meaning become bound together tighter and tighter and are recognized more quickly the more frequently the word is read *correctly*.

We think the word form integrative system works to build and support fluency in the following way: After a child has integrated the word components and *correctly* read a word several times, this integration process occurs more and more rapidly, so that the word is read virtually instantaneously. Subsequently, just seeing the word in print activates this primed process and the word is read fluently; that is, it all happens automatically, without conscious thought or effort.

Recognition of the neural processes underlying skilled, fluent reading has important implications for how reading is taught, especially to struggling readers. Currently phonology (sounding out words) is often taught to the relative exclusion of other elements of reading. With these new data supporting the interactive theory, educators should now give serious thought to enlarging their perspective early on in the primary grades to include early exposure to words in context by having the children begin to read as early and as much as possible and stressing vocabulary as essential rather than ancillary in teaching children to read.

Mapping out the neural pathways in good readers opened the door to understanding the nature of the difficulty in dyslexic readers. As I said earlier, imaging studies revealed markedly different brain activation patterns in dyslexic readers compared to those in good readers. As they read, good readers activate the back of the brain, and also to some extent the front of the brain. In contrast, dyslexic readers show a fault in the system—inefficient functioning of neural pathways in the back of the brain. It is here, within these neural systems, that dyslexic readers have difficulty in integrating the major word components, which in turn interferes with their becoming fluent readers. Even as they mature they remain slow and not fluent readers.

A NEURAL SIGNATURE FOR DYSLEXIA

Converging evidence from many laboratories around the world has now unequivocally demonstrated the "neural signature for dyslexia"—that is, an inefficient functioning of those reading systems in the back of the brain during reading (see Figure 23). This evidence from fMRI has for the first time made visible what previously was a hidden disability. For example, in one of the first studies of fMRI in dyslexia, we used fMRI to study 144 children, approximately half of whom had dyslexia and half of whom were typical readers. Results obtained during a task requiring phonologic analysis indicated significantly greater activation in posterior reading systems in typi-

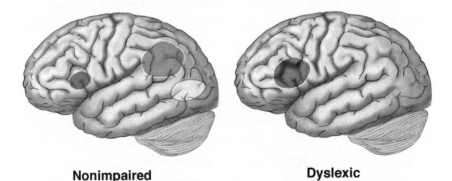

Nonimpaired   **Dyslexic**

Figure 23. A Neural Signature for Dyslexia: Inefficient Functioning of Neural Systems in the Back of the Brain

At left, nonimpaired readers activate neural systems that are mostly in the back of the left side of the brain (shaded areas); at right, dyslexic readers underactivate these reading systems in the back of the brain and tend to overactivate frontal areas.

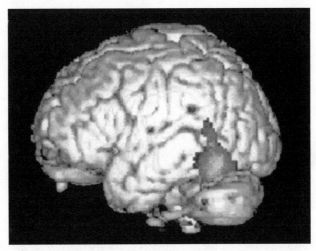

*Figure 24. Universal Inefficient Functioning of Posterior Reading
Systems in Dyslexia Found Around the World*

fMRI brain activations (shaded area) show left posterior brain
regions activated by French-, English-, and Italian-speaking university students who are good readers, but not by those with childhood histories of dyslexia who are now accurate but slow readers.

cal readers than in readers with dyslexia. These findings align with the classic
nineteenth-century reports by Dejerine I mentioned earlier, which describe
neural systems affected in adults who had a sudden loss of the ability to read
(acquired alexia), for example, following a stroke.

These data from fMRI studies in groups of children with dyslexia have
been replicated in reports from many investigators and show a failure of left
hemisphere posterior brain systems to function properly during reading,
particularly the systems in the left hemisphere occipito-temporal region.
(See Figure 24.) Similar findings have been reported in German and Italian
readers with dyslexia. Some studies in Chinese readers with dyslexia show
brain abnormalities in left occipito-temporal and anterior frontal regions
similar to those found in dyslexia in alphabetic writing systems.

These findings are helping us understand the roots of dyslexia. For example, we observed the inefficient functioning in young children indicating that
the wiring glitch is present from the *start of reading* and does not represent
the end result of years of poor reading. The identical posterior dysfunction
is observed in children and in adults, neurobiological proof that reading
problems don't go away. They are persistent, and now we better understand
why. We have also learned that dyslexic children and adults turn to alternate,

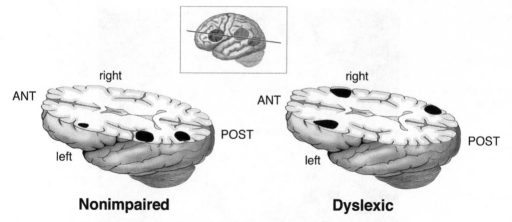

Figure 25. Dyslexic Readers Use Compensatory Systems to Read
The nonimpaired reader, on the left, activates neural systems that are mostly in the
back of the left side of the brain; the dyslexic reader, on the right, activates systems
on the right side and in the front of the brain on the left.

compensatory reading systems. Brain images recorded as dyslexic readers
try to sound out words show the posterior system on the left side of the
brain is not working efficiently; instead these slow but accurate readers are
relying on alternate, secondary pathways—not a repair but a different route
to reading. In addition to their greater reliance on frontal systems (Broca's
area), dyslexics are using other auxiliary systems for reading, ones located
on the *right* side and also in the *front* of the brain on the left—a functioning
system but, alas, not an automatic one (see Figure 25).

These findings now explain the previously puzzling picture of bright adult
dyslexic readers who improve in reading words accurately but for whom
reading remains slow and draining. The dysfunction in left posterior systems
prevents rapid, automatic word recognition; the development of ancillary
(right side frontal and posterior as well as left frontal) systems allows for
accurate, albeit very slow, reading. These dyslexic readers have to rely on a
manual rather than on an automatic system for reading.

Taken together, these studies from investigators in different laboratories
and examining different languages demonstrate unequivocally not only inefficient functioning of the occipito-temporal system for skilled reading, but
now for the first time provide the neurobiological explanation for the dyslexic reader's slow reading.

BRAIN IMAGING IN PREREADERS AND EARLY READERS

When does the neural signature for dyslexia develop? Do these changes in brain function and organization predate the poor reading, or are they the result of poor reading itself? In a series of experiments, Nadine Gaab at Boston Children's Hospital and John Gabrieli at the McGovern Institute at MIT addressed just these questions. While preschool children performed a phonological task, Dr. Gaab compared functional brain images in those with a family history of dyslexia to peers without a family history of dyslexia. Results showed that as a group, the preschoolers with a family history of dyslexia demonstrated reduced activation in posterior reading systems. In a separate study, Dr. Gabrieli and his associates examined the white-matter (nerve fiber) connections in kindergarten children who had not yet been taught to read. Results showed a larger volume and higher integrity of white matter in the left arcuate fasiculus (the large white-matter tract connecting two reading systems, the left anterior language system in Broca's area with posterior language systems in the temporal and parietal areas) in those children with the best phonological awareness. The findings of both studies suggest that the differences in brain function and structure are present even before the children learn to read.

> *CAUTION:* These findings must be viewed cautiously, because those conducting both studies did not know if the children would develop dyslexia or not. For example, in the study by Dr. Gaab, while children with a family history of dyslexia are at higher risk for dyslexia, many children with a family history never develop dyslexia, while some who lack any known family history are eventually identified as dyslexic. Future studies should clarify these issues.

BRAIN CHATTER: NEW DATA REVEAL HOW BRAIN SYSTEMS FOR READING COMMUNICATE WITH ONE ANOTHER

The initial challenge in the elucidation of the neural basis of reading and dyslexia was identifying the specific regions of the brain involved in reading. Once that was accomplished, we moved on to the next important question, one that involves *function:* How do the key systems in reading communicate with one another, how are they *connected,* and are there differences in connectivity between good and struggling readers?

Connectivity analyses of fMRI data represent the most recent evolution in characterizing brain networks in dyslexia. Measures of functional connectivity are designed to detect interesting differences in brain regions which, although they have *similar magnitudes of activation,* demonstrate *differences* in how this is synchronized with other brain systems in dyslexic and typical readers.

In fact, this synchrony between anatomically distinct regions might be equally as or more important for cognitive performance than the magnitude of activation in any single region. Although there have been some functional connectivity studies of dyslexia, these have examined connections between just a few specific regions and potentially fail to provide a full, complete picture of the connectivity profiles of dyslexic compared to typical readers. For example, one recent study examined differences between readers with dyslexia and typical readers in the connectivity of the visual word form area (VWFA) and other components of the reading system. In typical readers, the VWFA was connected to distant as well as adjacent reading systems in the left and right hemispheres. In contrast, functional connectivity in dys-

Connectivity to Skilled Reading Area

Typical readers   Dyslexic readers

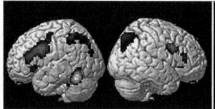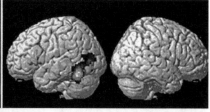

Connections from skilled reading area to phonology and semantic areas

Connections from skilled reading area limited to adjacent areas

Van der Mark et al., 2011

Figure 26. Differences in Connectivity Between Typical and Dyslexic Readers

lexic readers was significantly reduced to primarily adjacent areas in the left VWFA (see Figure 26).

Rather than examining connectivity between two or three arbitrarily chosen brain regions, we studied connectivity patterns in typical compared to dyslexic readers across the *entire* brain. Based on a large group of seventy-five children, we found that compared to typical readers, in dyslexic readers brain connectivity to the word form area (critical to reading fluency) is disrupted. Connectivity is also disrupted between neural systems for reading and those for attention.

The brain's reliance on patterns of connectivity may have particular relevance to the teaching of reading, since within these systems patterns of neural connections are continually reinforced and strengthened as a result of repeated reading practice and reading experiences. We can then imagine that each time a six-year-old is able to associate a letter with a particular sound, the neural pathways responsible for making this linkage are further reinforced and even more deeply imprinted within her brain.

With these studies, the science of dyslexia has taken a major leap forward, progressing from talking about the location of isolated processors in the brain to a much more dynamic and meaningful accounting of a neural network of interconnected, interactive brain systems for reading in typical readers compared to disruptions in the connectivity in dyslexic readers.

Attention Systems in Dyslexia

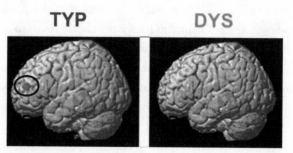

Left frontal attention systems activated in typical but not in dyslexic readers

Kovelman et al., 2012

Figure 27. Attention Systems in Dyslexia

ATTENTIONAL MECHANISMS IN READING AND DYSLEXIA

For almost two decades the central dogma in reading research has been that the generation of the phonological code from print is modular, that is, automatic, not attention-demanding and not requiring any other cognitive process. New findings, including recent imaging studies, now suggest that attentional mechanisms may also play an important role in reading and dyslexia. For example, compared to typical readers, children with dyslexia failed to recruit an area in the front of the brain (technically referred to as the left dorsolateral prefrontal cortex) during a phonological task (see Figure 27).

This region has long been known to play an important role in attention, and this study demonstrated that when asked to rhyme words they heard, typical children activated the region while children with dyslexia did not. Active work now taking place in our laboratory is examining the long-sought neurobiological relationship between reading and attention (to be discussed later).

QUICK—I WANT AN fMRI
TO DIAGNOSE MY (OR MY CHILD'S) DYSLEXIA

Think again—no you don't!

> *CAUTION*: As a scientist, I am truly excited to see the responsible neurons coming online and communicating with one another to produce reading. But this leads to the question that I am asked over and over again: "Can you image my child's brain to see if he is dyslexic?" I can even recall a physician who came to see me. "I must have my brain imaged," he blurted out with great anxiety. "The images are so compelling and I want to see what my brain looks like as I read. Do I have the usual changes seen in dyslexic people? It would be wonderful to see my brain in action and finally settle the question of my dyslexia. I have been diagnosed as dyslexic and I seem to have all the symptoms, but it still would be quite special to see my working brain read."

Yes, I agree it would be spectacular for that child or parent to be able to undergo functional brain imaging and see the source of the problem, but as a responsible neuroscientist and clinician, *I must recommend that you neither request nor undergo fMRI to diagnose your or your child's dyslexia.* This procedure is simply not ready to be used clinically. Why? In carrying out research, we image *groups* of individuals and analyze the fMRI activation patterns based on *group* results. Although we and others are working toward obtaining reliable results based on a single person, there is just too much variability within each individual and across individuals to provide a reliable interpretation based on a single person. My advice to you at this time is not to look to brain imaging to diagnose dyslexia, as tempting as it may be. Such a procedure has as much chance of providing incorrect information as it does of being correct. The good news, however, is that with the deep understanding of dyslexia that now exists, dyslexia can be reliably diagnosed clinically based on a synthesis of an individual's history, symptoms, observations of the person speaking and reading, and test measures.

PUTTING IT ALL TOGETHER

Because I've covered so much ground so far and some of it has been scientifically demanding, let me summarize what we've learned.

Reading is a code, and no matter who we are, each of us must somehow represent print as a neural code that the brain can decipher. Functional imaging makes this process transparent, allowing scientists to watch (and record) the neural systems at work as children and adults attempt to transcribe letters into sounds. For most people this process is incredibly quick and smooth; it's effortless. For others it's an entirely different story. Imaging provides the neurobiological—the physical—evidence of the difficulties dyslexic readers have in transforming this written code into the linguistic code that is the key to reading.

Seeing these images leaves no doubt that the core problem in dyslexia is phonologic—turning print into sound. Neuroscientists now have the holy grail that we have been searching for since Pringle Morgan first introduced us to Percy F.: the basic foundation for a neural explanation for the understanding of dyslexia. Now that we have the growing ability to pinpoint the complex neural networks that serve reading, our quest to understand dyslexia has taken a quantum leap forward. Having identified the neural networks engaged by reading and having pinpointed the anatomic location of the glitch in dyslexia, we are now probing deeper within these systems, examining the network of connections in typical compared to dyslexic readers.

As I said earlier, dyslexia is a "hidden disability," and skeptics have tried to explain it away. But at long last, thanks to functional imaging, dyslexic readers have the proof they have been seeking. These remarkable images provide concrete evidence of the physical reality of their reading difficulty. They also explain how adolescent and young adult dyslexic readers can read accurately but slowly. Meanwhile, it is now possible to use what we have learned about the basic phonologic weakness and growing evidence of the important role of attention, along with dyslexic readers' substantial strengths, to provide better solutions for them.

MAKING SCIENTIFIC DISCOVERIES WORK FOR YOU

The new science of reading has direct application to diagnosing and treating dyslexia: It allows us to spot early warning signs of a reading problem, to know what specific clues to look for at any age, and to understand which are the most scientifically sound approaches to teaching reading. The science happily dovetails with practical everyday needs. This knowledge can be used to answer virtually all the common questions concerning dyslexia. The possibilities for someone with dyslexia are just about limitless; the potential for success and for a happy, fulfilling life is greater than ever before. Applying all that we now know allows virtually every dyslexic child to dare to dream. The very good news is that our *knowledge gap is closing;* the frustrating news is that this has been replaced by *an action gap*—that is, we, especially schools and policymakers, are way behind in translating the existing scientific progress into policy and practice.

The greatest stumbling block preventing a dyslexic child from realizing her potential is widespread ignorance about the true nature of dyslexia. The remainder of this book is devoted to eradicating this ignorance by applying scientific knowledge to the diagnosis and management of dyslexia. We can do it. We must do it.

PART II

TRANSLATING SCIENCE INTO PRACTICE:
DIAGNOSING DYSLEXIA

8

DIAGNOSING DYSLEXIA—
WHY, WHAT, AND HOW

Given the solid science behind dyslexia, its diagnosis should be straightforward and readily obtainable. Yet this necessary first step in getting help for a struggling child (or adult) is far too often and unnecessarily elusive. Of the mountains of requests for help I receive, obtaining a valid diagnosis for a child reigns as the top concern. It need not be this way.

I want to share with you, from the broadest perspective, the newest information about how best to conceptualize dyslexia, the great value of obtaining a diagnosis of dyslexia, and how some of the sharpest dyslexic minds view dyslexia. Dyslexia will take on new meaning for you when you understand how the scientific discoveries pinpointing the origin of the difficulties it causes allow you to know which symptoms to look for and to understand why and how they cause your child's or your own difficulties. It will all fall into place as you read about the paradox of slow reading and quick thinking that is dyslexia.

Indeed, a very large and often misunderstood and overlooked segment of the dyslexic population is the child who is both bright—often gifted—and dyslexic (sometimes referred to as 2e, meaning twice exceptional). Can this be—a struggling reader and a swift thinker? Yes, it can. Learning about the latest scientific data should end any doubts about dyslexia as an *unexpected* difficulty characterized by the frequent co-occurrence of high IQ and surprisingly slow reading. For far too long, very bright children and young adults have been discriminated against in being denied a diagnosis of dyslexia and, later on, accommodations for high-stakes standardized tests. We now have the highest level of scientific evidence that should rightfully put an end to this misguided policy and support the diagnosis and accommodations that make such a difference to these boys and girls who are so full of potential. I will begin with the essentials making up the big picture or main idea of dyslexia and then fill in with more specific details.

DIAGNOSIS: THE ESSENTIAL FIRST STEP

A diagnosis is akin to finding the key to unlock a mystery. It puts the pieces together, explains so many puzzling aspects or seeming contradictions often noted in your child, and gives a direction. Once you know the diagnosis, you will know what you and educators can do to make the most positive difference. A diagnosis is specific and definitive, reflecting a synthesis of the best scientific knowledge and the characteristics of your child. Failure to receive a diagnosis tops the list of practices harmful to a dyslexic child. There should be no difficulty in obtaining one. But there are potential pitfalls.

> *CAUTION:* A great concern is the well-intentioned, albeit misinformed, recent tendency by some evaluators to list the child's performance on various neuropsychological tests but neglect to synthesize these, where appropriate, into a specific diagnosis of dyslexia.

It is important to understand the difference between a child receiving a specific diagnosis of dyslexia and getting a report that simply lists an array of your child's strengths and weaknesses. Some parents describe this experience as akin to reading a mystery novel in which the last chapter, the denouement—the resolution—has been omitted. A great many parents have asked me to look at their child's evaluation following such an approach. When I am handed a document of more than thirty pages without a diagnosis—it happens often—I feel for those parents, and I find such reports incredibly disappointing. It doesn't have to be that way.

Twelve-year-old Olivia was given a listing of the *pieces*—that is, the processes—in which she did well and those in which she struggled. When asked what she could surmise from the report, she could only regurgitate: "I have problems with auditory processing, processing speed, rapid naming, working memory, letter-word identification, pseudoword decoding, and reading fluency. Oh, and I do better in comprehension, long-term retrieval, fluid intelligence, and visual-spatial processing." Olivia was essentially saying that she had no idea what she was talking about.

But when these pieces were synthesized into a coherent whole, a diagnosis of dyslexia, her response represented notable progress: "I am dyslexic. Nothing about me has changed, only now I understand what is happening. It has a name and I know others who are also dyslexic." Olivia's relief was evident; she felt so much better. *A diagnosis of dyslexia puts all the pieces together so they are not floating, unattached fragments. They are coherently joined together in a pattern that makes sense and leads to a road map—an effective plan of action.*

Reliable scientific and clinical information in support of the diagnosis is vital. It will make clear what the child is experiencing—which brain systems are involved—and what effective treatment will look like. The diagnosis also allows the child to know she is not alone, as was the case with Olivia—that she is a member of a community of boys and girls who are dyslexic. Right before a patient of mine, Evan, popped the question to his girlfriend, Emma, he looked her directly in the eye, cleared his throat, and said, "Before we go any further, there is something about me you should know." "What's that?" asked Emma. "I'm dyslexic," Evan said. "Oh," laughed his fiancée-to-be, "don't worry about that. I'm dyslexic too."

A diagnosis also brings together the various adults who interact and make ongoing decisions impacting the child's daily life and her future. Having the symptoms coalesce into a meaningful diagnosis that has a name and explanatory power also provides a common language that facilitates communication with specialists in disparate disciplines who have an impact on the life of a child who is dyslexic.

Skye Lucas is a wonderfully happy and successful young woman now. However, when she was just starting school, she simply could not read. As Skye tells it,

> The thing that was really difficult for me was that I didn't realize what I had. So I would always think I was dumb, and it was tough, because kids in the class would always read about five or more books each week and I couldn't even get through the first chapter of any of them. I struggled a lot.

According to Skye's mom, Geralyn, "She seemed very anxious and then she would call me from the nurse's office and say she had a stomachache." At one point Skye said, "My brain is not working. I don't know what's going on." "And I think that when I saw Skye and how depressed she was," says Geralyn, "and how she thought she was dumb, and she didn't believe in herself, I just thought I was losing her at age six."

"I went to talk to the principal. I thought it would help me," says Skye. "And I asked him, 'Am I dumb? Is this not the right place for me?'"

With the powerful, indisputable scientific knowledge available today, no child should have to experience that terrifying and lonely feeling of isolation that comes with not knowing why you are suffering and why reading is so difficult for you. For many children, knowing you are not stupid—rather, like Skye, are quite smart—can be a game-changer.

DRILLING DOWN: WHY DYSLEXIA AND NOT A LEARNING DISABILITY

The terms *learning disability, dyslexia,* and *unexpected* have a strong relationship, but it is important to discuss their significant differences. But first let me offer a sobering reminder.

> *CAUTION:* For parents of a dyslexic child and for dyslexic young men and women, beware of such innocuous-sounding but essentially uninformative and unhelpful terms as *learning differences* and *learning disability*. A dyslexic individual learns little from this kind of label. Self-awareness is certainly not strengthened by these terms, which are vague and lack specificity.

What are the specific symptoms of a learning disability? What is the difficulty, what is the origin of the difficulty, and where in the brain does it reside? What are the proven treatments and what happens over time? The answer to each of these questions for terms such as *learning differences* and *learning disability* is unknown. These terms lack all the benefits of empowerment, knowledge, and specific direction provided by a precise diagnosis of dyslexia. When something can't be precisely defined, it often raises anxiety, whereas knowing the name and all the scientific knowledge that goes with a diagnosis of dyslexia is reassuring and empowering.

Earlier you read about Dr. Pringle Morgan's initial scientific report in

1896 of an unexpected difficulty in reading (developmental dyslexia), followed by many reports of similar cases of unexpected reading difficulties in children originating in Britain, Europe, and the United States. As I noted earlier, Scottish ophthalmologist James Hinshelwood reported many cases in which he emphasized the *unexpected* nature of the reading difficulty; it was due not to a generalized deficit in intelligence but to a "localized" problem affecting reading. Indeed, Dr. Hinshelwood pointed out that such children had strengths as well as a reading difficulty.

By the early twentieth century, clinicians had elaborated descriptions, possible causes, and potential treatments for developmental dyslexia. Perhaps the most cogent description came from the American physician Samuel Torrey Orton. In his 1937 monograph, *Reading, Writing and Speech Problems in Children,* Dr. Orton noted a number of characteristics of dyslexia, many of which remain true today. These include the observation that though some dyslexic individuals may learn to read, their reading remains slow compared to that of their peers in age and intelligence, a phenomenon we have come to refer to as a lack of reading fluency. He further observed that while their reading often could appear appropriate for their age, it was below what was expected for their intelligence.

Around this time, other astute clinicians were beginning to elaborate on what came to be known as the brain-injured child, which by 1962 had evolved into what American psychologist Sam Clements and American psychiatrist John Peters described as Minimal Brain Dysfunctions (MBD). The first finding mentioned was specific learning deficits, and the first symptoms noted were failure to read and poor spelling. In the same year American educators Samuel Kirk and Barbara Bateman first used the term *learning disability* (LD) and illustrated the evaluation of a learning disability with the steps in diagnosing a reading disability. Kirk and Bateman noted that the very first step in determining if the child has a learning disability is determining the child's reading *capacity*, a measure reflecting IQ. Then actual reading achievement is determined, and the "discrepancy between the capacity for reading and actual achievement in reading is examined." Prominent American psychologist Helmer Myklebust emphasized the hallmark of learning disability as a discrepancy between potential and actual success in learning resulting in significant underachievement.

In an effort to clarify the relationship between learning disability and dyslexia, Bennett and I wrote on our Yale Center for Dyslexia & Creativity website: "Learning disabilities, as described in the Individuals with Disabilities Education Act (IDEA), refer, in part, to such symptoms as 'imperfect ability . . . to think' which not only do not at all apply to dyslexia but could even be harmful to those who are dyslexic. *Dyslexics have a reading impairment, not a thinking impairment.* This is one reason why it is important to differentiate dyslexia from the general and more heterogeneous 'learning disabilities.' "

In September 2014, when I testified before the House Committee on Science, Space and Technology on the "Science of Dyslexia," I again addressed the distinction between dyslexia and learning disabilities.

> *Dyslexia is specific; learning disabilities are general.* Dyslexia differs markedly from all other learning disabilities. Dyslexia is very specific and scientifically validated: We know its prevalence, cognitive and neurobiological origins, symptoms, and effective, evidence-based interventions. *Learning disabilities* is a general term referring to a range of difficulties which have not yet been delineated or scientifically validated. Learning disabilities are comparable to what in medicine are referred to as "infectious" diseases [that is, learning disabilities are more like the general term i*nfection,* which can have many different causes along with many different symptoms and differing treatments depending on the specific infection], while dyslexia is akin to being diagnosed with a strep throat—a highly specific disorder in which the causative agent and evidence-based treatment are both known and validated.

I concluded my testimony with a plea "to bring education together with current scientific knowledge," followed by a list of specific recommendations, beginning with my number-one recommendation: "First and foremost, schools must not be allowed to ignore, fail to recognize, or deny the reality or diagnosis of dyslexia." A diagnosis doesn't make the child dyslexic; it clears the fog and reveals what is already present.

As I said earlier, school personnel often refuse to acknowledge or diagnose dyslexia. Typically, this reflects lack of knowledge—actually, ignorance—about dyslexia. At the same time, it may well be the child's parent who, for very different reasons, is in denial and avoids obtaining a firm diagnosis. Just as a school's failure to provide a diagnosis may do harm, a parent's fear can have a long-lasting negative impact. For all the moms and dads reading this, my message is, *Don't be afraid.* Perhaps the most common source of fear and anxiety is the unknown, and that is far too often the case with dyslexia. My insightful friend Karen refers to dyslexia as "the disorder that can't be named," sarcastically but sympathetically referring to parents' misplaced fears.

It's important to remember that if your child is dyslexic, it's going to impact her whether it is diagnosed or not, causing havoc in her life. Not identifying a problem does not make it go away. Specifically, diagnosing dyslexia is the necessary first step to taming the disorder and successfully managing it. In my decades of experience working with scores of parents and children, I do not know of one parent or child who regretted obtaining an accurate diagnosis of dyslexia.

Please remember as well that the diagnosis reveals true intelligence, a morale booster. Let your child in on the news. Yes, your child may struggle to decipher written words, but at the same time she is also likely to be an out-of-the-box, creative thinker. She will be happy to hear that.

UNLOCKING THE MYSTERY OF DYSLEXIA: WHAT IT IS AND HOW IT IMPACTS YOU

A major reason a diagnosis of dyslexia is able to pull together and make sense of what may at first blush appear to be a random array of symptoms is that science has uncovered the origin of the difficulty. As I said, the basic deficit in dyslexia is a problem getting to the individual sounds of spoken words. This scientific knowledge allows the clinician, educator, and especially concerned parents to recognize the symptoms of dyslexia—to know in a tangible way the *why, what,* and *how* of dyslexia. This allows a reliable diagnosis at the earliest possible moment. Recognition and diagnosis of dyslexia is clearly the essential first step in the development of an effective plan to support and treat the difficulty and to unburden both the child and his parents, as well as the teacher involved, from what is often a confusing and frustrating situation.

Once the mystery of the *why* is solved, confusion and apprehension are replaced by logic and confidence. Knowing the source, the *what,* the origins of what your child is experiencing is empowering. Once you know, you are in charge of determining what is best for your child. You are an active player, not merely a passive observer. A parent is the child's number-one champion, and once a diagnosis is made you have the critical information required to go forward with determination. As demonstrated in Figure 28, once you know the diagnosis is dyslexia, you also know the origin of the difficulties and how to deal with them to best help your child. Working from a diagnostic

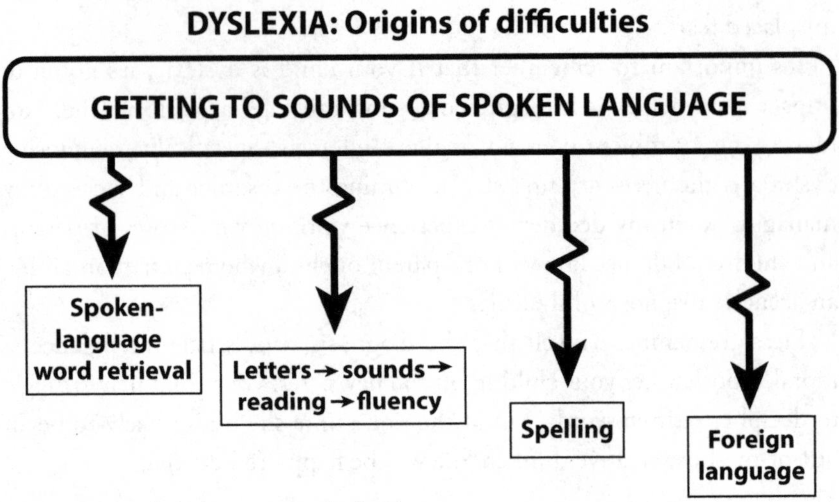

DYSLEXIA: Origins of difficulties

```
┌─────────────────────────────────────────────────┐
│        GETTING TO SOUNDS OF SPOKEN LANGUAGE       │
└─────────────────────────────────────────────────┘
```

Spoken-language word retrieval

Letters → sounds → reading → fluency

Spelling

Foreign language

Figure 28. Dyslexia: Origins of Difficulties

model makes so much sense, clarifying just what dyslexia is and why and what characteristics are associated with the condition.

As shown in this figure, if you have difficulties getting to the individual sounds of spoken language, you will likely have problems with

→ spoken language itself, both delayed language early on and word-retrieval difficulties later on.
→ attaching letters in a word to the sounds they represent.
→ decoding/reading difficulties, impacting both accuracy and fluency.
→ encoding difficulties, that is, transforming sounds into letters, impacting spelling.
→ learning the sound system of a foreign language.

Knowledge of the underlying deficit in dyslexia translates into a very useful clinical tool that brings reason and understanding to the diagnosis, an important step in the development of targeted, focused, and, most critically, successful school, home, and personal intervention plans. Once you understand that the diagnosis is dyslexia, you know that you are dealing with a problem that affects multiple areas, including speaking, spelling, writing, and learning a second language in addition to reading.

The importance of this awareness was brought home to me when I heard from Harold, whose son, Eli, was never diagnosed but was placed in a reading program in second grade that produced some modest results. At the time Harold was pleased that his son didn't have to go through an assessment process. Moreover, Eli's teacher seemed to be pleased with Eli's performance. However, today Harold very much regrets the lack of a diagnosis for his son and not having had the benefit of the knowledge that goes with the diagnosis. Now fourteen years old and in ninth grade, Eli "can't spell very well, hates writing, and is already frustrated in trying to learn a foreign language in school. How much frustration we would have been spared had Eli been diagnosed as dyslexic when he was in second grade and received the evidence-based interventions he required and the knowledge of who he is that goes with the diagnosis."

Harold's experience carries an important message for all parents: Don't settle for some limited support for your child. A true diagnosis is essential.

DYSLEXIA IS A PARADOX: SLOW READING, FAST THINKING

One of the most essential features to know about dyslexia is its paradoxical nature. Slow reading does not imply slow thinking. People often improperly confuse the two, a false notion that may limit the aspirations and options of a dyslexic child. Such a misconception harms adults, too. For instance, I have heard people say, "Oh, I would never want a dyslexic doctor to care for me when I am taken in to the emergency room."

Dr. Karen Santucci, professor of pediatrics and director of the very busy, fast-paced, and world-class emergency department at Yale Children's Hospital—and dyslexic—refutes that:

> It's not like you have to sit down with a textbook; rather, hearing things, seeing things, experiencing things, taking care of patients, listening to patients, identifying their signs and symptoms, putting the pieces of information together, I can assimilate that rapidly and without a moment's hesitation. I may not be able to read very quickly, but I can process the information quickly.

(If your doctor has to sit and read about your symptoms in a book when you are taken to the ER in the midst of a heart attack, blue and gasping for air, you are in deep trouble!)

Dyslexia is a true paradox, with weaknesses in concrete observable tasks like reading and spelling, which are noticed in the classroom and result in the student's being criticized. That same dyslexic student, however, also has significant strengths in big-picture thinking and reasoning—highly advantageous attributes that are not as easily noticed and are often overlooked. Should his or her life be defined by: high thinking ability and excellent reasoning or slow reading and poor spelling?

In an attempt to correct the prevalent misunderstanding of dyslexia, I have paired the symptoms of dyslexia to illustrate its paradoxical nature. Our conceptualization of dyslexia brings to life our Sea of Strengths model, discussed earlier, and how it expresses itself. I have placed this synthesis of dyslexia on a four-by-six yellow card titled, "ARE YOU A SLOW READER, AN OUT-OF-THE-BOX THINKER? COULD YOU BE DYSLEXIC?" I take this card with me whenever I travel and often share it with the person next to me on a plane or train. I continue to be amazed at the reactions I get: "Oh, I never knew that this is what dyslexia is," or, "Are you sure? Isn't dyslexia about reversing letters?" (No, it is not.) Or, "I can't believe it. I always thought my son might be dyslexic, yet this describes *me* to a T. I want to say, *How did you know? This is me. I am dyslexic. I never knew it.* This explains so much."

These pairings, a *weakness* and along with it a strength, capture the paradox of dyslexia: slow reading and creativity, two sides of the same coin.

YOU MAY BE DYSLEXIC IF YOU . . .
. . . *Read slowly and with much effort.*
. . . Are often the one to solve the problem.
. . . *Can't spell; have messy handwriting.*
. . . Show terrific imagination in your writing.
. . . *Have trouble remembering dates and names.*
. . . Think out-of-the-box; grasp the big picture.
. . . *Have difficulty retrieving and pronouncing spoken words.*
. . . Have excellent vocabulary and ideas.

(© S. Shaywitz, 2011)

From a more personal perspective, here is how two of the brightest, most insightful dyslexic minds have described this paradox:

Figure 29. David Boies, renowned attorney

Most annoying is when people equate dyslexia with a thinking disability. Dyslexia gives you the ability to see the entire picture and step back from it, and think. Because reading is hard [for a dyslexic], it forces you to rely more on thinking, and as you get out into the world it is thinking that is a lot more prized than reading. [David Boies]

Figure 30. Charles R. Schwab, groundbreaking financier

I always had great strengths in thinking. . . . Even though I couldn't read quickly, I could imagine things much faster than people who were stuck thinking sequentially. That helped in solving complex business problems. I could visualize how things would look at the end of the tunnel. . . . I intuitively get there much more quickly. [Charles R. Schwab]

That it is unexpected for a person who is highly intelligent to read at a much lower level defines a major aspect of dyslexia. It answers the question, can a very bright, even gifted child be dyslexic? Far too often parents are told, "Your son is too smart to be dyslexic," or "Her reading score is too high for her to be dyslexic." Without question, a highly intelligent person can be dyslexic.

THE TWENTY-FIRST-CENTURY DEFINITION OF DYSLEXIA

It has taken more than 120 years since dyslexia was first described, but we finally have a definition that is universal and incorporates all of our up-to-date knowledge: the science, the paradox, the educational implications, and the importance of a diagnosis for school and for personal life satisfaction. Given the importance of the definition of dyslexia to its diagnosis, a critical step has been its codification into federal law.

In this area, Louisiana senator Bill Cassidy's dedication and perseverance have been vitally important. The devoted father of a child who is dyslexic, Cassidy understands the impact of dyslexia along with the often overwhelming anxiety and worries of parents who are waiting for their child's school to provide their child with the benefits of a specific diagnosis. He is a passionate, unrelenting advocate for children who are dyslexic. Recently he spearheaded the inclusion of the scientific definition of dyslexia into the First Step Act, which was passed overwhelmingly by both houses of Congress and when signed into law by President Trump on December 21, 2018, became Public Law 115-391. The law reads: "The term 'dyslexia' means an unexpected difficulty in reading for an individual who has the intelligence to be a much better reader, most commonly caused by a difficulty in the phonological processing (the appreciation of the individual sounds of spoken language), which affects the ability of an individual to speak, read, and spell."

This definition applies to all those boys and girls, men and women affected throughout the world. More and more countries are recognizing their citizens who are dyslexic. I have been both surprised and gratified that the first edition of *Overcoming Dyslexia* has now been translated not only into the usual alphabetic languages, such as Portuguese, Dutch, Croatian, and Polish, but also, as I noted earlier, into Japanese, Chinese (Mandarin; see Figure 31), and Korean.

Japanese

Chinese

Figure 31. Dyslexia Impacts Logographic as Well as Alphabetic Languages

HOW DYSLEXIA IMPACTS BRIGHT CHILDREN

Especially for bright children and adults who struggle to read and for whom reading is a slow, laborious process, the concept of dyslexia as an unexpected difficulty has great relevance. These bright, often gifted individuals, through extremely hard work, much practice, and resilience, often learn how to read relatively accurately but almost never automatically. Although they are able to read or make good guesses at what a word is, reading remains a slow and tiring process. That said, I have come across quite a number of dyslexic men and women who are relatively fluent readers of materials in their area of expertise, such as those in finance who are daily readers of *The Wall Street Journal* or those in a particular area of medicine who regularly read articles in their specialty, including cardiology, psychiatry, and anesthesiology. The individual's focused interest in a topic draws her to reading and frequently rereading materials, which leads to a relative degree of fluency in that specific area.

What is a student to do when she knows how difficult reading is for her, how much slower she reads than her peers, and how often she tries to rush in her reading only to find that she has to reread the material? What is she to do when the disparity (often referred to as a discrepancy) between her intelligence and reading is disregarded by the school or testing agency? Now, finally, powerful new evidence validates the disparity between intelligence and reading as a hallmark of dyslexia.

As noted earlier, I am the principal investigator of the Connecticut Longitudinal Study, which has followed a large random sample of schoolchildren from kindergarten entry through to adulthood. While in school, each child was administered an individual IQ test, the Wechsler scales, in alternate years, and a Woodcock-Johnson reading test yearly. I was eager to examine the relationship between reading and intelligence in this group of students.

In our analyses, focusing on *typical* readers, we found that reading and intelligence are dynamically linked, meaning that not only do IQ and reading travel together over time, they influence each other as well. A child's IQ one year will influence his reading score the following year, while his reading score one year will influence his IQ in a subsequent year. Not surprisingly, and conforming to what most people would anticipate, if a typical person is very bright, that individual will most probably be a very good reader, and if someone is a very good reader, he will most likely also be quite intelligent.

In planning and carrying out experiments, a scientist proposes a theory (technically, a hypothesis), plans a series of studies to test this hypothesis, and analyzes the resulting data. In examining the data from the Connecticut study, we were thrilled by what these same analyses showed when they were applied to *dyslexic* readers. These data (see Figure 32) indicate that *in the case of dyslexia* (right panel), *IQ and reading are not linked at all.* The data show that dyslexics can have a very high IQ but still struggle to read; that is, they can think quickly and read slowly. To many of us, this should be self-evident, but as a researcher, I find empirical findings that validate logical but often questioned associations are necessary, and in this case rewarding. These data provide the long-sought-after empirical validation of the *unexpected* nature of dyslexia.

From a practical perspective, these findings are of great value to bright dyslexic students who require interventions and/or accommodations. They provide the critical evidence supporting the validity of a disparity (discrepancy) between intelligence and reading achievement scores as a core component of the definition of dyslexia. Figure 32, along with the definition of dyslexia in the recent First Step Act, will come in handy if you request intervention or accommodations and you are told your child is too bright.

It is not unusual for struggling dyslexic readers to come to doubt, or at least to question, their own intelligence. Seeing the information in Figure 32 should have a powerful and positive effect on them, reaffirming their

Typical   Dyslexia
IQ-Reading Linked   IQ-Reading Diverge

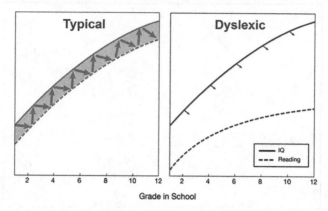

Scientific Validation of "Unexpected"
Ferrer, Shaywitz, et al., 2010

Figure 32. Unexpected Nature of Dyslexia

intelligence and going a long way to begin to repair any bruises to their self-concept. Every evaluator should take the time to visibly demonstrate to every tested child (I like to do this on a bell-shaped curve) where his or her reading lies and then where on this curve his or her intelligence score lies. When this information comes from an objective authority figure rather than the student's parents, it makes a powerful impact.

I can recall a very sweet, somewhat shy young man named Rob, who was dyslexic and about to enter college. He expressed great concern about his ability to succeed in what seemed like a very demanding new setting. He expressed feelings of inadequacy, of not being smart enough to succeed, yet he tested at the very superior range of intelligence, higher than ninety out of 100 peers who took the test. I placed in front of him a classic bell-shaped curve drawn on a piece of white paper, representing the distribution of IQ values in the general population. I pointed out to Rob that, as in Figure 33, the raised midpoint indicates the average IQ score; the left extreme end represents the very lowest IQ scores and the right extreme end the very highest IQ scores. I then asked him to place an *X* where he thought his IQ fell on the curve. As expected, Rob's line was to the left of the midpoint, slightly below average. "Well," I said to him, "would you like me to take the pencil and make a mark where your IQ really tested?" He mumbled a somewhat reluctant "Yeah."

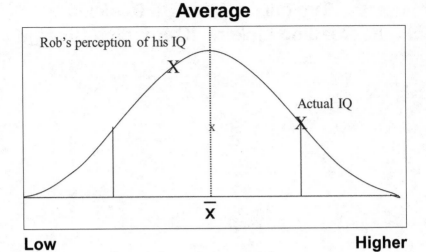

Figure 33. Difference Between Student's Judgment of His Intelligence and His Score on an Intelligence Test

When I put my pen down close to the extreme right side of the curve, indicating his exceptionally high score, he scowled as he uttered, "You're kidding me, right?" Rob was exceptionally bright. In previous assessments, no one had actually sat down with him and explained his performance and how very intelligent he was. As his dad said, "This changed Rob's life, a real turnaround for him. No one had ever told him he was really smart." He had based his self-evaluation on the reading struggles he had experienced over the years, on the criticism he had gotten from many of his teachers, and on his belief that reading and intelligence were linked.

The published empirical demonstration of a disparity between intelligence and reading has significant implications for the diagnosis of dyslexia.

→ First, it means that an IQ test is integral to the definition of and to the assessment for dyslexia. The results of the student's performance by this measurement are necessary information, especially for a very bright student.

→ Many schools try to take shortcuts and omit this important test as part of the evaluation of a student for dyslexia. Although different reasons are cited for this omission, I believe it usually reflects financial constraints. Such an omission is a disservice to the child and improper for the school that has the responsibility for his education.

→ Simply relying on reading tests alone is in essence placing a ceiling on a child's reading. Educators, especially those who are charged with overseeing a child's education in the fullest sense of the term—providing academic, social, and moral instruction—must not prevent their students from knowing fully and from receiving an accurate and helpful diagnosis that is linked to effective instruction.

CAUTION: It is important for you to be aware of several key pieces of information in considering the findings of your child's (or your own) IQ test. In the Appendix I focus on the most widely used measures of intelligence.

Awareness that you are dyslexic and that it is an unexpected difficulty can be a happy surprise at any age or stage of life. Let me introduce you to Pat, a highly patriotic and philanthropic citizen who because of her contributions reflecting both of these characteristics had been nominated for a high position serving our government. I met Pat a few years back and was pleased to speak to her when she called. As is often the case when people call me, I thought she was calling to ask me about one of her children or perhaps a grandchild. No, Pat was calling to ask me about herself. She told me she was unsure if she would or should accept this honor, because, she said, "I don't read very well. Most people, aside from my inner family, don't know this, and I don't want to embarrass our country. I worry that my poor reading will be problematic, especially if I have to read a speech in public. I would not be up to it." Could she come to New Haven and be evaluated for dyslexia? she asked. I said, *Of course.*

After she was evaluated, I told Pat she was, indeed, dyslexic—and incredibly smart. "How do you know?" she asked me. "Did you give me one of those dumbed-down tests for dyslexics?" "You were given the Wechsler Adult Intelligence Scales (WAIS), the gold standard IQ test," I told her. "The WAIS has no reading on it, it's all about thinking and not reading."

"But we give tests to our employees, you know," Pat interjected, "to help see which have the ability to move forward, to contribute to the company. These tests have lots of reading."

"Yes, Pat," I said. "The irony is that in your own company you are relying on tests that must be read, and in the process are missing perhaps the brightest men and women, who may not be great readers but are highly creative thinkers and problem-solvers. Ironically, your company would be missing a person like you."

Pat breathed a sigh of relief. She knew for certain for the first time that she was not fooling everyone. She was a truly smart woman, capable of making complex decisions, not in spite of her being dyslexic but because of it. Pat accepted and served with honor in the position she was offered. Bennett and I were invited to her swearing-in ceremony in Washington, DC. When it came my turn on the receiving line to shake her hand, she instead reached out and brought me close to her with a heartfelt hug. There were tears streaming down both our faces. "Thank you," she whispered, "for introducing me to who I am and for giving me the confidence and courage to accept this position." Later, at the reception, several people came up to Bennett and me, asking, "You received such a warm hug, one that seemed so very special. Are you family whom we have never met?"

"No," I responded, "just very good friends."

SPECIAL BUT NOT RARE SITUATIONS INFLUENCING THE DIAGNOSIS OF DYSLEXIA

A woman once said to me, "My grandson is very bright. He reads in what we have been told is the average range. Does this mean he cannot be dyslexic?" First it needs to be said that the so-called average range covers a wide territory. For example, a child might score as low as the ninth percentile (lower than 91 percent of the test-takers of the same age; if one hundred students took the test, ninety-one would have scored higher) and yet still be considered by some testing agencies and schools as performing in the average (albeit low average) range. Often when a student's reading is in that range and he scores well in reading comprehension, he is not properly diagnosed as dyslexic, which prevents much-needed effective intervention and accommodations. Reading in the average range should not be used to deny a diagnosis of dyslexia or to withhold needed services.

> → Reading scores of bright dyslexic boys and girls may be in the average range, but the *manner in which they read, the effort required, and the lack of automaticity observed as slow reading is not average.*

→ What matters most in diagnosing dyslexia is what the person must do to read—*the energy he expends in deciphering the printed word.* Legally and scientifically, no ceiling is placed on what level of reading is consistent with a diagnosis of dyslexia.*

→ Bright students can have a reading score in the average range and still be considered, and indeed are, dyslexic.

There is also a surprising relationship between dyslexia and reading comprehension. Interestingly, when reading connected text (sentences, paragraphs, chapters), bright dyslexics, by using their high level of reasoning ability, often excellent vocabulary, and strong general knowledge, are frequently able to glean the gist or meaning of the material read, even though they may misread words here and there. So it should come as no surprise that in a paper published several years ago, we reported that *at young adulthood there was no significant difference in reading comprehension scores in groups of students who were dyslexic compared to those who were typical readers.* The study found that while these dyslexic young men and women continued to struggle with reading individual words either accurately or rapidly, they were able to *comprehend* connected text better than they were able to read single words. The finding that a person comprehends what she reads does not negate the fact that she may be dyslexic and as a result has to exert a great amount of effort to read. Importantly, the reading comprehension tests were not timed.

I can think of no better example of a slow reader who is an out-of-the-box creative thinker than the brilliant and dyslexic filmmaker Steven Spielberg, who in an interview shared that his dyslexia still affects him. For example, it may take him longer than others to read a script, but his comprehension is superior to that of most others, including those who may have whizzed through the script.

→ *In dyslexia, reading comprehension is often superior to the accuracy or speed of reading words.*

→ *Reading comprehension is a reflection of more than a child's ability to decode. It also reflects his vocabulary, background*

---

* See page 471 for discussion of the newest (2016) interpretation, referred to as the Final Rule, of the Americans with Disabilities Amendments Act (ADAA).

knowledge, and reasoning. Even with less-than-stellar decoding ability, a child who has a large storehouse of words, is aware of the world around him, and possesses strong reasoning powers can be an excellent comprehender of what he reads.

→ *Given sufficient time, a dyslexic individual has the necessary reasoning ability, vocabulary, and knowledge to ably comprehend what he is reading. These well-functioning higher cognitive abilities are what make extra time on exams able to reveal a student's ability rather than reflect his disability.*

"I am not sure how much slower I am in reading than the normal person," says David Boies. "I'm noticeably slower—maybe half as fast. I absorb a lot by going slowly. And so when I'm reading for comprehension, even though I may read at a pace that's one-half or one-third the pace that other people may be reading at, at the end of that period I will have learned a lot—and absorbed a lot more than many of them [the fast readers] will."

CAUTION: There are myths about dyslexia that need to be shattered. Among them:

Dyslexia is a medical, not an educational, diagnosis. While dyslexia is neurobiologically based, it brings with it significant educational implications. Dyslexia originates from the brain, as does every other learning ability or disability. Fortunately, informed educators and parents know what symptoms to look for, what questions to ask, and what assessment measures to administer or to request.

Only neuropsychologists can administer a test of intelligence (an IQ test), such as the Wechsler Intelligence Scale for Children, fifth edition (WISC-V). According to Pearson, the company that developed and distributes the WISC-V, qualified individuals must have a "doctorate degree in psychology, education, or closely related field . . . or licensure or certification to practice in this state in a field related to the purchase . . . or certification by, or full active membership in, a professional organization that requires training and experience in the relevant area of assessment."

Only a neuropsychologist can make a diagnosis of dyslexia. The diagnosis of dyslexia can be made by a professional with a deep understanding of dyslexia, including an educator, a school psychologist, a physician, and a speech and language therapist. Dyslexia is a clinical diagnosis based on the individual's history, observation of him speaking and reading aloud, and test scores, as I said earlier.

It is important to know that you do not need to be the person who administers each test in order to be the one to make a diagnosis of dyslexia. In a similar way, if I or anyone else knowledgeable about dyslexia sees a child who may be dyslexic, I will take a history, listen to him speaking and reading aloud, and order a test of intelligence along with other relevant measures of reading. The same approach can be and is taken by the professionals I noted above.

At this point you have a broad overview of the essentials of dyslexia and know that an extremely intelligent child can also be dyslexic. You understand the rationale for and critical importance of obtaining a diagnosis and are knowledgeable about the key elements of a diagnosis of dyslexia. You are also fully equipped not to be fooled by claims often made about dyslexia by some schools in denying a diagnosis to deserving students.

In the discussion that follows you will find specific, detailed information that offers a closer, more practical, and more personal guide to the diagnosis of dyslexia.

When a dyslexic child enters school, there are two potential paths he can take, one leading to a life of satisfaction and self-fulfillment, the other leading to a dispirited feeling of "what might have been," of lost potential and missed opportunities. These possibilities are shown in Figure 34 on page 110, illustrating the different pathways and choices before a young child and their consequences. For example, to the left is a child who is dyslexic but does not receive a diagnosis, and as a result does not receive evidence-based interventions (EBI) or accommodations. This failure to diagnose dyslexia often leads to school failure, unrealized potential, and low self-esteem—a tremendous loss to the child, to his family, and to society. In contrast, the child who receives a diagnosis of dyslexia is on the pathway to receiving EBI and accommodations leading to academic success, self-knowledge, empowerment, and fulfillment. The child has a far better chance of realizing his potential and contributing to society.

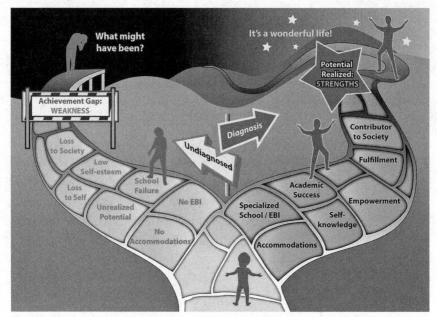

Figure 34. Dyslexia: Which Road Taken?

9

RECOGNIZING DYSLEXIA'S EARLY IMPACT: SPOKEN LANGUAGE

Now that you have a clear understanding of the scientific roots of dyslexia, I want to show you how to translate this knowledge into practice, beginning with how to detect the earliest clues to dyslexia. I believe the most helpful guide to accurate identification and effective treatment of dyslexia comes from Bennett's and my Sea of Strengths model: an isolated weakness in getting to the sounds of words surrounded by an array of higher-level strengths in thinking and reasoning. Using this simple model, many parents and teachers have become experts at recognizing the telltale signs of dyslexia, and you can, too. To begin, I want to describe exactly how a phonologic weakness impacts first spoken language and then written language.

Applying our model, the essential ingredients of a successful program for the dyslexic child are

→ *identifying* a weakness in getting to the sounds of words and strengths in thinking and reasoning, and then

→ *providing* early help for the weakness—self-awareness of how dyslexia impacts that individual and accommodations to help access the strengths.

I cannot emphasize enough the importance of focusing on the strengths as well as the weaknesses. The goal is to ensure that it is these strengths, and not the weaknesses, that define a child's life. This is how we apply the Sea of Strengths model to recognizing signs of dyslexia in our children.

LISTEN TO YOUR CHILD SPEAK

You will discover the earliest and perhaps the most important clues to a potential reading problem by *listening* to your child speak. Listening for subtle signs that your child's phonologic skills are not developing as they should is a lot easier than it sounds. That's because fuzzy phonemes leave their distinct mark on everyday language in highly predictable ways, influencing how well your child is able to pronounce certain words, sing nursery rhymes, and eventually learn the names and sounds of the letters of the alphabet.

→ The very first clue to dyslexia may be *a delay in speaking*. As a general rule, children say their first words at about one year and phrases by eighteen months to two years. Children who may be dyslexic might not begin saying their first words until fifteen months or so and may not speak in phrases until after their second birthday.

CAUTION: A seemingly innocent speech delay may be a very early warning signal of a future reading problem. Then again, some dyslexic children may not demonstrate a speech delay, or at least it may be subtle and go unnoticed.

→ Once a child begins to speak, *difficulties in pronunciation*— sometimes referred to as "baby talk"—that continue past the usual time may be another early warning. By five or six years of age, a child should have little problem saying most words correctly. Attempts to pronounce a new word for the first time or to say a long or complicated word can reveal problems with articulation. It is as if there were a jam in the articulatory machinery churning out spoken language and the phonemes were literally tripping over one another as they come out of the child's mouth. Typical mispronunciations involve either leaving off beginning sounds (such as saying *pisgetti* for *spaghetti, lephant* for *elephant*) or inverting the sounds within a word (*aminal* for *animal*). As part of one of our research studies, a group of six-year-old children with early signs of language difficulties were asked to repeat a series of words. Looking at the words and their pronunciations, you could just feel the phonologic mayhem that was occurring as each child tried to gather the appropriate phonemes and spew them out in some perceived order: physicist → *sysitist;* specific → *pacific, spestickit, pacissic, spastific.*

→ Even older children and adults are not immune to tripping over phonologic debris, as the medical student Gregory came to know as he tried to pronounce the long (anatomical) names of body parts, or as any dyslexic soon discovers as she attempts to say unfamiliar or complicated words. Gregory was having difficulties with word retrieval, as you read earlier, a common symptom of dyslexia.

Most preschoolers love to play games with sounds and with rhyming words, and indeed, much of the humor in books geared to this age level exploits the young child's fascination with rhyme. Three- and four-year-olds derive great glee from hearing and repeating sounds: for example, *Peter, Peter, pumpkin eater* or *Hickory, dickory dock, the mouse ran up the clock.*

→ Sensitivity to rhyme implies an awareness that words can be broken down into smaller segments of sound and that different words may have a sound in common; it is a very early indicator of getting ready to read. *Dyslexic children, on the other hand, have trouble penetrating the sound structure of words and pulling words apart. As a result, they are less sensitive to rhyme.* Children's familiarity with nursery rhymes turns out to be a strong predictor of their later success in reading. In England, researchers asked three- and four-year-olds, "Can you say [for example] 'Humpty-Dumpty' for me?" Regardless of intelligence or family circumstances, children who were the most familiar with nursery rhymes were the top readers three years later.

→ At age four, parents may notice that their son is still not able to recite popular nursery rhymes. He may *confuse words that sound alike* or may not even appreciate that certain words rhyme.

→ By age six or so, when most boys and girls are able to judge if two words rhyme, dyslexic children may *still not be able to demonstrate that they detect rhyme.* For example, when asked, "Do *food* and *foam* rhyme?" or "What about *talk* and *walk*?" dyslexic children are often unable to tell you which word pair rhymes. They are unable to focus on just one part of a word—in this example, *alk.*

The British researchers also compared a group of children with reading difficulties to a group of younger children who read well. Each child was asked to listen to a list of words such as *fun, pin, bun,* and *gun* and then to select which word did not belong. Here the reading-disabled child's troubles with rhyme became apparent. Although several years older, the poor readers experienced much more trouble choosing which word in the list did not rhyme. It is not a matter of maturation, just *an insensitivity to the sound structure of language.*

→ Not infrequently, *an incorrect phoneme is accessed.* As I mentioned earlier, a child like Amy looks at a picture of a volcano (which she has seen many times) and the word she pulls up is *tornado*—close in sound but not in meaning.

→ Because of difficulties in accessing the intended phonemes, a child may *talk around a word,* as Alex did when he could not retrieve the word *ocean,* or he may begin to say a word and end up with a series of *um*'s or "Um, um, um . . . I forgot."

→ Such children often *point instead of speaking* or *become tearful or angry* as they become increasingly frustrated at being unable to utter the word they have in mind.

→ As a child gets older, she may resort to *using words lacking in precision or specificity* to cover up her retrieval difficulties, such as constantly using vague words like *stuff* and *things* instead of the actual name of the object. Sometimes it may be hard to follow the conversation of a dyslexic because her sentences are so filled with pronouns or words lacking in specificity: "You know, I went and picked up the stuff and took it there. The things were all mixed up, but I got the stuff anyway." *It is important to remember that the problem is with word retrieval and not with thinking.* She knows exactly what she wants to say; the difficulty is with pulling the right word out. These word-retrieval language difficulties continue as the child matures. She *may be quiet, appear inarticulate, or in general experience difficulties expressing herself.* This is a frequent pattern, but it is not invariable, and in fact some dyslexic children may be quite articulate or fluent when they speak.

As with Alex and Gregory, the ideas are often advanced; it is the need to produce a word on command that is problematic for the dyslexic child. Parents and teachers may become irritated with a child because he seems so bright and they cannot understand why he spits out the wrong word. As a dyslexic child matures into adulthood, his speech continues to show evidence of his difficulties getting to the sound structure of words. *His speech is littered with hesitations;* sometimes there are many long pauses, or he may talk around a word, using many indirect words in place of the single word he can't seem to come up with (technically referred to as circumlocution). He is neither glib nor fluent in his spoken language.

Given a choice, the dyslexic can almost always *recognize* the correct word. For example, if asked whether a sudden ghostly appearance is an *apparition* or a *partition*, a dyslexic will invariably choose the correct response. However, when confronted and put on the spot to recall or come up with the word for a sudden ghostly appearance, the dyslexic may reach into his lexicon and pull out a word that sounds similar to the word he intended—in this instance, *partition* instead of *apparition*. For those who do not understand the phonologic basis of this confusion, it may appear that the child guessed or is totally confused.

→ Parents must always keep in mind (and remind their child's teacher) that these are phonologic slips and that the child most likely knows the meaning of the word but just can't get it out. Commonly the child (or adult) will say, "It was just on the tip of my tongue," or "I just can't get the word out. It's stuck."

→ It is also good to keep in mind, and to remind the child's teacher, that he understands many more words than he can articulate well. This is why it is well known that a dyslexic child's listening vocabulary (understanding words that he hears) is much larger than his spoken language vocabulary.

→ Finally, and perhaps most critically, fuzzy phonemes interfere with the beginning reader's *ability to learn the names and the sounds of the letters of the alphabet,* an important transition for the would-be reader. This *is* the beginning of reading. It is a necessary if not entirely sufficient accomplishment that must be in place in order to read.

Conversely, difficulty in acquiring these skills is an early warning signal that your child may have a reading problem. All the clues I have discussed in this section must be followed up quickly and seriously. I cannot overemphasize that.

As children mature, they develop a sensitivity to the parts of language. Initially spoken language is like a strip of movie film: It appears continuous and without any breaks. With time, children begin to appreciate the segmental nature of language. Just as strips of film on closer examination are composed of individual frames, ribbons of speech can be broken down into separate words; words in turn can be pulled apart into syllables, and syllables into particles—phonemes.

WORD → SYLLABLES → PHONEMES
Mistake → Mis—take → m / i / s—t / a / k

Children's increasing sensitivity to the sounds of language and to the segmental nature of language can be observed in the preschool child.

→ Beginning at close to three years of age, preschoolers love to learn and to sing their ABCs. At this stage the alphabet is really an undifferentiated string going from A to Z; children generally are not aware of the individual letters. If you stop a three-year-old child in the middle of singing the ABC song and ask what letter comes next, she most likely will have to start to sing the song all over again. This is perfectly normal. The ABC song helps a child become familiar with the names of the letters, and it is not expected that she will be able to differentiate individual letters until she is a little older.

→ A year or so later, by about age four, a child begins to recognize and to name individual letters; she is typically most interested in learning the letters of her own first name.

→ By the time she is ready to enter kindergarten, she will know the names of most, if not all, of the upper- and lowercase letters. Of course, the accomplishment of this critical milestone may vary, reflecting differing preschool and home experiences.

→ Certainly by the time a child has had a full year of instruction in kindergarten, she should be able to recognize and to name all the letters of the alphabet, both upper- and lowercase.

→ Learning letter sounds, which is critical for reading, is intimately linked to learning the names of letters. Mastering letter-sound relationships is the focus of intense activity during the kindergarten year, and children generally leave kindergarten knowing the sounds of most of the letters of the alphabet.

Gradually and sequentially, from the time children are toddlers through their first year of formal education, they are busily acquiring these raw materials for becoming a reader: knowledge of the alphabet, recognition of individual letters, and the ability to associate sounds with letters. Failure or delay in acquiring these skills is among the earliest clues to a potential reading problem.

DIGGING INTO YOUR GENETIC ROOTS

With the remarkable progress in genetics, the initial hope that dyslexia would be explained by one gene or just a few genes has not been fulfilled. Studies in identical twins have shown that although an individual's genetic makeup contributes to the risk of having dyslexia, other factors, such as the environment (including family, community, school, and health), likely play a major role as well. If dyslexia were entirely genetic, both members of a set of identical twins (sharing the exact genes) would have dyslexia. However, in only 60 to 70 percent of the cases are both twins dyslexic; in 30 to 40 percent one identical twin is dyslexic and the other reads well.

In addition to nongenetic factors playing a role in the risk for dyslexia, it appears that the genetic component itself is quite complex and not explainable by one or even a few genes. Much to the chagrin of scientists studying the genetics of dyslexia, genome-wide association studies (GWAS) found that a number of genes were involved, each producing a small effect. Recent advances in genetics have confirmed what the GWAS studies suggested. As articulated by the science writer Ed Yong, "If you told a modern geneticist that a complex trait—whether a physical characteristic like height or weight, or the risk of a disease like cancer or schizophrenia—was the work of just 15 genes, they'd probably laugh. It's now thought that such traits are the work of thousands of genetic variants, working in concert. The vast majority of them have only tiny effects." Distinguished geneticist Simon Fisher has commented on "the popular misconception of genes as abstract entities that directly specify traits, including key aspects of human cognition and behaviour. Actually, there's an enormous gap between DNA and the distal outcome [e.g., reading] . . . there is no one-to-one mapping between a single gene and a specific neural circuit."

> *CAUTION:* I must share with you my deep concern with the often thoughtless embrace of genetics based on ignorance of the actual science or misrepresentation of the scientific data. I feel compelled to emphasize that current evidence suggests "that common diseases involve thousands of genes and proteins interacting on complex pathways" and that similar to experience with other complex disorders (such as heart disease and diabetes), *it is unlikely that a single gene or even a few genes will reliably identify people with dyslexia. Rather, dyslexia is best explained by multiple genes, each contributing a small amount toward the expression of dyslexia.* The search for a gene to detect dyslexia has been very disappointing. Obviously, I would not look to genetics to reliably identify whether your child is dyslexic. *Parents should be very skeptical about any genetic testing that claims to diagnose dyslexia, not only in infancy but at any stage of life.* At the same time, parents should look to evidence-based screening and diagnostic assessments that do in fact reliably diagnose dyslexia.

As conceptualized by Dorothy Bishop, professor of developmental neuropsychology at the University of Oxford, even if one or more gene(s) are slightly more common in a dyslexic population, the gene or genes are still quite common in nondyslexics. Since there are so many more nondyslexics, the gene will appear in many more nondyslexics than it will in those who may be dyslexic. This means that a positive genetics test is not at all equivalent to a definitive determination of dyslexia and a negative test does not rule out dyslexia. Professor Bishop warns that "it would be ineffective to screen children for future dyslexia on the basis of these DNA variants . . . there are many failures to replicate findings of association in this field. . . ."

That does not mean that there is no genetic influence in dyslexia but rather that the influences are complex (perhaps many, many genes each contributing a small fraction of risk) and/or that other, nongenetic contributions play an important role. Like many common conditions (such as type 2 diabetes), dyslexia sometimes runs in families. Having a parent or a sibling who is dyslexic may increase the possibility that you are, too, but it doesn't mean that you are going to develop dyslexia. In fact, our Connecticut Longitudinal Study data led us to conclude that family history as a screening measure is unwarranted and potentially counterproductive.

Regardless of whether we can point to a familial or genetic influence on dyslexia, does this affect how we treat children with dyslexia? As if to emphasize this point, I am often asked if anything can be done to help if dyslexia is inherited. Of course it can. Many disorders that are inherited (therefore pointing to a genetic component) respond well to treatment (again, such as type 2 diabetes). Ironically, being aware of reading problems in a child may also promote increased awareness of one of his parents' own struggles with reading; the writer John Irving and the financier Charles Schwab both discovered they were dyslexic after their children were diagnosed. Many adults often wonder if they may be dyslexic. To determine if they are at risk for dyslexia, we have developed an adult version of our screener, the Shaywitz DyslexiaScreen™, Adolescent-Adult Form (see pages 406–407).

MYTHS AND MISUNDERSTANDINGS ABOUT DYSLEXIA

While being alert for clues to dyslexia, it is also important to be aware of some of the popular misconceptions and myths surrounding it. Sometimes a child's diagnosis is wrong or delayed because she fails to demonstrate one or more of these presumed "symptoms."

One of the most enduring misconceptions is that dyslexic children see letters and words backward and that reversals (writing letters and words backward) are an invariable sign of dyslexia. While it is true that dyslexic children have difficulties in attaching the appropriate labels or names to letters and words, there is no evidence that they actually *see* letters and words backward. In one study, primary-school children who had reading problems and a tendency to make reversals were asked to copy a series of letters and words. They had no difficulty doing so. What they did find troublesome was having to *name* the words accurately. For example, the children were able to copy the word *was* and to say the individual letters making up the word in the appropriate order. Then, typically, they would name the word as *saw*. This study confirmed that dyslexic children have problems in naming but not in copying letters. A related misconception is that mirror writing invariably accompanies dyslexia. In fact, backward writing and reversals of letters and words are common in the early stages of writing development among dyslexic and nondyslexic children alike.

Because these beliefs about dyslexia are so prevalent, many dyslexic children who do not make reversals are often misdiagnosed. In the words of one frustrated father who was trying to obtain help for his daughter, "They say she doesn't have a problem, she just has to work harder, and that she can't have dyslexia because she doesn't transpose letters or write backward."

Left-handedness, difficulties with right-left orientation, trouble tying shoelaces, and clumsiness have at times also been said to be associated with dyslexia. These are certainly not core findings that we would expect to observe in most people with dyslexia, but of course within the larger population of individuals with dyslexia, there are clusters of people who are also left-handed or who have difficulties with right-left orientation. Whatever subgroups of children with dyslexia may exist, it is clear that the vast majority of the dyslexic population share a common phonologic weakness (about 88 percent, our research team determined).

Other myths I will discuss later include:

→ Dyslexia is not real.
→ Dyslexia is just a developmental lag or is outgrown.
→ Dyslexia overwhelmingly affects boys.
→ Dyslexia occurs in only a few countries.

Now I want to move beyond the earliest clues to dyslexia and focus on the later clues which can alert you that your child is not progressing along the path to becoming a skilled reader and may need help to do so.

RECOGNIZING DYSLEXIA'S IMPACT
ON LEARNING TO READ

Breaking the reading code allows a child to cross the threshold into the world of reading. But learning to read marks only the beginning of a continuing process. Just as we now understand how a child first learns to read, a clear picture has emerged of how a child becomes a *skilled* reader. Surprisingly, it involves increasing focus on the details—the letters making up a word.

When a child first begins to read, the so-called *logographic stage,* he does not use knowledge of letter names or letter sounds to "read" a word. We are all familiar with four- and five-year-olds "reading" familiar signs, like STOP or MCDONALD's, or words on boxes of breakfast cereal. We now know that these very young children are simply relying on visual cues of various sorts, like the shape of the red stop sign or the famous golden arches. In one study, although researchers had pasted a *Coca-Cola* logo over a Rice Krispies box, the majority of preschoolers continued to "read" it as Rice Krispies. This level of reader will recognize only a few words, invariably associated with a highly distinctive visual cue, which they have memorized. Using such arbitrary cues is obviously of limited value. Children can memorize hundreds of words, but by the time a child reaches fifth grade he will come across as many as ten thousand new words as he reads during the school year. At that point, relying on memorization simply won't do. To make progress in reading, he must learn how the alphabetic code works. Linking letters to sounds and then sounding out words is the only guarantee of being able to decode the thousands of new words a reader will encounter. This is exactly what primary-school children do.

Generally during kindergarten, or as early as nursery school, as children learn letter names, a type of early or primitive reading becomes evident. If a child knows the letter names *j* and *l*, she can *read* the word *jail* by just saying the first letter name (*jay*) and the last (*ell*). This kind of reading is considered primitive because the reader is not paying attention to the full complement of letters in the word; she is relying on letter names, not letter sounds. To read effectively, a child needs to pay attention to all the letters in a written word so that she can link them to the sounds she hears in the spoken word and then decode the word. Otherwise she will confuse words that have the same initial and final consonants but differ in their interior vowels (such as *book* and *beak*). As children progress in learning to read, they rely more and more on trying to relate letters and groups of letters to the sounds they hear in spoken language.

One group of researchers discovered how important it is for early readers to try to match letters to sounds even if the children don't always succeed. They found that the kinds of errors children make while reading in the first grade offer important clues to their ability to use the phonologic code and ultimately to become proficient readers. Children whose reading errors reflected an attempt to match letters to sounds (for instance, reading *big* for *beg*) tended to be good readers at the end of the school year. Conversely, children whose errors indicated a lack of awareness of the relation of letters to sounds typically ended the year as poor readers. Such children might read *like* for *milk*, words that have some letters in common but do not sound the same. These children were not flexing their phonologic reading muscles. Parents should be concerned when their children do similar things. Above all else, the transition to being a skilled reader requires a child to focus her attention on the *internal* details of the word—on the specific letters making up the word and the sounds they represent.

But reading is more than associating letters with sounds. The aspiring reader must build his reading vocabulary so that eventually he can read complex, long, or unfamiliar words. Since he has created specific neural pathways linking letters to their sounds, he has accumulated within his brain a broad range of letter-sound representations. If our young reader were to stop here, his reading would be very slow and laborious, since he would have to read letter by letter. But as a child reads, he builds up his vocabulary and with it enriches the neural connections linking the sounds, the letters, and the meaning of more and more words, and now things really begin to accelerate. You will recall the interactive theory I presented earlier: that the neural mechanisms for the elements of reading (phonology, orthography, semantics) can be thought of as distributed patterns of neural activity representing each of these elements. In other words, multiple neural pathways must be activated and then joined together in order to recognize a word. The more the child reads, the more he develops each of these pathways, so that they are not only robust but can now rapidly come together to identify a word.

The child goes from linking images of individual letters associated with specific sounds to linking larger and larger chunks of printed material—common letters that frequently go together (-at, -gh, -th) and larger groups of letters that recur (-ight, -eight, -ought). With time, after the child has read many, many books and come across and successfully decoded hundreds of words again and again, he has created increasingly stronger linkages for more and more words. All the child needs to do now is to look at the printed word on the page and well-developed neural linkages are activated so that the sounds, letters, and meaning of the word are instantaneously integrated and the word identified. As I noted earlier, as a person becomes proficient in reading, the different kinds of relevant information—the word's spelling, its pronunciation, and its meaning—are more tightly linked together as part of the same resonating neural circuit. Once you plug into the word, the whole circuit lights up and the word is immediately recognized and understood. A skilled reader has a huge internal dictionary of words whose features are rapidly integrated so the word is identified.

The beauty of this process is that it allows the reader to decipher and read a word that she has never before encountered. She sees a word, scans all the letters, and in essence asks herself, *Do any of the letters fall into a familiar pattern? Do they resemble letter groups—parts of words—that I've encountered?* If so, she is able to take these letter patterns and connect them to a known pronunciation. For example, if she sees the unfamiliar printed word *architect*, she may know that the letters *t-e-c-t* go together and how they are pronounced. She may also know from experience that the letters *a-r-c-h* are often grouped together and that *arch* sounds either like the arch of your foot or like Noah's Ark. She tries to pronounce the unknown word both ways, *arch-itect* and *ar-ki-tect*, and uses the surrounding text to judge which pronunciation fits. From the context she realizes that the word is *architect*, meaning a designer of buildings, and is pronounced like *ark (ar-ki-tect)*. Once she has successfully decoded this word, it joins the other words in her lexicon. Each time she reads the word again, the connections between how the word sounds, how it looks, and what it means become more integrated and more quickly recognized. After correctly reading it several times, she will activate the linkages connecting its spelling, pronunciation, and meaning, instantly adding it to the growing list of words she reads fluently.

Fluency—reading words quickly, smoothly, and with good expression—is acquired by practice, by reading words over and over again. This is consistent with what we know about neural circuits that are reinforced and strengthened by repetition. A reader must have four or more successful encounters with a new word to be able to read it fluently. (As I will discuss later, I use the term *fluent* to describe how a skilled reader reads; I reserve the term *automatic* for describing the neural processes that underlie fluent reading.) Once a word can be read fluently, the reader no longer has any need to rely on context. Fluency does not describe a stage in which a reader is able to decode *all* words instantly; rather, we become fluent word by word. Studies in which the eye movements of readers are tracked have shown that a skilled reader pauses at between 50 and 80 percent of the words in a text. He needs to fixate on the word, essentially to scan the word in, but does so very, very quickly, because the words—their spelling patterns and pronunciations—are well known to him. If you meet someone on the street you don't know well, you may continue to stare at him until some shred of recognition appears; however, if you know someone well, a glance is enough.

Initially, small common words are the ones read instantly. As the reader progresses, larger and more complicated words, including words that may appear infrequently, join the words read at a glance. Although children's books may contain thousands of words, only a relative few of them make up most of the text. In the primary grades, about one hundred words make up 50 percent of the entire text, one thousand words 75 percent of the text, and five thousand words about 90 percent of the text. The remaining 10 percent of the words occur infrequently. Since as few as one hundred words make up half of the words in a typical book for schoolchildren, you can understand how boys and girls who are good readers can become fluent quite quickly and are soon reading books with relative ease.

In addition to reading words accurately and quickly, a skilled reader understands what he reads. Reading comprehension develops gradually, so that over time the balance tips from learning mostly from listening to learning through reading. Thus the beginning reader learns much more from what he hears than from what he reads in print. This gap narrows in the primary grades as the child becomes a better reader. By seventh grade the balance begins to favor reading; at this point the reader is said to have *mature* reading comprehension. Gradually the student gains more and more from reading, so that by the time he is in college most knowledge and vocabulary is acquired through reading.

A large vocabulary is a key element in facilitating reading comprehension; in turn, reading itself is a powerful influence in developing a child's vocabulary. Indeed, it is difficult to overestimate the importance of reading to a child's intellectual growth. Reading researchers Anne Cunningham and Keith Stanovich have emphasized how dependent a child's vocabulary is on reading. A child learns about seven new words a day, which amounts to a staggering three thousand words a year. To acquire a new word for his vocabulary, a child must scrutinize the inner details of the word and not gloss over it. For the most part, analyzing each letter and letter group in a word is the way an accurate representation is formed and the means of having a new written word become a part of the child's reading vocabulary.

Books offer almost three times as many interesting or complicated words—words outside the general vocabulary of a sixth-grader—compared to even the most educated speakers. Books for adult readers have about fifty rare words for every one thousand words; the spoken language of a college graduate has only about seventeen rare words per one thousand spoken words. Children's books, too, "have 50 percent more rare words in them than does the conversation of college graduates." So simply relying on alternatives, ranging from even the most sophisticated conversations to the very popular YouTube videos to increase vocabulary, falls short of what can be gained through reading.

The powerful influence of early reading on later reading and vocabulary growth was demonstrated when researchers had children keep diaries of how they spent their time when they were not in school. As shown in Figure 35, the very best readers, those children who scored better than 90 percent of their peers on reading tests, read for more than twenty minutes a day (about 1.8 million words a year), while those at the fiftieth percentile read only 4.6 minutes a day (282,000 words yearly). The poorest readers, those children reading below the tenth percentile, read less than *one minute* each day (a meager 8,000 words a year) and would require one year to read what the best readers read in two days.

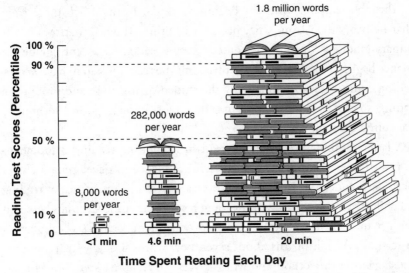

Figure 35. Good Readers Read Many Words Per Year; Poor Readers Read Very Few
Good readers spend more time reading each day, so that they read many more words
in a year compared to poor readers.

Given such worrisome consequences of poor reading, is it possible to identify early on a child who is not on track to becoming a skilled reader? Fortunately, the process of becoming a skilled reader is well mapped out. It consists of a series of discrete and discernible accomplishments, ones you can look for and monitor to determine if your child is on track. To help you do this, I've included a chart laying out the expected developmental sequence and timing of specific reading-related skills or benchmarks. Keep in mind, though, that these are general guidelines for the typical child, and they are certainly not written in stone. Each of these expected steps represents a single point. What we are looking for is a series of points that together, over time, can help to affirm that your child is on course or to alert you early on that there might be a problem. Because teaching and learning experiences influence phonologic skills, the child's level of schooling, rather than age, may at times be the more appropriate guide to expectations.

BENCHMARKS: A GUIDE TO THE DEVELOPMENT
OF READING-RELATED SKILLS
EARLY PRESCHOOL ACCOMPLISHMENTS (AGES 3–4)

→ Begins to develop awareness that, like a roll of perforated
   postage stamps, sentences and then words come apart

→ Shows an interest in the sounds of language: repeats and plays with sounds, especially rhymes; recites nursery rhymes (Humpty Dumpty, Jack and Jill)

→ Identifies ten alphabet letters, most likely from his or her own name

LATE PRESCHOOL ACCOMPLISHMENTS (AGES 4–5)

→ Breaks spoken words into syllables, such as *protect*→ *pro tect* (50 percent of children can count the number of syllables in a spoken word)

→ Recognizes and names a growing number of the letters of the alphabet

BEGINNING KINDERGARTEN ACCOMPLISHMENTS (AGE 5–5½)

→ Continues to progress in breaking spoken words into syllables (90 percent of children can count the number of syllables in a word)

→ Pulls apart spoken words into even smaller pieces of sound (50 percent of children can count the number of phonemes in a word)

→ Compares whether two spoken words rhyme: Do *cat* and *mat* rhyme? Do *hop* and *mat* rhyme?

→ Names a word that rhymes with a simple word like *cat* or *make*

→ Recognizes and names just about all upper- and lowercase letters

END OF KINDERGARTEN ACCOMPLISHMENTS (AGE 5½–6)

Spoken language:

→ Identifies which of three spoken words (or pictures) begins with the same sound as a given word (that is, when instructed, "Tell me which word begins with the same first sound as *car—mat, can,* or *dog*?" he answers *can*) or with a different sound from the other two (that is, when asked, "Which word begins with a different sound: *man, dog,* or *mud*?" he answers *dog*)

→ Identifies which of three spoken words (or pictures) ends with the same sound as a given word (for example, when instructed, "Tell me which word ends with the same sound as *bag—man, dig,* or *dad,*" he answers *dig*)

→ Counts the number of phonemes in a small word (for example, when asked to count the sounds he hears in *me,* he finds two)

→ Pronounces the beginning sound in a word (that is, when asked, "Can you say the first sound of the word *mat*?" he answers *mmmm*)

→ Blends (pushes together) parts (phonemes) of a word into the complete word (that is, when asked, "What word do these sounds make—*zzzz, oo*?" he answers *zoo*)

Print:

→ Names all the letters of the alphabet

→ Knows the sounds of almost all the letters of the alphabet

→ Masters the alphabetic principle; understands that the number and sequence of sounds within a written word represent the sequence of sounds heard in the spoken word

→ Begins to decode simple words

→ Recognizes a growing number of common words by sight (*you, my, are, is, the*)

→ Uses *invented spelling,* such as writing *krr* for *car*

→ Writes many uppercase and lowercase letters

→ Writes his or her own name (first and last) and names of family members or pets

FIRST-GRADE ACCOMPLISHMENTS (AGES 6–7)

Spoken language:

→ Phoneme awareness is in place (70 percent of children can correctly count the number of phonemes in a spoken word)

→ Counts the sounds in longer (three-phoneme) words (for example, when asked, "Can you count the sounds you hear in *same*?" he answers *three*)

→ Says what word remains if a given sound is taken away from the beginning or end of a three-phoneme word (that is, when instructed, "Say *bat* without saying the *b,*" he says *at*)

→ Blends the sounds in three-phoneme words (that is, when asked, "What word do the sounds *p aah n* form? he answers *pan*)

→ Breaks off and pronounces each sound in a two-phoneme word (that is, when asked, "Can you say *no* one sound at a time?" he says *nnnn, o*)

Print:
→ Reads aloud with accuracy and comprehension any text that is meant for first grade
→ Links letters to sounds to decode unknown words
→ Accurately decodes one-syllable words (real words like *sit* and nonsense words like *zat*)
→ Knows sounds of common letter groups or word families, such as *-ite* and *-ate*
→ Recognizes by sight common irregularly spelled words (which do not follow the pattern of a word family), such as *have, said, where, two*
→ Has a reading vocabulary of three hundred to five hundred words, including sight words and words that are easy to sound out
→ Monitors his own reading
→ Self-corrects if an incorrectly identified word does not fit with cues provided by the letters in the word or with cues provided by the context surrounding the word
→ Reads simple instructions such as "Open your book"
→ Reads regularly
→ Begins to spell short, easy words accurately

SECOND-GRADE ACCOMPLISHMENTS (AGES 7–8)
Spoken language:
→ Blends and segments longer words, with three or four phonemes in a word
→ Can take out a sound inside a word and recombine the remaining sounds (for example, when asked, "What word remains if you take the *t* sound from *stay*?" he answers *say*). This is called *phoneme deletion.*

Print:
→ Routinely links letters to sounds to decode unknown words
→ Accurately decodes some multisyllable real and nonsense words, such as *Kalamazoo*

→ Begins to read more and more words with fluency—reads smoothly, rapidly, and with inflection

→ Reads and comprehends fiction and nonfiction meant for second grade

→ Represents the complete sound of a word when spelling

→ Reads on his own voluntarily

THIRD-GRADE ACCOMPLISHMENTS (AGES 8–9)

Print:

→ Reads aloud with fluency and comprehension any text meant for third grade

→ Uses knowledge of prefixes, suffixes, and roots to infer meanings of words

→ Reads longer fiction selections and chapter books

→ Summarizes the main points from readings

→ Correctly spells previously studied words

→ Uses a dictionary to learn the meaning of unknown words

FOURTH-GRADE AND ABOVE ACCOMPLISHMENTS (AGE 9 AND OLDER)

→ Reads to learn

→ Reads for pleasure and for information

DYSLEXIC READERS

For dyslexic readers the process of learning to read and of becoming a skilled reader is torturously slow. Benchmarks are significantly delayed. At the beginning, difficulties linking letters to sounds interfere with learning to read. Over time, as the dyslexic learns to read, she, too, begins to build up her neural pathways linking the way the word looks and sounds and what it means. Unfortunately, the dyslexic reader may correctly connect only a few of the letters in a word to their sounds. As a result, the neural connections for that word are imperfect and incomplete. Later on, when she comes across that printed word again, she struggles to recognize it.

As I said, part of the process of becoming a skilled reader is creating more detailed and robust linkages representing each of the elements making up a word. Dyslexic readers generally require many more exposures to a printed word over a much longer period of time before the neural pathways are well developed and come together each time they encounter that word. In some instances, these connections continue to be less than robust, impeding the ready retrieval of words. As a result, even when dyslexic readers are able to decode words accurately, they are still not quick in their reading of these words.

As a further consequence of their imperfect connections linking each of the elements of a word (the way the word looks, the way it sounds, and what it means), dyslexic readers often have to continue to rely on context to get to a word's meaning. This means that the identification of the word is limited to that particular context and cannot be generalized to other situations. Because a dyslexic reader often gets to the meaning without having first fully decoded the word, the linkages between the way the word looks, the way it sounds, and what it means are not reinforced and remain imperfect. Next time he comes across that word, it is often as if he has never seen it before, and he will have to go through the same exercise of using context to get to the word's meaning.

Context also explains a common puzzling symptom of dyslexia. Many dyslexic readers complain of difficulties in reading the little words, such as *in, on, the, that,* and *an.* The parents of a nine-year-old dyslexic boy, Noah, described his peculiar difficulty with reading small words while at the same time "reading without apparent hesitation longer, more difficult words, such as *museum* and *Metropolitan Opera* and shorter words representing *things,* such as *tree, bat,* etc. Our son, a baseball fan, when in third grade could read *Metropolitan Stadium,* but the word *on* gave him real difficulties."

Since dyslexic readers rely so much on context, it is often difficult to figure out a small, so-called *function* word, whose meaning cannot be gleaned from the context. For example, a ball could be *on, over,* or *under* the table, which makes it difficult to decide which of these choices is the one the author intended. For the same reason, a dyslexic might be able to read words such as *tree* and *bat* because they represent concrete objects that can be predicted from the text as well as visualized. The small function words are so neutral that it is difficult for the dyslexic child to find something in the text to help him anchor and remember the word.

Finally, the ability of a child—say, one who is an avid baseball fan—to read the name of a local baseball stadium and not simpler words argues for the importance of reading material that is relevant and meaningful to the reader. Rather than having to be pushed, he finds the content so inherently interesting that he is pulled into the reading material. Whether it's science

Figure 36. Left Panel Illustrates Push, *right Panel Illustrates* Pull.

or baseball, even those with the most severe dyslexia can learn to read, not only decoding isolated words but comprehending the materials—for after all, this is the pull. After Charles Schwab and I gave a joint talk on dyslexia at the Commonwealth Club in San Francisco, he responded to a question about how he reads. To paraphrase Mr. Schwab, "I take a book, *Shogun*, with me when I go on vacation. I haven't made much progress getting through it. You give me *The Wall Street Journal* and I eagerly go through it cover to cover, and pretty quickly."

JUDGING YOUR CHILD'S PROGRESS

Knowing how skilled reading develops and when to expect the successive milestones characteristic of skilled reading demystifies reading; it allows you to observe your child's reading in a new, informed way. You know when to expect your child to walk and to talk, and now you will be able to know the milestones of reading—what to anticipate in your child's reading development and when to expect it. You will also learn how to recognize the danger signs—the specific clues that should alert you that reading is not progressing smoothly.

Just as a parent would not think of ignoring her child's scheduled physical with his pediatrician, every parent should regularly observe her child reading. Given the high prevalence of dyslexia, it is much more likely for your child to have a reading problem than almost any other physical problem for which he is being checked. I recommend reading with your child as often as possible, at least several evenings a week. Part of that time should be devoted to hearing your child read to you. Listening to your child read prevents a problem from developing without your knowledge; moreover, it's fun. I have wonderful memories of reading with my children, and although they are now adults, they often surprise me with details of stories we read together years ago.

FIRST GRADE

As your child begins to read, listen carefully.

→ Is there evidence that he is trying, even imperfectly, to link letters with sounds?

→ Has he taken that important first step of matching the initial letter of a word with its sound?

As the year progresses, you should continue to listen to him read. Some parents begin reading with their child for a short while, and if he seems to be on track, they no longer read with him. That is a mistake. Children may start off following the milestones, but as the words become more complicated they begin to struggle and lose interest in reading.

→ Notice if he is matching sounds to letters in each position in a small word (beginning, end, and middle).

→ He should also be recognizing common letter groups (such as *-ate, -at, -ick*) and associating them with their sounds.

→ He should leave first grade reading.

SECOND GRADE

By second grade his basic tools for reading should be in place.

→ You should see the emergence of the child's ability to read easy multisyllabic words (such as *animal* and *sleepy*). This important step involves paying attention to the individual parts *within* the word. He is not only matching the first and last parts of the word but recognizing the inner details of longer words as well.

→ Follow along as he reads aloud and listen carefully. You should be concerned if your second-grader is not yet

sounding out words, is taking wild stabs at words, is not
able to read new or unfamiliar grade-level words, has not
yet penetrated the inside of a word when he is reading,
cannot decode most single- or some multisyllabic words,
is not building a vocabulary of words that he can read
fluently, and doesn't seem to enjoy reading.

While most children with dyslexia are not identified until third grade, the
ideal time to identify and help a child is during the very first few years of
school, which is why I have concentrated so much on clues that are easy to
observe during kindergarten and grades one and two. As I have already noted
and will elaborate on, good data show that an achievement gap between
dyslexic and typical readers is already present and detectable by the end of
first grade. This adds even more incentive to screen for and identify dyslexia
as early as possible. I developed the Shaywitz DyslexiaScreen™ in response
to the finding that the achievement gap is present so early, which brought
home to me the critical need to determine if a child is at-risk for dyslexia.
Given the availability of a rapid, evidence-based screener and the high cost
of not identifying dyslexia early, waiting is no longer acceptable. Not only is
the achievement gap evident as early as first grade, but reading intervention
studies have documented over and over again that while improvement in
decoding can occur, there is little improvement in reading fluency the longer
one waits to begin intervention. The message for parents is: If you suspect
your child has dyslexia, act now.

THIRD GRADE AND ABOVE

As your child progresses through third and higher grades, your focus shifts
from wondering if she is learning to read to wanting to know if she is learning to read a critical core of words fluently. Reading is changing in character
now. Words are more complicated, and there are many more of them. In
class the emphasis is less on teaching reading than it is on using reading to
gain information. So it is easy to understand why dyslexia is so often noticed
and diagnosed for the first time in third grade.

Since dyslexic readers often do not use a decoding strategy to identify a
word and instead rely heavily on the surrounding context to figure out its
meaning, you should notice if your daughter uses word substitutions. These
replacement words make sense in the context of the passage but do not
resemble the pronunciation of the original word. For example, a child might

read *car* for *automobile*. Making repeated substitutions is a sure sign that the reader is using context to guess at the meaning of words she has been unable to decode. Because these substitutions often make literal sense, you may find it helpful to obtain a second copy of the books your child is reading so that you can look at each word as she reads it. Otherwise you may not always notice that she is substituting one word for another.

Pay attention to the overall rhythm of her reading—is it smooth or hesitant? She should be reading most of the words on the page fluently, with appropriate expression. Does she tend to trip over the words instead? Slow, choppy oral reading with words omitted, substituted, or misspoken are important clues that a third-grader is not on track for becoming a skilled reader. Such children are in terror of being called on to read aloud and often go to great lengths to avoid such embarrassment; they often stay home, ask to leave the room, or act out and have to be sent out of the room. I've heard from more than a few dyslexic students that they would rather be sent to the principal's office for bad behavior than have to stand up and struggle through reading a passage aloud and face embarrassment and the taunts of classmates.

Poor spelling is very often a sign of dyslexia. Spelling and reading are intimately linked. To spell correctly, a child relies on the neural connections he has formed between the word's letters and sounds, and, as you have read, these are imperfect in dyslexia. Spelling difficulties may be an indication that the child is not forming robust connections between the letters and sounds of that word. As a child who is dyslexic goes on in school, difficulties with spelling persist. In fact, spelling errors may remain long after a dyslexic child (or adult) has learned to decode most words accurately. Analysis of spelling errors often reveals omissions of entire blocks of sounds or confusion of the order of sounds. Mark, a thirtysomething Canadian, reflects the experience of many dyslexics when he says, "My spelling is atrocious. . . . So often I compensate by using a smaller word or words [often a poor compromise]. Spell-checkers are a great help, but not always. Often the spelling is mangled so much that the spell-checker can't find a match."

Children who are dyslexic frequently have abominable handwriting—a problem that continues into adulthood. Word processing is often a relief to dyslexic children and adults alike (perhaps it is the freedom of not actually having to form the letters that is so liberating). Bruce Miller, who is dyslexic and head of his own computer programming company, describes his experiences with the keyboard: "None of my friends type as fast as I do. I think it's because I memorize the words, not the parts. So I'm just reproducing words. I guess I'm really just memorizing all the words and using them as tokens both for input and output." I think Bruce is absolutely right. He is not interacting with the word, nor is he sounding it out; he is acting as a pass-through—just lifting the word as a whole (think of a pdf) and placing it onto the page via the keyboard.

The inability to read fluently leaves its mark on young adults who are dyslexic. By adolescence, good readers have created strong neural connections linking the elements making up each of thousands of words, allowing them to read each of these words fluently. The connections are there and ready to go. Reading is now a pleasure. Decoding is not a problem for them, so their energies can be devoted to thinking about what they are reading. Bright dyslexic adolescents love to think, but for them it's hard to take in the raw material—the printed words—that serve as the source of inspiration for new ideas. They must devote their full concentration to decoding words. As noted previously, many are surprised that in dyslexia, comprehension is typically good. (In fact, some children may have adequate decoding but impaired comprehension; this is not dyslexia but rather a separate entity, Specific-Reading Comprehension Deficits [S-RCD].)

However, accessing comprehension may be quite hard work, since so much of the dyslexic reader's attention is being siphoned off to decode the individual words. Reflecting the lack of fluency, they read slowly—a hallmark of dyslexia. Good readers are fluent; that is, they can read rapidly and automatically. This means that these readers do not require or need to dip into their attention resources. Let's say that each of us has a bucket of attention available. For typical readers this bucket remains relatively untouched and full during reading. In contrast, the dyslexic reader does not read automatically but rather *manually*, which means she has to continually dip into her bucket of attention to support her lack of automatic reading. All the effort of reading required of her is very fatiguing, and eventually her bucket of attention runs out.

The lack of fluency causes significant problems for dyslexic adolescents as they try to cope with large volumes of written work. Homework assignments take a great deal of time to complete and often go unfinished. Fluency is what binds a reader to the text. If a child cannot effortlessly decode a critical mass of words on the page, it is difficult for him to engage with the text. He will be at odds with it. I frequently hear, "My son is not paying attention to his homework." If you cannot penetrate and decode enough words on a page, reading is like gliding over ice: You may never get into the words and, of course, their meanings. Why would someone continue to pay attention to words that she is unable to penetrate? Would you continue to read what was in essence a foreign text that you could not decipher? After a while a dyslexic reader may consume much of her available attention and start daydreaming, staring out the window, and then giving up.

It is important for the child's teacher and her parents to be aware of what I refer to as the *empty bucket* syndrome. The child has not decided to willfully stop paying attention but rather has simply and involuntarily run out of attention. Just as a car that runs out of fuel on a highway comes to a halt, so a dyslexic reader who runs out of attention has little or no choice but to stop reading. Far too often, unfortunately, she is chastised and even punished for "not paying attention."

Given this new understanding, as you notice your student or child starting to lose attention, give her a break from reading. Wait ten or twenty minutes and then try going back to reading. Each child is different and will require her own amount of time to refuel her attention supply. Depending on the specific situation, it might be an optimal time to talk about the story she has just tried to read, its theme, characters, and what she thinks will happen next. At other times it may be best to have her listen to soft, soothing music.

The necessity of devoting all his attention to decoding the words on the page makes a dyslexic reader extremely vulnerable to any distraction, such as noises or movements. For him, reading is fragile, and the process can be disrupted at any moment. Any little sound that draws his attention away from the page is a threat to his ability to maintain his reading. He needs all his attention to try to decipher the printed words. (In contrast, a fluent reader has attention to spare, so room noises are not likely to interfere with his reading.) The practical consequences of this fragility is that dyslexic readers often require an extremely quiet room in which to do their reading or to take tests. Music can be on, as long as it has no lyrics; speech always distracts us. On the other hand, music without words can be supportive.

The difficulties a dyslexic reader has in gaining command of the phonology of his primary language are exacerbated when he tries to learn a new language. Persistent difficulties in learning a foreign language provide an important clue that a student may be dyslexic.

There is one final clue to dyslexia in children and adults alike: the fact that they say they are in pain. Dyslexia inflicts pain. It represents a major assault on self-esteem. In grade-school children, this may express itself as a reluctance to attend school, or moodiness, or spoken expressions such as "I'm dumb" or "I get teased a lot." Adolescents may develop feelings of shame and work hard to hide their dyslexia by avoiding school, pretending to have forgotten assignments, doing anything not to read aloud in class. Adult dyslexics invariably harbor deep pain and sadness reflecting years of assaults to their sense of self-worth.

At the same time, I want to make clear to the dyslexic reader that in spite of these difficulties, he shouldn't be discouraged. The work of one reading researcher, in fact, offers great encouragement. Rosalie Fink has studied a very accomplished group of dyslexic men and women whose honors include a Nobel Prize and election to the National Academy of Sciences and the National Academy of Medicine. Many of these people, even as adults, continue to struggle with getting to the sounds of words and continue to rely on context to get to meaning. The late Baruj Benacerraf, a 1980 Nobel laureate in medicine or physiology and a dyslexic, commented: "Even today, when I can't figure out a word, I guess from the context. Yes, I guess what makes sense." His daughter, Beryl Benacerraf, a physician and a highly respected professor of radiology at Harvard, is also dyslexic. Clearly, dyslexic men and women are able to make significant, even profound contributions.

11

SHOULD MY CHILD BE
EVALUATED FOR DYSLEXIA?

Having described how spoken language and skilled reading develop, and how a phonologic weakness might impact each, I want to combine all the clues along with new scientific data into an early-warning system for recognizing dyslexia. This will help you answer the specific question, should my son or daughter (or I) be evaluated for dyslexia?

No one wants to be an alarmist and put her child through an evaluation for trivial or transient bumps along the road to reading. Evaluations can take time, and those carried out privately can be expensive. But our children are precious, and we must always bear in mind that if a child is not evaluated and later proves to have dyslexia, we have robbed him of precious time. The human brain is resilient, but there is no question that early intervention and treatment bring about more positive change at a faster pace than an intervention provided to an older child. There is also the erosion of self-esteem that accrues over the years as a child struggles to read.

Childhood is a time for learning. A child who delays breaking the phonetic code will miss much of the reading practice that is essential to building fluency. As a consequence, he will fall further and further behind his peers, begin to see himself as not belonging in school, and question his own intelligence. To see this happen to a child is sad, all the more because it is preventable.

"Once children fall behind in the growth of critical word-reading skills," says Joe Torgesen, a reading researcher who has carried out many of the critical studies on intervention, "it may require very intensive interventions to bring them back up to adequate levels of reading accuracy, and reading fluency may be even more difficult to restore because of the large amount of reading practice that is lost by children each month and year that they remain poor readers."

Most parents and teachers delay evaluating a child with reading difficulties because they believe the problems are temporary, that they will be outgrown. This is simply not true. As the participants in the Connecticut Longitudinal Study have demonstrated, *at least three out of four children who read poorly in third grade continue to have reading problems in high school and beyond.* What may seem to be tolerable and overlooked in a third-grader certainly won't be in a high schooler or young adult. Without a specific diagnosis of dyslexia, along with proven interventions, virtually all children who have reading difficulties early on will struggle with reading when they are adults and, worse still, often feel like failures in life.

Groundbreaking new longitudinal data make explicitly clear why you cannot wait or delay in having your child evaluated for dyslexia. The earlier the diagnosis, the sooner the evidence-based reading interventions can begin. These data, published in the *Journal of Pediatrics* in November 2015, demonstrate not only that most changes in reading take place from grades one to three, but that the dyslexic child is already behind by first grade. These data find that the achievement gap persists.

As you can see in Figure 37 on the next page, the reading trajectories of dyslexic readers never catch up with those of typical readers. The clear takeaway is that effective reading interventions must be implemented as early as possible, when children are still developing the basics for reading acquisition, such as phonological awareness, understanding letter-sound relationships, and building a lexicon. *Later interventions may minimally decrease or stop the gap from widening but will not overcome the already existing differences in early grades.* Interventions must begin no later than first grade.

IMPLEMENTING THE SEA OF STRENGTHS MODEL

Luckily, parents can play an active role in the early identification of a reading problem. All that is required is an observant parent who knows what she

Achievement Gap Between Typical & Dyslexic Readers Occurs as Early as First Grade

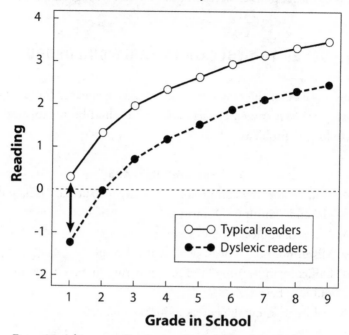

Figure 37. Achievement Gap Between Typical & Dyslexic Readers Occurs as Early as First Grade

is looking for and who is willing to spend time listening to her child speak and read.

The specific signs of dyslexia, both deficits and strengths, in any one individual will vary according to the age and educational level of that person. The four-year-old who can't quite learn his letters becomes the six-year-old who can't match sounds to letters and then the fourteen-year-old who dreads reading out loud and finally the twenty-four-year-old who reads excruciatingly slowly. The threads persist throughout a person's life. The key is knowing how to recognize them at different periods during development.

Therefore, as promised, I have gathered the clues together to provide three distinct portraits of dyslexia: first, in early childhood from the preschool period through first grade; next, in school-age children from second grade on; and last, in young adults and adults. Examine these clues carefully, think about them, and determine if any of them apply to you, your child, or someone else you are close to. Look for clues in the weaknesses *and* strengths. Identifying the weaknesses makes it possible to spot dyslexia in children before they are expected to read and in adults after they have developed some degree of reading accuracy but when they are continuing to show the remnants of earlier problems, reading slowly and with great effort.

CLUES TO DYSLEXIA IN EARLY CHILDHOOD

Naturally, the earliest clues involve mostly spoken language. *The very first clue to dyslexia may be delayed language.* Once your child begins to speak, look for the following problems.

THE PRESCHOOL YEARS

→ Trouble learning common nursery rhymes such as "Jack and Jill" and "Humpty-Dumpty"
→ A lack of appreciation of rhymes
→ Mispronounced words; persistent baby talk
→ Difficulty in learning (and remembering) names of letters and numbers
→ Failure to know the letters in his own name

KINDERGARTEN AND FIRST GRADE

→ Failure to understand that words come apart; for example, that *batboy* can be pulled apart into *bat* and *boy,* and, later on, that the word *bat* can be broken down still further and sounded out as: *b, aah, t*
→ Inability to learn to associate letters with sounds; for example, being unable to connect the letter *b* with the *b* sound
→ Reading errors that show no connection to the sounds of the letters; for example, the word *big* is read as *goat*
→ The inability to read common one-syllable words or to sound out even the simplest of words, such as *dog, cat, hop, nap*
→ Complaints about how hard reading is, or running away or hiding when it is time to read
→ A history of reading problems in parents or siblings

SIGNS OF STRENGTHS IN HIGHER-LEVEL THINKING PROCESSES EARLY ON

→ Curiosity
→ A great imagination

→ The ability to figure things out
→ Eager embrace of new ideas
→ Getting the gist of things
→ A good understanding of new concepts
→ Surprising maturity
→ A large vocabulary for age
→ Enjoying solving puzzles
→ Talent at building models

CLUES TO DYSLEXIA FROM SECOND GRADE ON

PROBLEMS IN SPEAKING

→ Mispronunciation of long, unfamiliar, or complicated words; the *fracturing* of words—leaving out parts of words or confusing the order of the parts of words; for example, *aluminum* becomes *amulium*

→ Speech that is not fluent—pausing or hesitating often when speaking, lots of *um*'s during speech, no glibness

→ The use of imprecise language, such as vague references to *stuff* or *things* instead of the proper name of an object

→ Not being able to find the exact word, such as confusing words that sound alike: saying *tornado* instead of *volcano*, substituting *lotion* for *ocean* or *humanity* for *humidity*

→ The need for time to summon an oral response or the inability to come up with a verbal response quickly when questioned

→ Difficulty in remembering isolated pieces of verbal information—trouble remembering dates, names, telephone numbers, random lists; difficulty with rote memory; remembers concepts better than isolated facts

PROBLEMS IN READING

→ Very slow progress in acquiring reading skills
→ The lack of a strategy to read new words
→ Trouble reading *unknown* (new, unfamiliar) words that must be sounded out; making wild stabs or guesses at reading a word; failure to systematically sound out words

→ The inability to read small function words such as *that, an, in*

→ Stumbling on reading multisyllable words, or failing to come close to sounding out the full word

→ Omitting parts of words when reading; failing to decode parts within a word, as if someone had chewed a hole in the middle of the word: for example, reading *convertible* as *conible*

→ A terrific fear of reading out loud; the avoidance of oral reading

→ Oral reading full of substitutions, omissions, and mispronunciations

→ Oral reading that is choppy and labored, not smooth or fluent

→ Oral reading that lacks inflection and sounds like the reading of a foreign language

→ A reliance on context to discern the meaning of what is read

→ A better ability to understand words *in context* than to read *isolated* single words

→ Disproportionately poor performance on multiple-choice tests

→ The inability to finish tests on time

→ The substitution of words with the same meaning for words in the text he can't pronounce, such as *car* for *automobile*

→ Disastrous spelling, with words not resembling true spelling; strange spellings may be missed by spell-check

→ Trouble reading mathematics *word* problems

→ Reading that is very slow and tiring

→ Homework that never seems to end; parents who are often recruited as readers

→ Messy handwriting despite what may be an excellent facility at word processing—nimble fingers

→ Extreme difficulty learning foreign languages

→ Trouble reading anything but memorized words

→ A lack of enjoyment in reading, the avoidance of reading books or *even a sentence*

→ The avoidance of reading for pleasure, which seems too exhausting

→ Reading accuracy improves over time, though it continues to lack fluency and is laborious

→ Lowered self-esteem, with pain that is not always visible to others

→ A history of reading, spelling, and foreign-language problems in some family members

CLUES IN OTHER AREAS

→ Difficulty memorizing multiplication tables, impacting math calculation and carrying out basic math operations such as multiplication and division

→ Good understanding of math concepts but a tendency to solve math problems in her head without showing on paper how she arrived at the answer

→ Problems with directionality, often getting lost within a building or while walking and, later, driving to a destination

→ Difficulty in proofing what she has written

→ Poor spelling that overshadows great ideas and imagination

SIGNS OF STRENGTHS IN HIGHER-LEVEL THINKING PROCESSES IN THE OLDER CHILD

→ Excellent thinking skills: conceptualization, reasoning, imagination, abstraction

→ Learning that is accomplished best through meaning rather than rote memorization

→ Ability to get the "big picture"

→ A high level of understanding of what is read *to* her

→ The ability to read and to understand at a high level overlearned (that is, highly practiced) words in a special area of interest; for example, if his hobby is restoring cars, he may be drawn to and able to read auto mechanics magazines

→ Improvement as an area of interest becomes more specialized and focused when he develops a *miniature* vocabulary that he can decode

→ A surprisingly sophisticated listening vocabulary

→ Excellence in areas not dependent on reading—math, computers, visual arts—or excellence in more conceptual

(versus factoid-driven) subjects like philosophy, biology, social studies, neuroscience, creative writing
→ Often, exceptionally empathetic

CLUES TO DYSLEXIA IN YOUNG ADULTS AND ADULTS

PROBLEMS IN SPEAKING

→ Persistence of earlier oral language difficulties
→ The mispronunciation of the names of people and places; tripping over parts of words
→ Difficulty remembering names of people and places and the confusion of names that sound alike
→ A struggle to retrieve words: "It was on the tip of my tongue"
→ Lack of glibness, especially if put on the spot
→ Some anxiety when called upon to speak publicly
→ Spoken vocabulary that is smaller than listening vocabulary, hesitation to say aloud words that might be mispronounced
→ Difficulty learning a foreign language

PROBLEMS IN READING

→ A childhood history of reading and spelling difficulties
→ Word reading becomes more accurate over time but continues to require great effort
→ Lack of fluency
→ Embarrassment caused by oral reading: the avoidance of Bible study groups, reading the Haggadah at Passover seders, or delivering a speech
→ Trouble reading and pronouncing uncommon, strange, or unique words such as people's names, street or location names, food dishes on a menu (often resorting to asking the waiter, "What's your special of the day?" or resorting to saying, "I'll have what he's having," to avoid the embarrassment of not being able to read the menu)
→ Persistent reading problems
→ The substitution of made-up words during reading for words that cannot be pronounced; for example,

metropolitan becomes *mitan,* and the word *metropolitan* is not recognized when it is seen again or heard in a lecture the next day
→ Extreme fatigue from reading
→ Slow reading of most materials—books, manuals, subtitles in foreign films, teleprompter
→ Penalized by multiple-choice tests, leading to very bad results
→ Unusually long hours spent reading school- or work-related materials
→ Frequent sacrifice of social life for studying
→ A preference for books with figures, charts, or graphics
→ A preference for books with fewer words per page or with lots of white showing on a page
→ Disinclination to read for pleasure
→ Spelling that remains disastrous, and a preference for less complicated words in writing that are easier to spell
→ Particularly poor performance on rote mechanical or clerical tasks that require minimal thinking or reasoning

SIGNS OF STRENGTHS IN HIGHER-LEVEL THINKING PROCESSES IN YOUNG ADULTS AND ADULTS

→ The maintenance of strengths noted in the school-age period
→ A high learning capacity
→ A noticeable improvement when given additional time on multiple-choice examinations
→ A noticeable excellence when focused on a highly specialized area such as medicine, law, public policy, finance, literature, or basic science
→ Excellence in writing if content and not form is important
→ A noticeable articulateness in the expression of ideas and feelings, and exceptional empathy and warmth, feeling for others
→ Successes in areas not dependent on rote memory
→ A talent for high-level conceptualization and the ability to come up with original insights
→ Big-picture thinking

→ Inclination to think out of the box

→ A noticeable resilience and ability to adapt

→ Ability to get the point, often almost instantly, leaping over others who are stuck thinking sequentially

Parents, educators, and the dyslexic person herself may not perceive how strong her intelligence is due to the preoccupation with the impact of dyslexia in schools, where dyslexia is often neither acknowledged nor understood. Here's what Charles Schwab has to say about how he solves problems: "I can sort of move ahead in a more radical way, to visualize what the end zone might be, where other people have to go through a sequence of various things to get there. And I know that's frustrating to some of my cohorts here at the company. But I think in some ways that is a processing thing and so maybe I'm not burdened by this sequential thinking that so many of the people are."

If you think you or your child has some of the problems you just read about, it is important to note how frequently they manifest themselves and how many there are. You don't need to worry about isolated clues or ones that appear very rarely. For you to be concerned, the symptoms must be persistent. Anyone can mispronounce a word now and then or confuse similar-sounding words occasionally. What you are looking for is a persistent pattern—the occurrence of a number of these symptoms over a prolonged period of time. That represents a likelihood of dyslexia.

GETTING HELP

Identification of a problem is, of course, the key to getting help. The sooner a diagnosis is made, the quicker your child can get help, and the more likely you are to prevent secondary blows to her self-esteem. Here are some cautionary notes along with helpful hints to guide you as you pursue early identification of possible dyslexia in your child.

PRESCHOOL

If your preschool child struggles with language, particularly with rhymes and with pronouncing words, you should not keep your worries to yourself. You need to seek help.

A preschooler can be seen by her *general pediatrician,* who can consider her history and listen to her speak, judge her ability to appreciate rhymes and her ability to identify pictures or objects beginning with a particular sound, and then make a referral for further evaluation. Since the focus of the evaluation of such a young child is on her spoken (rather than her written) language, I often turn to a *speech-language pathologist* to carry out this type of assessment. These specialists are quite knowledgeable about early language development and are often extremely helpful in assessing phonologic skills in young children. Parents can call or visit the website of the American Speech-Language-Hearing Association (800-638-8255; www.asha.org /profind) for the names of certified speech-language pathologists in their area.

KINDERGARTEN

Parents (and teachers) must closely monitor a child's progress in learning to read, hopefully no later than kindergarten. As I discussed earlier, a large gap between the reading of dyslexic and typical readers is already present by the end of first grade. Kindergarten is definitely a time to act.

Kindergarten is in many ways a watershed in identifying children vulnerable to dyslexia. For the first time, a child is in a public environment where he is exposed to a formal curriculum geared to teaching those skills necessary for reading and he is surrounded by his peers. He is now a student, and there are expectations for what must be learned. Even acknowledging that children come from different backgrounds and may intrinsically differ in their pace of learning, the clues I have listed are important signals that reading is not progressing; they should not be ignored. The cost to your child and to you will be too great. I have yet to meet a family who feels they acted too soon.

> *HELPFUL HINT:* If you observe these signs, I urge you to speak to your child's teacher. Before doing so, it often helps to list your observations and your concerns. Parents are often so nervous when speaking to their child's teacher that they forget what they want to say. The teacher will appreciate having such a list as well.

Many of my patients have found this brief checklist helpful in setting up a meeting with their child's teacher:

→ Set up a specific time to speak to your child's teacher; don't catch her on the run.

→ Take a deep breath and try to be positive and a good listener.

→ Find out how your child is progressing in reading; you want specifics, not generalities or euphemisms (you can check these against the benchmarks listed on pages 126–130).

→ Inquire as to what type of reading program your child is receiving. Is it more focused on the whole word, or are the specific components of reading taught systemically and explicitly?

→ Pin down exactly how her reading progress is being measured.

→ Ask what reading group she is in and what level reader that group represents.

→ Ask how she compares to others in her class and in her grade.

→ Ask what the expectations are for her by year's end.

→ In very specific terms, ask what help she is receiving (type of reading program, size of group, and minutes per day, days per week).

→ Ask what specifically is being done to ensure that your child increases her reading progress. This is a vitally important question. I cannot emphasize enough that scientific data show that reading problems and an achievement gap in reading are already present by the end of grade one and are persistent; they do not represent a temporary lag in development.

→ If you can, visit the class and observe your child along with her classmates during a reading lesson.

→ Ask if she has been screened for dyslexia. If she has not, suggest she be screened as at-risk for dyslexia (for example, by using the Shaywitz DyslexiaScreen™).

Don't delay. It is wishful thinking to believe there will be a sudden, magical improvement. Schools often have a tendency to want to wait, and sometimes parents have this same tendency. (*Why don't we see how things go until after Thanksgiving?* Or, *We can't spoil his Christmas.*) You don't want to be like the mother of one of my patients, who was surprised to be told by her son's teacher on the very last day of school, "I really hope he has a better year next year."

As a knowledgeable parent you must not accept allusions to a temporary or developmental lag, or that "some children are just slow readers," or that girls are not dyslexic. Nor should you accept that there is "no such thing as dyslexia." Anyone who says this needs to be educated. Every parent should be aware that in 2018 a key step toward major policy changes occurred that (should) positively impact schools' approach to dyslexia.

→ For the first time, federal law provides the scientific definition of dyslexia as part of the First Step Act, passed in December 2018 as Public Law 115-391.

→ In the fall of 2015, Assistant Secretary of Education Michael Yudin issued a letter unequivocally stating that when appropriate, the term *dyslexia* should be used in diagnosis and educational planning. The word *dyslexia* is used for the first time in the Every Student Succeeds Act (ESSA), the latest version of the nation's main K–12 law and the successor to the No Child Left Behind Act.

→ Together, these developments pave the way for the schools to identify and provide effective needed services to dyslexic students.

If you are unhappy after your discussion with your child's teacher, speak to the principal or to the reading specialist (if there is one). Your concern is more likely to be taken seriously if it is expressed in writing to the school principal, along with a request for a screening and evaluation.

Be persistent. Parents are often fearful of antagonizing school personnel, and justifiably so. However, the alternative—waiting—can be even more harmful to your child. Your daughter's classroom teacher may have been worrying, too, and your interest may be just enough to jump-start the process of screening, diagnosis, and effective intervention. The most dangerous thing you can do if you suspect a reading problem is to do nothing.

Under no circumstances should you allow yourself to be talked into waiting—that is not an option. Set a deadline for a response. Remember that you do not have to depend solely on your school for an evaluation. Your pediatrician can refer you for testing as well. If you would like more information, check the Yale Center for Dyslexia & Creativity website (www .dyslexia.yale.edu), the LKA Dyslexia Resource Center website (www .dyslexia1n5.com), the International Dyslexia Association website (https:// dyslexiaida.org), or the parent-based Decoding Dyslexia website (www

.decodingdyslexia.net). Scottish Rite Learning Centers (a branch of Freemasonry) can be helpful, both by offering evaluations and by providing interventions, while the Neuhaus Education Center in Houston offers a range of very useful support services to both educators and parents.

Don't be intimidated. Act, and act quickly. Trust your instincts—you know your child best. With the information you have now absorbed, you are an informed parent who is an effective advocate for your child. Your child is depending on you. It is a good feeling to be in control and to know what is and what is not acceptable.

Parents often ask who is capable of carrying out an informed, trustworthy, helpful examination for dyslexia and whom they should turn to for an evaluation. A major requirement is that any assessment be carried out by an individual with expertise in diagnosing dyslexia. That individual must be up-to-date on twenty-first-century scientific knowledge of dyslexia and its expression at different developmental levels from preschool through adulthood and in different contexts, including home, school, and the workplace. An evaluation is not only about administering tests; it's about understanding dyslexia. So obtaining an in-depth history looking for the relevant signs of dyslexia across development is vitally important.

Furthermore, such an evaluator should be knowledgeable about schools, have experience working with schools, understand what schools must do to support their dyslexic students, and be knowledgeable about both federal and state laws regarding dyslexia. You must ask any evaluator if he or she has diagnosed dyslexia, and how often. You will want to know what tests he or she uses to make a diagnosis. A basic test battery for dyslexia should include a test of cognitive ability (IQ); tests of reading, including real and made-up (nonsense) words; tests of reading fluency; oral reading of both single words and connected texts; tests of phonological processing; tests of vocabulary and reading comprehension; math tests, both calculation and word problems; and perhaps other tests, depending on the child's history. I will return to this subject shortly.

As I've said, such testing can be administered by a psychologist, a neuropsychologist, a learning-disability specialist, a reading specialist, or, especially for children under five years, a speech and language therapist. The major requirement is for the individual to have a deep understanding of dyslexia as well as training, experience with, and expertise in the measures used.

Administration of the most complex tests, such as the Wechsler Intelligence Scale for Children-V (WISC-V) requires specific degrees, licensure, or certification.

Be aware that dyslexia can often go unrecognized even in the very best school and home circumstances. The following story, Sarah's, is about how dyslexia existed but went undetected for several years in one of the best schools in the country, with excellent, caring teachers and under the watchful eyes of highly involved, bright, doting parents who are, in fact, active in education. Sarah's parents are sharing their story to raise your awareness, so that you will be alert to the clues, not rationalize them away, and act to have your child tested and diagnosed. Sarah's years of struggle and self-doubt and the enduring guilt of her parents could have been avoided by early recognition and detection of signs of dyslexia.

Sarah entered school anointed with the halo left behind by her two older brothers, ideal students, excelling in every subject. She was exceptionally bright, worked very hard, and was a pleasure to have in class. What everyone seemed to miss was that as bright as she is, Sarah is also dyslexic. In her mom's words:

> Looking back at all the things I missed, this is sort of a heartbreaking memory for me, because I consider myself an aware mom and there were all these clues before our eyes but ones that we didn't allow ourselves to see.
>
> Because Sarah was so strong in many areas, we'd just sort of make excuses for her. There was this paradox: She was very slow to learn to read and yet she was great at memorizing—she memorized everything. And so it went for the first several years of school. Sarah worked very hard. For example, even as far back as kindergarten, at parent events where children got up on a stage and read aloud, you would not know Sarah had any problem in reading. She stood up and read to all the other kids and the parents.

Unknown to everyone listening was that Sarah had memorized everything. "And so she would do these very good readings and I thought we were right where we were supposed to be and she compensated for many, many years."

Sarah was not only keeping her reading difficulty from her parents and teachers but struggling emotionally, keeping this from herself as well. The result was many requests to stay home and avoid going to school. I have heard this now from so many parents that I think frequent school avoidance should set off some alarms. Be wary of a child's repeated pattern of complaining of being ill and asking to stay home when she appears quite well, with no fever or overt signs of illness. If the child is sent to school and you receive frequent calls from the school nurse that the child claims to feel sick but doesn't show any signs, you should also be wary. Repeated pleas to remain at or return home are a sign of the child's overwhelming anxiety. Although her parents couldn't understand "how someone this smart was having this much trouble," Sarah knew. In fact, she reached a point of concern when she noticed that her classmates who were always part of her academic group doing comparable work were now ahead of her and that "something was wrong."

Finally Sarah said, "I think I need to see Pearl. There's something that's not right." (Pearl, who performed evaluations for possible learning difficulties, was Sarah's step-grandmother.) Indeed, something was not right. Results showed a large discrepancy between Sarah's high intelligence, in the ninety-fifth percentile, and her reading skills, at only the twentieth percentile. Without any question Sarah had a glaring, unexpected difficulty in reading, and yet neither her parents nor her school had picked up on it. During the first three years of school, reading consists of relatively few words and much repetition of the same words, and that is convenient for memorization. But from fourth grade up there are many more words and less repetition, making memorization much more difficult to accomplish.

Schools and parents must raise their antennae not to overlook students like Sarah, who despite significant reading difficulties force themselves to work, work, and work some more, so that homework begins when the child arrives home from school and goes on well into the night. This pattern results in major spillover into other aspects of the child's life, so that there is little, or often no, time for socializing or participating in sports or other after-school activities.

Sarah's parents were overwhelmed with guilt because they had missed so many clues, and then they saw how much difference it made for their daughter to know she was dyslexic. As her dad describes the transformation following her diagnosis in ninth grade, "It was like the weight of the world came off Sarah's shoulders. And once we knew, we all jumped into action with her and with the school."

"I should have acted sooner on things that were right there," says Sarah's mom. "I think, too, it's not stopping but going to the next step. For example, asking the teacher the next question when he said she needs a tutor, going to the next step and asking if Sarah should be tested to determine the source of her difficulty."

Happily, Sarah's parents have much to say on all the positive things that have come with Sarah's identification as dyslexic.

> What started to pick Sarah up off the floor is the idea that she can manage this. Once her dyslexia was diagnosed she took charge of it.
>
> We missed it, but what I would say to parents is don't be afraid of it. Become empathetic, learn about it, do the best you can to help your kid understand, also, that they're going to be fine. This isn't the end of the world.
>
> *Not only is it not the end of the world, it's a great opportunity.* Sarah in high school had to learn not only how to strategize for herself but also how to advocate for herself, which is good—a lot of kids get through high school without ever having to really take charge of where they are in the whole system. She would talk to teachers. She would explain—if her friends asked why she was staying later—and so she had to do all of this. So she was ahead of the game, I think, in learning how to take care of herself. And when she applied to college, she wasn't even going to apply to very competitive ones. She was afraid she was not smart enough to get in. And then she decided to do it and she got in. She didn't want to put herself in a situation where she was not successful. It takes her longer to do her work—she's spending all this time in the library—but she's doing it.

Her mother smiled as she told me: "But she's doing it and she's doing wonderfully." Sarah's proud dad added, "She has grit and resilience in addition to a brilliant mind. I think back at how many times Sarah would have to tell the teacher, 'I need a half hour more. You have to give me a half hour more.' She would have to do that for herself. Clearly, that's the kind of resilience many people never learn."

> *HELPFUL HINT:* Parents, too, have to think of ways to make the days go better. Here, as well, Sarah's mom worked out a plan to focus her on the positive.

Every day in fifth grade I'd pick Sarah up and she was just like mopey and depressed and upset. And she'd get in the car and I said, "Okay, the first three things out of your mouth now are what went well today." Then she had to worry about that and focus on it during the day and that just got her through—it's like one of my brilliant moments.

Now that you are aware of the warning signals and the urgent need to obtain an evaluation for dyslexia, I want to tell you about the evaluation itself: how we go about making a diagnosis of dyslexia.

12

DIAGNOSING DYSLEXIA IN THE SCHOOL-AGE CHILD

As common as reading problems are, and as much as we have learned about them, the diagnosis of dyslexia is often missed. Take the experience of one of my colleagues, Juliet, in her own words, talking about her son.

> I would say that from the beginning we were always expecting Andrew to read. When he was four, he would sit with us for hours at a time and listen to us read *Alice in Wonderland* and even show a glimmer of interest in *Harry Potter*. We knew he was interested in books and he had a great attention span for listening to books. So we just knew as soon as he got to school, reading was going to take off. That didn't happen. Actually, in kindergarten other kids were reading and Andrew wasn't. And in first grade I'd meet with his teacher, and she'd say, "Well, he's not where we expect him to be, but it clicks in for everyone at a different time, and it'll click in for him." She was a veteran teacher and well respected. So I thought, *What do I know?* But Andrew knew. And I said, "You read much better than you did in kindergarten." And he said, "But I read like a kindergartner." He knew that I was trying to put one over on him. The school did some cursory testing and they said,

"Yeah, somewhat behind in the reading, but not enough to worry about. He'll be fine." Andrew was very bright, and he was reading very poorly, especially in relation to his very superior intelligence. Things did not improve as he went from one grade to another. But he had dyslexia, which they didn't get at all; they told me that they didn't believe in dyslexia. What frustrated him the most was that no matter how hard he tried, he couldn't do what the other kids were doing. And so Andrew, who couldn't wait to start school and become a reader, instead became one of those class clown kind of kids.

There were enough clues in Andrew's early reading struggles, as well as the demonstration of a large gap between his high intelligence and his poor reading, to tell us that Andrew warranted a full evaluation for dyslexia. Alas, tens of thousands of children who are struggling to read and have clinical symptoms of dyslexia are going undiagnosed.

Today's evaluation for dyslexia puts into practice all that we have learned about reading and dyslexia. It is focused, it follows a structure, and it reflects a scientific rationale.

DYSLEXIA IS "UNEXPECTED"

For the last hundred years, perhaps the most consistent and enduring core of the definition of dyslexia is the concept of *unexpected* underachievement. The construct of unexpected underachievement refers to a child whose reading is unexpectedly low in relation to his or her intelligence. This unexpected nature of dyslexia, as you'll remember, was originally described by Dr. Pringle Morgan in his patient Percy F., the first published description of dyslexia. The unexpected nature of dyslexia has been scientifically validated in a 2010 report published in the journal *Psychological Science,* and that scientific finding has now been translated into federal law—specifically, in the First Step Act (Public Law 115-391), which I referred to earlier.

DYSLEXIA IS A CLINICAL DIAGNOSIS

It's important to consider each of the components of the diagnostic puzzle: what each component contributes and how to synthesize all the pieces to arrive at a diagnosis. Current practice recognizes that dyslexia is a clinical diagnosis, based on clinical judgment, made by an experienced clinician who considers both the individual's history and symptoms and his or her learning profile, including cognitive and achievement abilities that may point to an unexpected difficulty in reading as well as the reasons for this difficulty. In this schema, reading below intellectual ability, sometimes referred to as an ability-achievement discrepancy, is just one component of an overall clinical assessment, one that considers, for example, a person's history and observation of the person speaking and reading aloud as well as the results of testing.

While tests are helpful, they are only one component, albeit an important one, of the evaluation process. Supreme Court Justice Sonia Sotomayor understands this. Prior to her tenure on the Supreme Court, Justice Sotomayor served as a U.S. District Court judge for the Southern District of New York State. In this role, presiding over a trial involving a dyslexic attorney's suit charging the New York State Bar with discrimination for denying the plaintiff's request for accommodations, Justice Sotomayor astutely made these remarks concerning the diagnosis of dyslexia: "By its very nature, diagnosing a learning disability requires clinical judgment . . . not quantifiable merely by test scores." Test scores from psychometric measures are not necessarily sufficient to establish the presence or absence of dyslexia. Instead, knowledge of the individual's history, clinical observations, behavioral observations, and judgment are essential to a diagnosis.* Dyslexia is neither diagnosed nor accurately represented by a single score on a test; it is identified best by consideration of a broader clinical picture conforming to the known characteristics of the disorder. This makes it imperative that educators and evaluators become more deeply knowledgeable about dyslexia.

To repeat, the most consistent and enduring core of the definition is the concept of dyslexia as *unexpected* underachievement. The evaluation follows the universal definition of dyslexia, which describes its typical impact on several areas, including speaking (word retrieval), reading (accuracy and fluency), spelling, and learning a second language.

Today's scientifically based evaluation of dyslexia is *organic*. It reflects the actual experiences dyslexic children and their families have endured. Wise clinicians appreciate that tests are only approximations of the reality that is the individual's life experience. In fact, the tests that are ultimately selected must be chosen with great care. An evaluation must be tailored to

---

* See page 475 for more details on this case and Justice Sotomayor's ruling.

the individual so that it reflects the expression of problems appropriate to that person's age and education.

The diagnosis of dyslexia is basically no different than that for any other condition. Guided by knowledge of the underlying pathophysiology, the clinician seeks to determine through history, observation, and psychometric assessment if there is evidence to support a diagnosis of dyslexia. The three steps of the evaluation process are to

→ establish a reading problem according to age and/or education.

→ gather evidence supporting its "unexpectedness" (high learning capability may be determined on the basis of an intelligence test).

→ demonstrate evidence of a phonologic weakness, with other higher-level language functions relatively unaffected.

There is no one single test score that makes the diagnosis of dyslexia. As with any other scientifically based diagnosis, the diagnosis of dyslexia should reflect a thoughtful synthesis of all the clinical data available. What the clinician is seeking is converging evidence of a phonologically based reading disability as indicated by a disparity between the individual's reading and phonologic skills, on the one hand, and his intellectual capabilities or age, on the other. Dyslexia is distinguished from other disorders that may prominently feature reading difficulties by the unique, circumscribed nature of the phonologic deficit, one not generally intruding into other linguistic or cognitive domains.

Finding reading scores that are unexpectedly low for age or intelligence is the first step in the diagnosis of dyslexia. Reading tests are easy to understand once you recall that there are two major components of reading: *decoding* (identifying words) and *comprehension* (understanding what is read). Accordingly, the assessment focuses on how well the child reads words and on how well she understands what she has read.

In the school-age child, reading is assessed by measuring accuracy, fluency, and comprehension. While accuracy is critical early on, the ability to read fluently gains in importance as the child matures. *A child who reads accurately but not fluently is dyslexic.* One important element of the evaluation is how accurately the child can decode words (that is, read single words). This is measured with standardized untimed tests of single real word and pseudoword reading, such as the Woodcock-Johnson Test of Achievement-IV (WJ-IV). In addition, the Wechsler Individual Achievement Test-III (WIAT-III) and the Kaufman Test of Educational Achievement III (KTEA-III) are reliable, valid, and comprehensive measures of academic achievement. Each test includes multiple tasks designed to evaluate early reading skills, such as letter identification, word reading, pseudoword decoding, fluency, and comprehension.

READING REAL SINGLE WORDS

In a young school-age child like Andrew, the reading evaluation first determines how accurately he can decode words—that is, read single words in isolation. For example, Andrew was asked to read increasingly difficult words, going from *go, the,* and *me* to words like *pioneer, inquire,* and *wealth,* and finally to the most complex: *epigraphist, facetious,* and *shillegah.*

READING PSEUDOWORDS

Even more central to the diagnosis of dyslexia is how Andrew reads so-called pseudowords or made-up words—words he has never seen before and could not have memorized. We might begin with *ree, ip,* and *din.* The words increase in difficulty to *rejune, depine,* and *viv,* and then to the most challenging of all, *pnir, ceisminadolt,* and *byrcal.* The point of these strange but pronounceable words is that they test a child's ability to *sound out words*—to map letters to sounds. Each word can be sounded out if you have acquired *phonologic decoding* ability. *The ability to read nonsense words is one of the best measures of phonologic decoding skill in children.* Reading tests often refer to this skill as *word attack.* The reader literally has to penetrate the sound structure of the word and sound it out, phoneme by phoneme; there is no other way. Most children generally reach their full capacity to sound out nonsense words by adolescence.

READING FLUENCY

Reading fluency, or the ability to read accurately, rapidly, and with good prosody, is of critical importance because it allows for the automatic, attention-free recognition of words. The Test of Word Reading Efficiency, second edition (TOWRE-2), assesses how accurately and how quickly someone reads single words and pseudowords. This is helpful, but it does not measure true fluency, which is based on reading *connected* text aloud and demonstrating *understanding* of what is being read.

Tests of oral reading fluency—reading passages aloud—are often particularly helpful in identifying any uncertainties a child may have in decoding a word. By its very nature, oral reading forces the reader to pronounce each word. Listening to a dyslexic reader as she painfully tries to decode one word after another leaves no doubt about her reading ability. We can observe how much effort it requires for her to pronounce each syllable of each word; we can hear words that are mangled or newly made up; we can note that some words that should be there are not; we can note the lack of cadence or inflection in her reading. We can thus identify a struggling reader who still has not fully mastered the connection between letters and sounds. As you know by now, such labored oral reading can be a telltale sign of dyslexia in an otherwise extremely gifted and accomplished person.

Oral reading fluency is generally assessed by asking the child to read *aloud* using the Gray Oral Reading Test fifth edition (GORT-5). This test consists of increasingly difficult passages, each followed by comprehension questions; scores are awarded for accuracy, rate, fluency, and comprehension. As you listen to the dyslexic child read and then review his scores, you will often see that as he attempts to read more quickly, his accuracy dips. This is because he is a manual rather than an automatic reader.

While the ability to read words accurately and rapidly is often the focus of measures of fluency, it is critical to understand that a major, if not the most important, component of the definition of fluency is *prosody*—that is, reading with *understanding*. So listening to how the person reads connected text out loud—listening for how she modifies her voice reflecting the content or emotion of what is being read—is an essential part of the assessment of fluency.

SPELLING

Children who have difficulty reading typically struggle with spelling—encoding the words, that is, converting the sounds into letters. Helpful spelling tests include the Test of Written Spelling-5, the WIAT-III, KTEA-III, and WJ-IV, which includes spelling as one of the skills measured on its Written Language subtest.

COGNITIVE ABILITY

Because dyslexia is defined as an unexpected difficulty in reading in a child or adult in relation to intelligence, it is not surprising that a measure of intelligence such as the Wechsler Intelligence Scale for Children fifth edition (WISC-V) is an important component of a comprehensive assessment of the child or adult with dyslexia. Very often an IQ test can reveal areas of *strength*, particularly in areas of abstract thinking and reasoning, which are very reassuring to parents and especially to the child herself. It also indicates that the reading difficulty is isolated and not reflective of a general lack of learning ability.

> *HELPFUL HINT:* I want to reemphasize that following completion of the testing and scoring of a test of intelligence, *it is essential to engage in a dialogue with the child or adult about his ability, asking him where on the bell-shaped curve he believes his intelligence lies and then indicating to him where it actually lies—typically significantly higher than he indicated.*

READING COMPREHENSION

In contrast to decoding and fluency, reading comprehension, as Andrew demonstrated, is often minimally affected in bright children with dyslexia, if at all. Tests of reading comprehension are generally, but not always, read silently. Accordingly, someone's score on such a test depends less on the accurate pronunciation of each word than on her being able to infer the meaning of the passage in order to answer questions based on it. She can use the context to guess at the meaning of some words and still answer the comprehension questions correctly. For example, if Andrew cannot read the word *giraffe* or pronounce it, he can still guess at its meaning by reading the words preceding it: "A tall animal with a very long neck is called a giraffe." Tests of reading comprehension are included on the Woodcock-Johnson Test of Achievement, Kaufman Test of Educational Achievement, and the Test of Reading Comprehension fourth edition (c). As a result, dyslexic readers (permitted sufficient time) often do better on tests of reading comprehension than on measures that ask them to decode isolated single words.

There are instances reflecting a range of issues, including anxiety or extremely poor decoding, when a child will perform poorly on a test of comprehension. Yet when comprehension is examined in more depth—for example, when the child is given a test such as the TORC-4—he performs well on many of the subtests of comprehension and would not be considered to have poor comprehension.

> *CAUTION:* If a child has major problems in reading comprehension, he or she should be diagnosed not as dyslexic but rather as having a reading comprehension problem—Specific-Reading Comprehension Deficit (S-RCD).

After a child is evaluated, we know exactly what pattern to look for in the test results to diagnose dyslexia (we look for both accuracy and effort required by reading):

→ difficulty reading single words
→ poor phonological skills
→ particular difficulty decoding nonsense or unfamiliar words
→ particular difficulty in reading single words when speed is required
→ lack of fluency as demonstrated by inaccurate, slow, and labored oral reading and a lack of inflection (prosody) reflecting the content
→ trouble reading short function words—*that, is, an, for*
→ poor spelling
→ reading comprehension often superior to decoding individual words
→ average or even superior intelligence
→ disparity between intelligence and reading

I mentioned earlier how a phonologic weakness and strengths in thinking and reasoning can be recognized. For example, parents can often observe problems with rhyming or with pronouncing words, or, as with Andrew, difficulty in retrieving words, having words on the tip of his tongue and not being able to pull them out.

Phonologic skills can be directly and reliably measured in a school-age child. The Comprehensive Test of Phonological Processing, second edition (CTOPP-2), assesses a particularly broad array. One specific kind of phonologic test that is quite sensitive to dyslexia in children asks a child to pull apart a word and then to delete a specific phoneme. Not surprisingly, this is called a phoneme deletion (or elision) test. The examiner asks the child, "Can you say *sour* without the *sss*?" (*Our.*) "Can you say *crane* without the *r*?" (*Cane.*) Children who cannot break apart spoken words into phonemes will not be able to link letters to sounds.

VOCABULARY

A child's vocabulary—his familiarity with the meanings of a range of common and uncommon words—provides a good index of his general language skills. In the most commonly used test, the Peabody Picture Vocabulary Test, fifth edition (PPVT-V), he is asked to point to which one of four pictures shows, for example, a *giant,* a *canoe,* a *mammoth,* or an *equestrian.* This test, which does not require a verbal response, is often especially informative for dyslexic children, who may instantly point to the correct picture but have difficulty retrieving the specific name if a spoken response is called for.

THE SYNTHESIS

Once Andrew had completed his evaluation, we examined all the information: what his parents (and teachers and Andrew himself) had told us about his development and school history; what his evaluator observed during her interactions with him; and what his pattern of performance on tests of phonology, reading, spelling, vocabulary, and intelligence told us about how Andrew reads. In each of these elements we looked for evidence of a reading problem, of intact higher-level thinking and language skills, and of a phonologic difficulty.

Andrew struggled to read words. He experienced the most problems in trying to sound out nonsense words. He was able to pronounce only the first few words, performing at the level of a first-grader. In addition to his difficulties deciphering isolated words, Andrew labored over reading passages out loud. He read the stories very slowly and haltingly as he tried to decode the words. Sometimes he would skip two lines and not seem to notice. In oral reading, Andrew scored dismally. Clearly, he was not a fluent reader. In contrast, as his mom had predicted, he scored much higher on a test of reading comprehension based on silent reading. Andrew's high learning capability was evident.

His parents (and grandmother) told us about his development: They described a bright, highly inquisitive, creative child who constantly wanted to know *why,* who put together two-hundred-piece puzzles himself, who spent hours looking at the globe and identifying different countries by their shapes, and who especially loved to be read to. Andrew could listen to *Harry Potter* and stories from Greek mythology over and over again. His teachers corroborated these accounts with their observations of his performance in class, in listening to stories and absorbing even the most abstract and subtle points. Andrew's ability to understand what he heard and his ability to read were, according to his teacher, "like day and night."

Although not a good reader, Andrew was apparently an excellent thinker.

This was clearly evident in the results of his IQ test, where he scored at the ninety-fifth percentile. His performance on tests to get to the sounds of words was the exact opposite. On the phoneme deletion test Andrew could not seem to pull apart the sticky phonemes in the test words. To Andrew, a word was a solid, impenetrable whole. There was no question that he had a phonologic weakness.

On the PPVT-V test, where he could point to pictures of words, Andrew performed in the superior range. Dyslexics often have surprisingly large vocabularies. Andrew was able to demonstrate this. Since he did not have to summon up the word itself and say it, he was not penalized by his phonologic difficulty in retrieving words.

Timed and untimed measures of reading, including tests of reading accuracy (regular and nonsense words) and fluency, reading comprehension, spelling, phonology, vocabulary, arithmetic, and cognitive ability represent a basic core battery for the diagnosis of dyslexia in school-age children. Of course, additional tests of academic achievement in arithmetic, language, writing, or memory may be administered as part of a more comprehensive evaluation. If a child performs much better on a test of math concepts, say, than on reading, that adds to the impression of an isolated reading deficit in the context of other well-developed academic abilities.

Keep in mind that at times dyslexic children experience difficulty with remembering the names of numbers, in memorizing multiplication tables, or sometimes in reading word problems. This should not be confused with the child having a basic problem in mathematical ability. Later you will read about a highly regarded dyslexic economist who experienced these difficulties in her early school years and then went on to develop widely admired complex economic mathematical models.

> CAUTION: There is no one single test score that ensures a diagnosis of dyslexia. It is the overall picture that matters. An extremely bright child who has a reading score in the average range but who struggles and cannot learn to read fluently, and who has all the signs just described for dyslexia, has dyslexia.

This pattern is so consistent and so replicable among dyslexic readers that the approach I have described should ensure that each struggling reader is identified before he endures years of frustration and failure. Keep in mind, as Justice Sotomayor said, that dyslexia is a clinical diagnosis and there is no ceiling score, a number below which a person must score to be considered dyslexic. As I will describe later, as the Americans with Disabilities Act Amendment Act of 2008 regulations indicate, the critical determinant is *how* the person reads, technically referred to as the *condition, manner, or duration* of *how* he or she reads. Is it with great effort, stumbling, skipping words, mispronouncing, or reading incredibly slowly? These factors of *how* the person reads must be given primacy.

> *CAUTION:* There are infrequent instances when older children, often very bright boys and girls who have received large doses of intensive code-based reading interventions, such as programs using the Orton-Gillingham approach (see pages 287–289), will perform surprisingly well on, for example, tests of phonological processing such as the CTOPP-2 but are still not fluent readers. Most likely the lack of fluency reflects lack of practice applying their newly learned phonologic skills through reading lots of material. In other words, these children have the basic skills but haven't honed them through reading practice.

Disorders other than dyslexia may affect reading.

In *developmental dyslexia,* the phonologic weakness is primary; other components of the language system are virtually intact, and the reading impairment is at the level of decoding the single word, initially accurately and later reading connected text fluently. Intelligence is not affected and may be in the superior or gifted range. The disorder is present from birth and not acquired.

In *language-learning disability,* the primary deficit involves multiple aspects of language, including both the sounds and the meanings of words. Here the reading difficulty is at the level of both decoding and comprehension, and language difficulties of all sorts are prominent. Measures of verbal intelligence are significantly affected by language deficits, and intelligence may be in the subaverage range. People are born with the disorder.[*]

In *acquired alexia,* there is a loss or a diminution of reading ability resulting from brain trauma, a tumor, or a stroke affecting the brain systems necessary for reading. There may be other accompanying symptoms, such as loss of speech or weakness on the right side of the body, depending on the exact region affected by the injury to the brain. In contrast to developmental

---

[*] Some speak of dyslexia as a language-learning disability. We differentiate dyslexia involving specifically the phonological component of language from language-learning disabilities affecting multiple components of language.

dyslexia, this disorder is acquired and is most often observed in middle-aged or elderly men and women.

Hyperlexia is a relatively rare disorder whose cause is unknown. In many ways it is the mirror image of dyslexia. Children who are born hyperlexic learn to decode words very early, sometimes even as toddlers or preschoolers. They show an early and intense interest in words and letters, and their exceptional word-recognition ability is often apparent by the age of five. In contrast, they have extremely poor reading comprehension. There are deficits in reasoning and abstract problem-solving. Not infrequently, affected children have difficulties forming relationships with other children as well as adults.

It is also important in diagnosing dyslexia to eliminate other potential contributing factors in school-age children who struggle to read. Children should be evaluated for hearing or vision problems. At one time it was thought that subtle hearing problems resulting from chronic ear infections might interfere with the acquisition of language and later with reading, but the most recent evidence indicates that is not true. There is also a scarcity of hard scientific evidence to support the notion that problems relating to the ability of eye muscles to converge cause reading difficulty.

There are some tests that are *unnecessary* in the evaluation of dyslexia— for example, laboratory measures such as imaging studies (MRIs, CT scans, X-rays), electroencephalograms (EEGs), and genetic studies. These tests should be prescribed only if there are specific clinical indications *other* than dyslexia.

Many people confuse attention deficit/hyperactivity disorder (ADHD) and dyslexia. Some even use the two terms interchangeably. ADHD and dyslexia are two distinct conditions. (See pages 419–429 for more details.) Dyslexia, as you know by now, is a disorder reflecting a difficulty accessing the sounds of spoken language. ADHD is a problem reflecting difficulties allocating, focusing, and sustaining attention. Lack of automatic (fluent) reading may drain attentional resources, resulting in the dyslexic reader appearing inattentive.

Dyslexia is one of a small group of public health problems that we have the ability to reliably screen for and detect early on and effectively treat. We must ensure that each child who is not learning to read in the first year or so of school is identified and treated. It is now possible to minimize the negative impact of dyslexia, but in order to do so such children must first be identified. As I've said, the earlier the diagnosis is made, the better the results are. This is why I now want to turn our attention to the question of evaluating younger children for the very first signs of a reading problem.

13

IDENTIFYING THE AT-RISK CHILD

As you read, Andrew was diagnosed as dyslexic after he had suffered years of frustration and had fallen further and further behind his peers. This destructive holding pattern—failing in reading, but also failing to have it recognized—is absolutely unnecessary for children with dyslexia. Today it is possible to reliably identify boys and girls at high risk for dyslexia *before* they fall behind.

Earlier I discussed the general clues to dyslexia. The focus here is the specific assessment approaches that help to identify potential dyslexia in a young child—the *screening* instruments designed to identify children most *at-risk* for dyslexia. The goal, of course, is to catch the potential difficulties before they become full-fledged problems and begin to harm the child.

Over the last two decades reading researchers have been trying to develop fast, efficient screeners to be able to identify at-risk children as early as kindergarten or first grade. A screener is not a diagnostic instrument; its role is to quickly and efficiently identify those children *at-risk* for dyslexia. Screeners are designed to be given to all children in the earliest grades, a process referred to as universal screening. In order to target those children who need extra help as early as possible, the screeners must be able to pinpoint the children who actually will develop dyslexia (referred to as "true positives") while designating which children definitely will not develop dyslexia (referred to as "true negatives").

Researchers like myself have recognized that one source of screening information that has been ignored is the *child's teacher's insights and judgment about the child's reading*. In the beginning grades teachers have ample opportunity to observe their students over a substantial period of time as the students engage in relevant learning contexts. They are ideally situated to assess the early language and academic behaviors indicative of high risk for dyslexia in their students. Given the often vastly differing preschool experiences of children and where they are on the reading continuum when they first enter school, it is best to wait until the second semester before screening. At this point the teacher's insights and judgment about the child's reading are an important source of information. It is the child's teacher who has spent months closely working with him and observing the daily ups and downs, the frightening frustrations, and the satisfied smiles that are all part of the process of becoming a reader. These experiences are reflected in the teacher's responses on the screening measure. Too often his or her unique insights and depth of experience are underutilized. This is unfortunate, and we hope it will change.

We found that using the teachers' responses to a small subset of questions (comprising ten items from kindergarten, twelve items from first grade, ten items from second grade, and ten items from third grade) from the Multigrade Inventory for Teachers (MIT) developed at our Yale Center for Dyslexia & Creativity allowed them and us to predict children at high risk for dyslexia with a high degree of accuracy. This screener, the Shaywitz DyslexiaScreen™ (Pearson, 2016; Figure 38), has a number of significant advantages compared to other screening measures: It engages the teacher, captures his or her insights about the child, is extremely efficient, is very inexpensive, is completed by the teacher who knows the child, and has very good psychometric characteristics. The screener is completed on a tablet, which reveals the results—"at-risk for dyslexia" or "not-at-risk for dyslexia"—almost immediately. Most important, this screener is evidence-based—it has been demonstrated to work. Furthermore, it does not take children away from precious

Figure 38. The Shaywitz DyslexiaScreen™, a reliable, user-friendly screening tool for K–3 students who may be at-risk for dyslexia

instructional time. Indeed, the Shaywitz DyslexiaScreen™ is enthusiastically being used nationally in diverse locations, including Palo Alto, California, the state of Missouri, and Brooklyn, New York.

For the child identified as *at-risk*, there are two possible pathways: (1) immediately being placed in an intense early intervention program, or (2) having a full diagnostic assessment. I strongly recommend the latter. A diagnosis is critical for ensuring that the child receives the most appropriate intervention. A diagnosis also allows the student, his family, and his teacher to understand and be able to address his needs as a dyslexic child. For those children who are screened and identified as not being at-risk, parents and teachers can follow what should be an evidence-based reading curriculum, monitoring their child's progress.

An evaluation of the skills (especially phonologic) known to be related to reading success is appropriate for preschool, kindergarten, and first-grade children who are identified as being at-risk for dyslexia and is carried out on an individual basis by a professional who is knowledgeable about reading and dyslexia: a speech and language pathologist, a learning disabilities specialist, a reading specialist, an experienced psychologist, or even a teacher trained to recognize and understand dyslexia and to administer the tests.

Pardon my repetition of something I said earlier, but it is terribly important: New data from the Connecticut Longitudinal Study indicate that dyslexic readers are reading far below their typically reading peers as early as first grade, and that this gap persists into adolescence and most likely beyond. If we are to close the reading gap between dyslexic and typical readers, we must implement effective reading programs as early as possible, ideally in kindergarten and no later than first grade.

Research has opened our eyes to how important it is to have phonologic skills in place during the very first years of school. To repeat: By the end of first grade, most children have just about mastered their basic phonological skills. From second grade on, a child's development of these competencies is more a matter of refining and gaining efficiency or automaticity in the phonological skills he previously acquired.

Phonological sensitivity refers to the ability to focus on the sounds rather than on the meaning of spoken words. The child can tell you what word rhymes with *cat* rather than simply whether a cat is a kind of animal. Beginning at about age four, phonologic skills develop gradually over time and in a predictable, logical sequence. Awareness of this sequence and its timing makes it possible to recognize when a child is veering off course and get her back on track before any serious damage is done.

In describing a child's phonologic skills, two terms are often used. *Phono-logic* awareness is a general and more inclusive term that includes all levels of awareness of the sound structure of words. It is also used to refer to the earliest stages of developing an awareness of the parts of words, including sensitivity to rhyme and noticing larger parts of words, such as syllables. *Phonemic* awareness is a much more specific term and refers to the more advanced ability to notice, identify, think about, and manipulate the smallest particles of language making up a word: phonemes. Phonemic awareness has the strongest relationship to later reading, and most tests focus on it.

Most clinicians and researchers have come to depend on the Comprehensive Test of Phonological Processing to test for the full range of phonological skills. The second edition now covers the four- to six-year-old range, an improvement over the first edition, which was normed down only to age five.

Though not as robust as tests of phonological awareness in predicting whether the young child is at-risk for dyslexia, a child's knowledge of letter names and sounds may also serve as a helpful guide to how ready to read she is. Testing letter knowledge is straightforward; it can be done informally by asking the child to name letters presented one at a time on a card. Similarly, knowledge of letters and sounds is tested by asking the child, "Can you tell me the sound(s) this letter makes in words?" Using a reading test that contains a letter-identification section—for example, the letter-word subtest on the Woodcock-Johnson Test of Achievement-IV—is more formal.

To reiterate, teachers are critical in identifying who may be dyslexic. As I noted, using the Shaywitz DyslexiaScreen™, we found that the end-of-year kindergarten and end-of-year first-grade teachers' ratings were able to identify with a high degree of accuracy which children would later be diagnosed as having dyslexia.

WHAT A PARENT CAN DO

Here is what I believe to be the most scientifically sound and sensible approach to identifying young children at-risk for dyslexia before they experience reading failure:

→ Observe your child's language development. Be on the alert
   for problems in rhyming, pronunciation, and word finding
   (detailed on pages 111–117).

→ Know your family history: Be alert to problems in speaking, reading, writing, spelling, or learning a foreign language. Some families with more than the average complement of dyslexics seem to have an abundance of photographers, artists, engineers, architects, scientists, and physicians (psychiatrists and radiologists). Somewhat less frequent but still impressive are the large number of families sprinkled with great writers, entrepreneurs, and jurists who are dyslexic.

→ If there are clues to problems with spoken language, have your child tested.

→ Make sure your child's school carries out universal screening to identify those children at-risk for dyslexia.

CAUTION: I am often asked about delaying kindergarten for a child with indications that he may have reading difficulties. *I strongly recommend not delaying kindergarten entry;* this will only delay needed help. While many parents believe that a child will respond better to the demands of kindergarten if he is a year older than his classmates, the evidence says otherwise. A group of researchers took advantage of what was in essence a natural experiment to test whether it was biological maturity or exposure to reading instruction that most influenced early reading. The researchers compared two groups of young children who were hovering on either side of the cutoff date for kindergarten entry. One group had birth dates that allowed them to begin kindergarten. A second group was similar in every way except they just missed the cutoff for kindergarten entry. They were a month younger and had to wait a year before starting. When tested later, those in the group that entered kindergarten were reading better than the delayed-entry group. They were also better off academically and emotionally. (In first grade, early starters do better, too.) Staying back did not help the children in their learning and seemed to carry an additional negative psychological burden.

This should come as no surprise. Earlier I discussed data indicating that there is no such thing as a developmental lag. If a child is experiencing a problem in reading, it is not going to get better by allowing more time to pass before he receives the appropriate help; he needs an evidence-based intervention. *Instruction that fails a child once is not going to help him a second time around.* So the evidence comes out strongly against retention. Keep in mind that if dyslexia is caught early and the child receives effective intervention, she can begin to catch up. In contrast, a child who is delayed in receiving such instruction has great difficulty in narrowing the gap, not to mention seeing herself as a success in school.

The evidence seems clear that delaying a child's entry to school does not help him become a better reader. On the other hand, early identification, when linked to effective programs of intervention, can make a difference. Such intervention can help to insure that the vast majority of children at-risk for dyslexia will never have to experience the trauma of prolonged reading failure. *Rather than age or "maturity," it is reading instruction that leads to better reading.* Good help is available now as never before.

14

DIAGNOSING BRIGHT YOUNG ADULTS

Almost three decades ago I received a call from the dean of undergraduate studies at Yale University. He wondered if I could help him with an unusual request from Kayla, an undergraduate about to enter her senior year, who wanted to do what was unheard of at Yale at that time: spend an extra, ninth semester at Yale. She said she was dyslexic and required the extra semester. The dean was astonished: "I've reviewed her records and she is absolutely brilliant!" Having had little experience with dyslexic young adults, I, too, was puzzled. So I started where every diagnostic journey must begin: with the history. This was my introduction to the world of bright young adults with dyslexia.

Kayla struggled early on and since age eight had attended specialized schools where she was able to receive close attention and specific help for her dyslexia. An evaluation from third grade seemed to reflect the comments of many of her teachers: "Her oral reading is characterized by frequent hesitations, pauses, and repetitions. Although she is working very hard to master reading subskills, she is not yet successful in using them to figure out unfamiliar words. Comprehension, however, is quite good."

The indicators were clear. I knew Kayla was dyslexic. I was impressed that a dyslexic could be successful in a college where there was such a heavy demand for reading, and I wondered how many students like Kayla attended top-flight colleges. Although Kayla's standardized test scores were not stratospheric, her qualities as "a creative thinker, an inquisitive scholar, and an independent leader," as cited by one of her high school teachers, must have caught the eye of an admissions staffer.

Kayla explained that she had completed her first three years "at a great cost." Reading very slowly and with great effort, she had devoted all her time and energy just to keeping up. Now that she was approaching her senior year, she did not want to see the quality of her education diminished because of lack of time:

> Though my grades in [my] courses are not bad, I feel that to the degree to which grades can reflect any truth at all, they reflect a level of mediocrity inappropriate to my intellectual aptitude, earnest scholarship and industriousness. . . . I am requesting a ninth term of study in which to complete these courses in a manner befitting my intellectual aptitude and earnest studentship. . . . The point is not that I am almost inarticulate, nor that I can barely struggle through a page of text. . . . I want the stuff of a liberal education, for whatever it's worth, to carry with me in my life. I want to possess, to understand, to challenge and to recreate the ideas, knowledge and visions of the liberal tradition.

Kayla was given permission to extend her studies for one semester and graduated from Yale the following year. She is now an accomplished screenwriter with an impressive record of television and movie credits.

Although no one is ever cured of dyslexia, highly intelligent, exceptionally hardworking, and strongly motivated young men and women with dyslexia can attend rigorous colleges and successfully complete their education, as hundreds have at Yale, Brown, Harvard, Stanford, Washington University, Duke, the University of Pennsylvania, Williams, Amherst, Middlebury, and other excellent colleges and professional schools. I know that at Yale our dyslexic students have performed well, reflecting the full range of abilities and personalities of all our students. Admissions officers are clearly attuned to the competencies of these students; they appreciate that the qualities they are looking for go far beyond what test scores have to say about a person. The University of Chicago is, too, having announced in June 2018 that standardized tests such as the SAT and ACT are now *optional* rather than required for admission.

From my thirty years of working directly with dyslexic students at Yale and other universities, and from my research and that of others, the sea of strengths interspersed with isolated islands of weakness model has held up. In particular, the students at Yale have shown me what it means to be bright and dyslexic. I consult about issues relating to dyslexia in students and faculty at Yale, and in the process I read through their records—essentially mini-biographies. As I peruse these documents, I read of little boys and girls who struggled to read or couldn't quite form their letters or print their names neatly, who could not spell, who stuttered and stammered as they mispronounced words, who puzzled their teachers and parents, and who were poked and prodded to try to understand why. I think of their mothers and fathers and all that they and their children endured. I cannot help but think of the worries each parent must have had. Reading one report after another, I imagine how much parents (and children, too) must have worried and how gratifying it must be to see their dyslexic child now as a successful student at such a high level.

I have come to know dozens and dozens of brilliant, extraordinary students and faculty at all levels, met with them, talked with them, problem-solved with them, and in great detail have reviewed their testing records. I have learned that at the level of college and beyond, standardized and other tests tell us very little of a dyslexic individual's potential for academic success and for future contributions to society.

> *CAUTION:* Be aware that many tests that are labeled as "cognitive" measures may not be fully tapping into and measuring some of the essential qualities that converge to bring about brilliance, creativity, or perseverance.

In a dyslexic student, so-called average scores on reading tests cannot measure the extraordinary effort that went into reading each word, to laboriously pronouncing it syllable by syllable, to rereading it over and over again until it began to sound right and make sense. What a juxtaposition: the tremendous cognitive strengths and conceptual ability of the bright dyslexic young adult on one level, and on another her ongoing struggle, at the most basic level of language, to try to decode and identify the printed word. Here she is, all geared up, her powerful intellectual equipment waiting to be fed its nourishment of words so that she can assimilate and feast on the ideas and the thoughts they contain, but to get to the basic substrate, the printed word, she must first decode and identify that word. What a frustration. What perseverance. A new study, the Yale Outcome Study (see pages 399–400), which I will share with you later, presents heartening and reassuring data about what happens when bright, hardworking, persevering dyslexic students meet up with an enlightened university.

For students attending competitive colleges or enrolled in graduate degree programs or professional schools, the texture of their lives and their accomplishments will tell us more than their performance on timed tests. If a student is thriving at a highly competitive college, am I to believe that a timed test is a better measure of her potential for learning than her insightful comments and thoughtful responses in the classroom, her ability to appreciate the big picture and get the gist or main point of what she has read or heard in a class discussion, or her compelling essays or performance on untimed exams? As comedian Richard Pryor said, "Who are you gonna believe, me or your lying eyes?"

DIAGNOSIS

In 1996, in celebration of the hundredth anniversary of the first description of developmental dyslexia, I was invited to write an article for *Scientific American*, "Dyslexia," in which I discuss Gregory, the medical student whom we met earlier, who typifies many bright young adults with dyslexia. He was extremely bright, understood the most difficult concepts, and yet had trouble pronouncing anatomical names and memorizing isolated facts. Following the publication of that article, I received many letters and emails from students and parents of students. Some of the students had already been diagnosed as dyslexic, while a substantial minority had never been evaluated for dyslexia. For example, Brandon Rogers, a philosophy major in a liberal arts college in the Midwest, wrote:

> I have always done well in school. . . . If I didn't get something, I would study it and study it until I did. I would revise papers over and over again until they were pretty much perfect. If it meant staying up much (or even most) of the night, I would put in whatever work was necessary to do well. Until recently, I never thought anything of this. I figured that the problems I had learning that made me have to put in so much overtime were the same problems everyone had. No one ever thought I might have a learning

disability, much less dyslexia. I always did well in school. I was the smart one. I got A's. Therefore, why bother testing me for something I clearly didn't have? . . . When I first heard about dyslexia, it was one of those meaningless terms thrown around, and seen as an excuse for people who don't like to work hard. But the more I have learned about it, the more I feel it might be affecting me. The pronunciation problems I have, the insane spelling errors, the abysmally slow pace of my reading, the needing to read things over and over again, the problems getting down what the professor says— I thought that maybe I had figured out what has been causing some of the issues I've always had. I just want to know for sure.

Brandon's questions echo those I hear from many college students and other young adults concerned about their reading. Yes, it is both possible and, I think, worthwhile to obtain a specific diagnosis of your reading difficulties. No, Brandon's constellation of problems is not normal, but it is exactly what we would expect to see in a bright young man or woman who is dyslexic.

Following scores of dyslexic children as they matured into young adults has given us a picture of the evolution of dyslexia over time. This clinical information is now enhanced by research studies of reading in adults (both dyslexic and typical readers) and by real-time studies of brain function during reading. These research findings have important implications for the diagnosis of dyslexia in young adults.

As children mature, bones elongate, muscles strengthen, voices deepen, fingers grow steadier, strides lengthen, knowledge accumulates, and skills sharpen. Yet the core phonologic deficit in dyslexia remains, even in the brightest and hardest-working children. Numerous studies conducted on college students and other adults document the persistence of the phonologic deficit.

The implication of all this for the diagnosis of dyslexia in adults is that, as Johns Hopkins researcher Maggie Bruck noted, phonologic deficits are constant through the lives of men and women who are dyslexic. This means that data gathered from a dyslexic as a child that demonstrate phonologic difficulties will align with data from his or her adult life that continue to give indications of problems in phonologic processing. Phonologic deficits persist—even though they may improve a bit—in virtually all those diagnosed as dyslexic in childhood. Of course, dyslexic readers can learn to read more words accurately and typically read with good comprehension. But the persistent phonologic deficit prevents fluency. Difficulties during childhood evolve into slow, labored reading in adulthood. Thankfully, it is our cognitive capability and not our phonology that allows us, as humans, to reason, to analyze, and to solve problems at the very highest levels—and to be capable of great accomplishment.

Researchers discovered that dyslexic college students take a very long time to identify words, even words they can eventually identify correctly. If dyslexic readers attempt to (or are made to) read too quickly, they simply will not be able to identify words accurately. This may be misinterpreted as a lack of understanding. Just how much of an issue speed is for dyslexic adults was demonstrated in what is referred to by researchers as a reading match study. College students with histories of dyslexia were compared to sixth-grade children reading at the same level of accuracy. *The dyslexic college students were the slower readers.*

In order to understand how college-age dyslexics read, we must separate the number of words a person can read from the *manner* in which he goes about reading those words. Two college students, one dyslexic and the other a typical reader, may read the same word accurately but via two entirely different neural routes.

Earlier I talked about evidence from brain research that offers tantalizing clues to how dyslexic and typically reading adults diverge in their reading pathways. Recall that in skilled readers, the fast-paced occipito-temporal system automatically recognizes a word, bringing together the core components (the phonology, the orthography, and the semantics), and the word is read rapidly and effortlessly. In a dyslexic reader, an inefficient occipito-temporal system prevents automatic reading. Instead he is dependent on secondary reading pathways, often those in the front and right side of the brain. Dependence on these secondary routes to reading has important implications for how he reads. First they allow him to read a large number of words, albeit very slowly. Second, rather than the sound-based route to reading, the dyslexic reader relies on her higher cognitive abilities to infer the meaning of the unknown word from the surrounding context. This is why the bright dyslexic reader does so much better in reading words in context compared to words in isolation or with scanty contexts.

Slow reading is the price the dyslexic pays for this more indirect route to getting to the meaning of an unknown word. At the same time, dyslexic young adults, with much perseverance and sufficient time, are well able to comprehend college-level materials. Studies have confirmed the universal nature of this faulty wiring; dyslexic students worldwide, speaking different languages, share a disruption of the primary neural systems for reading.

Understanding that the dyslexic reader relies on neural systems that permit accurate but not fluent reading helps explain why the kinds of tests and approaches used for diagnosing dyslexia in young children may not be suitable or sufficient for older children and young adults. Scores on traditional (untimed) tests of word recognition may not differentiate accomplished young adults with dyslexia from peers of a similar age who are good readers. Tests of reading fluency, which are often timed, are more valid measures of dyslexia in these young adults.

It is vital that scientific data is translated into everyday clinical practice so that decisions made about dyslexic young men and women are based on science and not on misconceptions or outmoded views. The latest scientific data provide strong validation of the unexpected nature of the reading difficulty in bright young adults who are dyslexic. Such individuals may score very high on a test of intelligence, especially on verbal reasoning components, and, in contrast to typical readers, for whom intelligence and reading are tightly linked, score quite poorly (much lower) on a test of reading, especially reading fluency. Tests of intelligence, happily, provide tangible evidence to the individual that he or she is intelligent. The difference this knowledge makes for that individual is incredibly powerful.

For the most part, though, the heart of a diagnosis of dyslexia in a bright young adult lies in her history. At this stage of life, dyslexic readers have accumulated enough life experiences to document a trail of phonologic-based reading difficulties. For some, like Kayla, it is a history of reading difficulties leading to evaluation, tutoring, or special schooling. For others, like Brandon, often because they are so bright, the diagnosis of dyslexia is delayed, and symptoms are brushed aside with the explanation, "Oh, but he's smart. Let's wait." This is the kind of child who falls between the cracks and escapes diagnosis until he is an adult.

I remember Aaron's mother's frustration as she told me how she knew that there was a problem when she noticed Aaron's agonizingly slow reading and abysmal spelling and how he would do his homework well into the night. She recalls how she worried about his bar mitzvah preparation, only to be told by his secular and religious instructors virtually the same things: not to worry, Aaron was so very, very intelligent that it would come. When she shared her concerns with his Hebrew tutor, he, too, was dismissive, again noting Aaron's superior intellect. However, after several lessons, the tutor sheepishly contacted Aaron's parents and agreed that despite his high intelligence, the boy had some serious reading issues, along with significant difficulties learning a second language, Hebrew. Aaron was finally diagnosed as dyslexic many years after he had endured painful frustration, anxiety, and self-doubt. (I would be remiss if I didn't mention, too, the very many bar and bat mitzvahs of dyslexic boys and girls I have attended, where, with appropriate support and tutoring, the children shone on their special day.)

> *CAUTION*: A child's high thinking and reasoning ability can work against him. As I have said, in dyslexia, intelligence and reading are not related. Do not allow a child's high intelligence to be used to delay diagnosis, help, and self-understanding.

We look for a pattern of findings pointing to a phonologic deficit permeating almost every aspect of a young adult's life. Each case will differ as to the specific ingredients of that pattern; *there is no single event, symptom, or test score that in itself is diagnostic.* The phonologic weakness is virtually always present, but we must know how to recognize it in different settings and during different life stages. A developmental history of difficulties with language, particularly phonologically based components of language, often provides the clearest and most reliable indication of a reading disorder. Slow reading is the primary symptom of dyslexia in an accomplished young adult. As I just explained, in these bright individuals, standard assessment procedures used in children may be inadequate and often misleading. So for now, the history of word-retrieval difficulties, laborious reading and writing, poor spelling, and requiring additional time in reading and in taking tests often represents the most sensitive and accurate indicator of dyslexia in young adults. The diagnosis of dyslexia at any age is not about formulas or magical cutoff scores; it is about understanding how a phonologic deficit affects people's lives. Theory provides the template, but true understanding comes from developing an intimate sense of the disorder as it is expressed day in and day out over a span of years, and even decades.

Our approach to the diagnosis in young adults comes naturally from all that we have learned about reading and the brain. Above all, it calls upon clinical judgment. It is a thinking man's (or woman's) approach, not based on unproven formulas or arbitrary decisions. Rather, cutting-edge science and good old-fashioned common sense drive the process. If one understands the basic nature of dyslexia, the diagnosis unfolds easily and naturally. It is critical therefore for evaluators to understand dyslexia fully so that they can function as *diagnosticians* rather than as rigid adherents to unproven theories.

Young men and women who are dyslexic may manifest what seems to be a confusing array of contradictory findings—honors and failures, commendations and warnings, the ninety-ninth percentile in some areas and the ninth percentile in others, brilliant thinker and hopelessly slow reader—all in the same person. Dyslexic readers experience awesome highs and devastating lows. One day they are praised for incisive thinking, so that they feel they must be very smart, and the next day a disappointing performance on a timed multiple-choice test makes them feel "really dumb" and leads them to wonder how they "could have fooled so many people for so long."

MAX'S STORY: IN SCHOOL, AND NOW IN THE REAL WORLD

For all those who continue to feel very nervous if called on to read and who to this day, as an adult, "can still hear my classmates' laughter and hurtful remarks," please know that there can be a happy ending, which is why I want to share Max's story with you. I met Max when we were both invited to speak at the September 18, 2014, congressional Committee on Science, Space and Technology hearing on "The Science of Dyslexia." Max gave testimony on the personal impact of having dyslexia. We went to dinner the evening before the hearing and Max shared many of his experiences growing up. He described having to read aloud a passage about John F. Kennedy to the class, recalling how he stuttered and stumbled while the whole class laughed at him. The experience was devastating, so that "it pretty much killed reading from a prepared text for the rest of my life." Sure enough, decades later, when testifying before Congress, Max began by stating, "I was asked here to talk about what it was like being dyslexic. I will say for me, first of all, let's just say that I didn't prepare a statement because the last time I tried to read aloud from a prepared statement, my whole fifth-grade class laughed at me. . . . So I am just going to be brief and speak from the heart."

And he did. Max is Max Brooks, the best-selling author of *The Zombie Survival Guide: Complete Protection from the Living Dead* and *World War Z: An Oral History of the Zombie War,* which was made into a motion picture with Brad Pitt. Most recently, Max wrote a novel called *Minecraft: The Island.* His books have sold millions of copies, and he is besieged by autograph seekers.

Thinking of so many of the dyslexic young men and women I have come to know, including Kayla and Brandon, I asked Max, "When you were a little boy, did you ever dream of hordes of people, even one person, wanting your autograph?"

"Never, never, never, never, never. I never thought I would be successful at anything. That's how low my self-esteem was."

When you feel you are up against a brick wall or unable to read aloud without hearing snickering from your classmates, please keep in mind the arc of Max's life—rising from shame, anxiety, and low self-esteem to a successful and happy adult life.

Science and common sense say that if a person reads extremely slowly and yet is able to comprehend materials at a college level or beyond, he most likely has a phonologic deficit as well as significant strengths. Together, this strongly suggests he is dyslexic. In my experience, the diagnosis becomes straightforward once the diagnostician understands the basic nature of dyslexia. This knowledge can now be applied to the three-step process we established earlier for the diagnosis of dyslexia in young children.

→ *Step 1. Establish a reading problem according to age, education, or intelligence.* A major step in establishing a reading problem in an accomplished adult comes from information gathered from a careful history. The most consistent and telling sign is slow and laborious reading and writing. The failure to recognize or to measure the lack of fluency in reading is probably the most common error in the diagnosis of dyslexia in bright young adults.

Keep in mind that a simple word-identification task often will not identify dyslexia in someone accomplished enough to graduate from college and attend law, medical, business, or any other graduate school. A dyslexic young adult may score well reading real words in an untimed setting, reflecting that she may previously have come across those words in her readings and committed the words to memory. When she sees one of those words on a word-identification task, she retrieves the stored word from memory and identifies it without having ever learned to decode it. So a dyslexic young adult's score when asked to read a list of real words may be deceptively high. Tests relying solely on the accuracy of word identification may therefore be inadequate for diagnosing dyslexia in these individuals.

Tests of reading single nonsense words (word attack) are more helpful, because to be read, these words *must* be decoded. Since they are made-up words, they could not have been previously encountered, stored, and then retrieved from memory.

→ To establish lack of fluency in accomplished adults, someone can read timed paragraphs and respond to questions on tests such as the Nelson-Denny Reading Test (NDRT, for ages fourteen to sixty-five years), or read timed connected texts aloud and answer questions based on the text on tests such as the Gray Oral Reading Test, fifth edition (GORT-5, for ages six to twenty-three years, eleven months).

CAUTION: An article that appeared in the *Journal of Learning Disabilities* in 2013 reported that reading comprehension test items on the NDRT can be answered correctly *without reading the associated passage.* Strange as it seems, test-takers were asked to respond to questions based on paragraphs they had never read. Results indicated that higher IQ, verbal comprehension, and broad reading skills were significantly associated with higher NDRT passageless scores; based on the question itself, test-takers were able to make reasonable guesses of the multiple-choice answers. This finding resonates with my and others' experience with bright young adults who lack fluency and must read with great effort and then go on to score high on the NDRT comprehension test (both time-limited and with additional time). It is often the very brightest and least fluent young man or woman who obtains such a score.

This is important to a young adult seeking a diagnosis of dyslexia, because it means that a person can score high on the comprehension components of the NDRT without ever reading the paragraphs. Seeing this score, an examiner may *incorrectly* conclude that the individual, scoring high on a timed reading comprehension measure, is not in need of extra time. The brighter you are, the more likely it is that you will be able to score well on the NDRT, even without ever having read the passage. A person can score high on the *timed* comprehension component of the NDRT and still require extra time on tests.

→ *Step 2. Gather evidence supporting dyslexia's unexpectedness.* High learning capability may be directly determined by administering a test of intelligence, such as the Wechsler Adult Intelligence Test (WAIS, ages sixteen to ninety) or, for those a bit younger, the WISC-V (ages six to sixteen years, eleven months). Alternatively, or in addition, high learning capability may be inferred on the basis of an educational or professional level of attainment. An accomplished adult's educational level or professional status provides a very good indication of learning capability. Graduation from a competitive college or status as a graduate, medical, business, or law student indicates high capability.

A lack of fluency (slow reading) in such an adult signifies an *unexpected* reading problem or dyslexia. Any scores obtained on testing must be considered relative to those of peers with the same degree of education or professional training. I prefer that the individual also be given an IQ test, such as the WAIS. As I've said, viewing their objective scores on the WAIS allows individuals to see how intelligent they truly are.

→ *Step 3. Demonstrate evidence of an isolated phonologic weakness, with higher-level language functions such as vocabulary and comprehension relatively unaffected.* An ongoing pattern of slow reading, word-retrieval difficulties, poor spelling, and difficulties in learning a second language in the face of high levels of educational or professional achievement provides evidence of an encapsulated phonologic weakness. Similarly, strong conceptual and reasoning abilities on the WAIS, or strengths in reading comprehension or other nonphonologically-based domains, such as higher-level math, indicate a sea of strengths.

Following these three basic steps and applying knowledge of the developmental pattern of reading and dyslexia results in a precise and scientifically valid diagnosis of dyslexia in an accomplished young adult or adult. If a person is dyslexic, he is dyslexic for life. There is no rational reason for college students to be retested if they have been evaluated and diagnosed in adolescence or later, unless there has been an extraordinary change in their symptoms or life circumstances. Retesting may actually be harmful, placing unnecessary psychological and financial burdens on an already overwhelmed dyslexic student.

This was legally affirmed in a 2014 settlement between the Department of Justice and the Law School Admission Council (LSAC). If a person has received accommodations on a high-stakes standardized test and is now requesting similar accommodations on a Law School Admission Test (LSAT), that individual should receive these accommodations without reassessment. In 1997, in a case brought by students with learning disabilities at Boston University, the judge ruled that there was no basis for requiring retesting of college students every three years, noting that "defendants have produced no peer-reviewed literature or scientific testimony that provides evidence for the idea that a person's learning disability will show any change after adulthood, or that a student's test scores will show substantial change during their college career." This legal opinion reflects scientific fact.

The often-repeated notion that some students *pretend* to be dyslexic is utter nonsense. In my experience with now literally hundreds of students, I have yet to encounter a young man or woman who falsely claimed to be dyslexic. For those who understand dyslexia and its tremendous costs to the individual, the very idea that someone would willingly seek such a diagnosis is absurd. As Sally Esposito, a former coordinator of disability services at Yale, explains, "The students just don't want to identify themselves. They don't want to be labeled. I guess they simply don't want to be viewed as different or disabled. They are so embarrassed by the stigma. They panic when they think their professors or classmates will find out. One of my biggest problems is getting students to identify themselves and come to the office for the services that they are entitled to. Often they just won't do it. They say the cost is just too high."

Similarly, I frequently hear that the diagnosis of dyslexia is somehow vague or lacking precision. That, too, is simply untrue. Each time I take my turn as attending physician on our acute in-patient pediatric wards, I am humbled by the often unavoidable lack of precision—even in this day of molecular medicine—in making even the seemingly simplest and most common diagnoses. What many do not appreciate is the huge variability inherent in human biology. In the case of dyslexia, though, our knowledge of the phonologic weakness and its impact across the lifespan now allows us to make a remarkably informed clinical judgment with a high degree of confidence.

15

RECOGNIZING DYSLEXIA IN ADULT MIDLIFE WOMEN

Science has provided new insights suggesting a relationship between dyslexia and certain difficulties experienced by many midlife women. These findings describe how signs of dyslexia may surprise women of a certain age and stage of life, particularly around the menopausal years. Menopause is due to the greatly reduced function of the ovaries, resulting in severely diminishing levels of ovarian hormones, especially estrogen. The time prior to menopause is termed perimenopause, a time of transition when the ovaries gradually make less and less estrogen. This process occurs over time, typically during a decade before menopause, when a woman is in her forties or even late thirties. During the perimenopausal time of life, a woman has greatly reduced estrogen available to her. The average age at menopause is about fifty years and is diagnosed after a woman has not experienced her period for about a year, at which point she enters postmenopause.

I want to focus on the relationship between the symptoms of dyslexia and a woman's estrogen loss beginning with perimenopause and extending to menopause and postmenopause. Let's start with Louise, who sent this email to me:

> My dyslexia appears to be worsening as I get older. I am approach-
> ing menopause and wonder if this has anything to do with what I

am experiencing. All the symptoms I had when younger seem to be increasing at a steady and concerning pace. The word finding is rearing its embarrassing head now more than ever, and it is taking me forever to read, whether it is work-related materials or even the newspaper. I am an attorney and you can imagine how troublesome and yes, even frightening this is to me.

Louise's concern is typical of many women throughout the stages of menopause who worry that they are having increasing trouble retrieving words. These worries often occur in highly successful women in their fifth to sixth decades of life; that is, they are entering or are in menopause. The symptoms of dyslexia may appear to be less prominent during middle adulthood. However, as these women mature, they often feel that their symptoms are becoming more prominent and troublesome. They worry that they are losing their intelligence or have the first signs of Alzheimer's. As I speak to these women, I am no longer surprised when I hear in a hushed semiwhisper, "I haven't told many people this, but I am also dyslexic." A distinguished colleague of mine complained of word-retrieval problems like Louise's: "It was on the tip of my tongue, but I just couldn't get the right words out." We spoke, and I asked if she had ever been diagnosed with dyslexia. "Oh yes, when I was in school," she replied. Soon afterward we were both attending the same lecture at a conference when she raised her hand to make a comment and couldn't find the words. She looked back at me in the audience and winked as she mouthed, "Yup, now I know, this is my dyslexia showing again." Afterward she shared with me how reassuring our conversation had been: Otherwise she would have been terrified when she could not get the words out during the lecture.

These women may or may not be surprised to learn that there may be a strong connection not only between their word-retrieval difficulties and their dyslexia but also between female hormones and reading and speaking.

SEX DIFFERENCES IN THE ORGANIZATION OF THE BRAIN FOR LANGUAGE

For decades dyslexia was thought to affect only males. As a consequence, only boys and men were invited to participate in research studies on dyslexia, and psychologists and educators tended to identify only males as having dyslexia. When we began our brain imaging studies of dyslexia, I insisted that women as well as men be included. As I said earlier, our first study was to use functional imaging to map out the language areas in the brain, and we discovered significant differences in the brain activation patterns of the women compared to the men. To remind you, women activated both left and right sides of their brains while men activated primarily the left hemisphere (see Figure 20, page 73). These findings spawned an international discussion of sex differences affecting cognition and language, including a series of cartoons (see figures 39 and 40).

What did these findings tell us? I began to delve into the subject, especially after I was appointed to serve on the National Research Council Committee on Exploring the Biological Contributions to Human Health: Does Sex Matter? For this committee I was responsible for writing about how a person's sex affects his or her cognitive function. I reviewed the cognitive areas in which males and females differed and looked at the list of functions in which women's performance exceeded that of men, particularly in the verbal domain. As I read through each of the functions, I was struck by how many of the functions that favored women were the very same functions strongly influencing reading and often reflecting phonological processing. Specifi-

Figure 39

"It isn't that I don't love you. It's just that I've evolved and you haven't."

Figure 40

cally, women show an advantage in verbal fluency, speech production, ability to decode a language, spelling, and articulatory skills.

As noted in the report, "the term verbal abilities is not a unitary concept," which means that sex differences do occur but not for every verbal function. Indeed, women do not generally excel over men in other areas of verbal function, such as comprehension and vocabulary. It's interesting that these two areas of verbal function are precisely those that typically are *not* affected by dyslexia. These and similar findings led me to postulate that the female sex hormone, estrogen, may have an impact on phonological function and, with it, reading and speaking.

To further understand the relationship of estrogen, cognitive function, and phonologically related skills, we designed the first studies to examine the potential impact of estrogen on reading-related skills. The initial study was a functional brain imaging (fMRI) experiment in which we examined the effects of estrogen on neural systems in the brain in postmenopausal women. We observed that the inferior parietal lobule, a brain region specifically activated by phonologic tasks, shows increased activation in postmenopausal women taking estrogen. This study led us to hypothesize that, through its actions on phonologic processing, estrogen would exert a positive influence on reading in postmenopausal women. It also suggests that in women like Louise, the reemergence of dyslexia symptoms during menopause may be related to their diminished estrogen levels.

In our next study, a state-of-the-art randomized, double-blind, placebo-controlled clinical trial (RCT, the gold standard in treatment studies), we found that, as we reported in the journal *Menopause,* in post-menopausal women *estrogen treatment was associated with significantly better performance on oral reading and verbal memory.* These findings should encourage those physicians caring for women in the menopausal years to take seriously concerns about difficulties in reading, speaking, and memory experienced by those women and to understand that these problems have a biologic component and that there is a possible effective treatment: estrogen.

> *CAUTION:* It is important to point out that while estrogen may be helpful to dyslexic women when they are at or approaching menopause, it also may have serious consequences for certain groups of women. As with any medication, you must first check with your physician.

If you are a woman approaching menopause and find yourself misspeaking or unable to utter the word you intend to say, experiencing reading difficulty—especially finding reading effortful and slow—and/or suddenly making spelling errors, think about dyslexia as a possible cause. Louise, who had been diagnosed as dyslexic, recognized these symptoms for what they were. There are many women who experience these symptoms who have never been diagnosed as dyslexic and worry about what they may imply. Never having been diagnosed as dyslexic does not mean you are not dyslexic. As we've seen, schools have not always done very well in identifying dyslexia in children. If these symptoms keep recurring, whatever your age, seek out an evaluation. There are highly effective steps we can take to teach reading at any age. It is to that subject I now turn.

PART III

HELPING YOUR CHILD BECOME A READER

16

ALL CHILDREN CAN BE TAUGHT TO READ

One of the most rewarding experiences I can imagine is seeing a child who once was sad and defeated in school transformed into one who glows with an eagerness to learn. The new scientific breakthroughs in the teaching of reading are responsible for exactly that, in case after case. Talia Matthews is one of these happy children.

When I first got to know Talia, she was a bright, bubbly, red-haired nine-year-old who radiated self-confidence. It had not always been that way. She was identified as dyslexic in third grade as part of one of our research projects. "Before, no one had said anything—not one word to even suggest that Talia had a reading problem," her mother told me. "But the other children knew. They'd say, 'Loser. You can't read.' "

I asked Ms. Matthews if she had spoken to the school or if the school had spoken to her.

Oh, yes. They told me, "We think it would be good if she buckled down and tried harder. . . . Maybe she just isn't as smart as you think she is." I can still hear them say, "Parents tend to think all kids are geniuses—when they are not." And then after her testing I was told, "Your daughter is extremely smart. Her IQ is very high.

Talia has a reading problem." When I heard that, I cried. I said, "Oh, my God." And they said they had a possible way they could work with her and teach her.

Talia was identified as dyslexic and placed in a reading program for one hour a day, four days a week for a year.

> I noticed a real change in Talia. I saw her being interested in coming home and wanting to do her homework. It just came together. They taught her a new way.
>
> She did exactly what they were showing her. It was completely different from what they were teaching her in school. And if she didn't know a word, she didn't want any help from anyone. And that was a big difference. Before, she would beg someone else, "Please just do it for me."
>
> It's unbelievable how this has affected our family. Before, we felt so much guilt for how we had been treating her. I would say, "What's wrong with you? Your brother has his homework done. And we can't even get you to do two answers." We were grilling her, you know, like in her face. So my husband and I had a lot of guilt about how we were treating her until she was diagnosed with dyslexia and went into the reading program.
>
> Life is better in our family; it's normal. And [Talia] has friends and she feels better about herself. She has girls call her, and she says, "Mom, they think I'm really cool."
>
> She tells everyone, "I couldn't read. And I was tested. And there was actually a problem with me, and they taught me how to fix it."

I asked Ms. Matthews what she wanted to tell other parents whose children might have dyslexia.

> There is a way to teach your child. You just have to find the right method. Reading can be the most important thing in the world because it affects everything—everything. I would say don't give up. Try to find someone who can evaluate your child and see what way she needs to be taught.
>
> Now Talia can do anything. She already has a vision of what she wants to do with her life. She wants to be a veterinarian.

Your dyslexic child can become a reader just like Talia did. Rather than a magic wand, such a transformation requires early screening, which leads to early diagnosis *and* effective treatment. I've already told you how to identify a reading problem early; now I want to focus on what to do once a child is identified. I will provide you with the basic principles to guide you and allow you to determine what is best for your child. They are the same principles that are helpful in identifying dyslexia; now you can apply these to obtain the most effective treatment.

PRINCIPLES TO GUIDE YOU

An effective program is tailored to a child's specific developmental needs. As you now know, what works best for a six-year-old is not going to be the most helpful approach for a sixteen-year-old. Accordingly, the first step in designing an effective reading program is to determine where your child is on the developmental continuum. Within such a framework, there has been a major change in emphasis. To repeat:

→ New data demonstrating that the achievement gap is already present at grade one mandate that schools must adopt an early identification model.
→ Young children *must* be screened in kindergarten and/or first grade so that those who are at-risk can have further assessment, be identified as dyslexic, and receive early intervention.
→ It is important that those children identified as dyslexic receive an evidence-based intervention.

CAUTION: Strong scientific data show that Response to Intervention (RTI) is not an effective approach. As I will discuss in detail later, a large federal study released in 2015 indicated that *first-grade children who received intervention using an RTI approach actually read less well than their peers who did not receive RTI.*

→ As the child progresses in school and reaches middle school and then high school—particularly those who have the potential to participate in demanding academic programs—the focus expands to ensuring that necessary accommodations are provided.

To soften the impact of the phonologic weakness, focus on fluency and vocabulary and access the child's higher-level thinking and reasoning strengths (through accommodations). This is important because it places emphasis not only on the child's reading difficulty but on his strengths. It reminds everyone that the isolated phonologic weakness is only one small part of a much larger picture. Far too often the focus is only on the weakness, and the child's strong capabilities (and potential) are overlooked. Whatever those strengths are—the ability to reason, to analyze, to conceptualize, to be creative, to have empathy, to visualize, to imagine, or to think in novel ways—it is imperative that these strengths be identified, nurtured, and allowed to define that child. Accommodations represent the difference between academic success and failure, between having a test result reflect ability rather than slow reading because of dyslexia, between a growing sense of self-confidence and an enduring sense of defeat.

There are some other important factors that can significantly influence your child's eventual success in reading and in life. They can be seen in Figure 41.

Considering solely the *child's qualities*, his long-term outlook will mainly

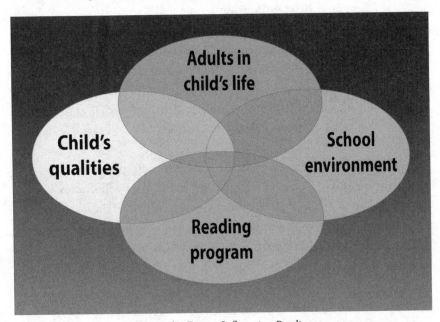

Figure 41. Factors Influencing Reading

reflect the severity of his phonologic weakness and the degree of his cognitive strengths. Not surprisingly, children with milder phonologic difficulties and higher intellectual capacities tend to do better. There is now evidence that children all along the continuum of dyslexia can benefit from *evidence-based* reading programs. In addition, the presence, severity, and effectiveness of approaches to associated conditions such as anxiety and ADHD will affect his future.

A dyslexic's unique blend of strengths and weaknesses reflects her biologic endowment and her experiences growing up. At this point, you cannot change the past. You can, however, brighten her future by maximizing the factors over which you have some control:

→ How early she is diagnosed as dyslexic
→ How she is treated by the adults around her
→ The reading program your child receives (which I will address soon)
→ The school environment

The significant adults in a dyslexic child's life—typically her parents and teachers—play an enormous role in determining her future outlook. While this would seem to be self-evident, I have witnessed far too many instances where the most loving and well-intentioned parents or teachers have taken an uncharacteristically passive stance and assumed that somehow "things will work out." It is not easy to be the parent of a vulnerable child, and particularly a child with a hidden disability that can escape diagnosis, as Talia's mother so poignantly demonstrated. But the reality is that in the overwhelming number of cases, the only way things will improve is if a knowledgeable and caring adult, a patient, persevering, and positive parent, takes the lead and *actively* creates change.

A child with dyslexia is in need of a champion, someone who will be his support and his unflinching advocate, his cheerleader when things are not going well, his friend and confidante when others tease and shame him, his supporter who by actions and comments expresses optimism for his future. Perhaps most important, the struggling reader needs someone who will not only believe in him and take positive action but who understands the nature of his reading problem and then relentlessly works to ensure that he receives the reading help and other support he needs.

Experience has shown me that if a child receives such help, she will succeed. David and Phyllis Lapin of Los Angeles, who have worked incredibly hard and been persistent in trying to understand dyslexia and to use that knowledge to support their very bright dyslexic daughter, Emma, can serve as models. Their strong belief in their daughter and desire to ensure she reaches her potential and feels good about herself have influenced virtually everything they do as parents.

→ They encouraged her to be able to express what it feels like to have difficulty with reading.

→ They investigated and found an effective reading program for her.

→ When they read in the first edition of this book how important it is to find something outside of school that your child is good at, David and Phyllis did just that. Although not "horse people," her family embraced Emma's desire to have a horse when, following a visit to a dude ranch, they saw how much confidence her riding skill gave her and how it helped her maintain her sense of self.

→ Emma met with a tutor regularly; she was designated as the tutor's "senior research assistant" and her father the "junior research assistant" who was to attend all the lessons. As a result, David was able to practice with his child on the days she was not scheduled to see her tutor, so rather than receiving one lesson per week, she received three.

→ Early on, when their daughter had a great deal of difficulty reading, her parents read most books before she did, often taking notes.

Emma attended a rigorous independent school for girls. As a senior there, she gained early admission to Washington University, from which she graduated with honors and a 3.85 GPA. She is now a proud alum entering Washington University School of Law and looking forward to a career as an attorney.

A child who has such champions will not only succeed academically but will maintain her sense of self and the possibilities for a happy future. This book is meant to give parents who want to be champions for their children the confidence and know-how that will allow them to make a real difference in their child's life—because they can. Not only parents but other adults as well can be champions who make a significant difference in a struggling dyslexic child's life.

ALIGNING TEACHING READING WITH
TWENTY-FIRST-CENTURY SCIENCE

Our knowledge of what works in teaching reading has come to us through science. My goal is to align teaching reading with twenty-first-century science. This new knowledge can be put to use to provide a sensible and successful approach to teaching children to read—children without reading difficulties, children who are at-risk for reading problems, and especially children who are known to be dyslexic. Constantly bombarded by information about different approaches to teaching reading and about new "miracle" treatments for dyslexia, some parents (and teachers, too) are confused about reading instruction. How do you know how to determine what is the best reading program for your son or daughter? The good news is that data coming in from laboratories all over the world are providing an increasingly deeper understanding of the process of learning to read. At long last, scientific evidence is replacing anecdotes, myths, and ingrained belief systems lacking in evidence.

For the first time ever, there is now an evidence-based guide to what elements are necessary in teaching children to read. By *evidence-based*, I mean that in a *clinical trial* comparing two different methods of teaching reading, one method has actually *proven* to be superior. This means that there is actual evidence of the effectiveness of this method. Evidence-based interventions differ substantially from *research-based* interventions, those that in theory should work.

Evidence-based means that the intervention has proven to be effective in a rigorous clinical trial, the gold standard for which is a randomized, double-blind, placebo-controlled trial in which the potential intervention is tested head to head with another intervention. In medicinal trials we test against a placebo; in education trials we test against what the school has typically used.

Randomized means that the children participating are randomly assigned to the treatments. This ensures that the researchers prevent potential bias that could occur if more good (or more struggling) readers were intentionally assigned to one group rather than another.

In medical trials, *double-blind* refers to the fact that neither the doctors testing a medication or treatment nor the patients receiving the medication know whether the patients are receiving the medication or a placebo. In trials of potential reading interventions, it is very difficult to keep everyone—the teachers, the school administrators, the children, the parents—in the dark about which intervention they are receiving, so the investigator modifies the experimental design, which is referred to as *quasi-experimental*. The key point is that the intervention is compared to another one in the same exact time frame.

THE REPORT OF THE NATIONAL READING PANEL

The National Reading Panel emerged as an outgrowth of a grassroots concern that while a substantial number of children were failing to learn to read, little was available to help parents and teachers make important choices among different approaches to reading instruction. So in 1997, as I mentioned earlier, Congress directed the creation of the National Reading Panel of experts to objectively and comprehensively review the existing research relating to the teaching of reading. As a member of the panel, I can attest to how seriously the group took its mission, traveling across the country listening to parents, teachers, and others; developing rigorous criteria to review the existing research; and identifying thousands of studies and then subjecting them to careful analysis. On April 13, 2000, after more than two years of effort, the panel released its findings in the *Report of the National Reading Panel: Teaching Children to Read: An Evidence-Based Assessment of the Scientific Research Literature on Reading and Its Implications for Reading Instruction.* It was with great pride that I along with the panel chair, Dr. Donald Langenberg, and then director of the National Institute of Child Health and Human Development, Dr. Duane Alexander, presented the report to Congress. The report, the most thorough ever to be undertaken in American education, provides a road map to guide parents and teachers to the most effective, scientifically proven components for teaching reading.

The findings of the panel signal a critical first step in a revolution in how children are taught to read, ensuring that every child is provided with a reading program that works. It is an important step in our goal of aligning education with science. The National Reading Panel made a major contribution by employing a rigorous scientific process to identify these essential components in any effective program to teach reading:

→ phonemic awareness
→ phonics
→ fluency
→ vocabulary
→ comprehension

Now we need to determine through rigorous studies what the optimal combination and timing of these elements is.

The establishment of the What Works Clearinghouse (WWC) within the Institute of Education Sciences (IES) of the Department of Education reflects an important positive development. Within medicine, the Food and Drug Administration (FDA) serves to protect the consumer by ensuring that medications and medical devices must have evidence of safety and efficacy before they are approved for sale. There is no equivalent organization in trials of potential reading interventions. Though the WWC has no enforcement power, it rates evidence behind various programs and textbooks, using criteria similar to the ones researchers use to assess the effectiveness of medical treatments. This represents a significant potential aid to educators and parents in selecting effective programs. Though the WWC is a major step forward, it remains a work in progress.

Caution is indicated. In a *New York Times* story about the WWC, Gina Kolata wrote in 2013 that there has been a paucity of "solid answers [about what works in education], just guesses and hunches, marketing hype and extrapolations from small pilot studies." She further pointed out that the selection of materials such as reading programs "can affect achievement as profoundly as teachers themselves; a poor choice of materials is at least as bad as a terrible teacher."

All who care about reading and dyslexia should be aware of the caveats noted in this piece. According to Kolata, this new-to-education scientific approach and the evidence-based data it brings with it are unknown to most educators, including principals, superintendents, and curriculum supervisors. Says Kolata, educators don't know that the data exist, "much less what they mean." Often claims of effectiveness made by program developers, when carefully examined, have "no evidence behind them. And when rigorous studies were done, as many as 90 percent of programs that seemed promising in small, unscientific studies had no effect on achievement or actually made achievement scores worse."

In programs intending to teach teachers how to improve their students' reading achievement, data may indicate that the teachers' knowledge of the subject has improved but *without* a comparable improvement in their students' achievement. During the time I served on the board of directors at the Institute for Education Sciences, I listened to a review of a highly touted professional development program for teaching reading and heard how teachers improved but their students did not. This taught me a lesson about the strong need for large trials before accepting that a program can help the students it is meant to help. According to Jon Baron, vice president of evidence-based policy at Arnold Ventures, "Most programs claim to be evidence-based but have no good evidence that they work." As Robert Slavin, director of the Center for Research and Reform in Education at Johns Hopkins University, says, educators "too often . . . are swayed by marketing or anecdotes or the latest fad. And invariably, folks trying to sell a program will say there is evidence behind it," when the evidence is not rigorous at all, or in fact is lacking.

In the discussion that follows of what approaches to teaching reading are most effective at different points along the reading continuum, you may notice that, based on the newest data available, I have removed some of the programs mentioned in the first edition. I also question some of the recommendations of the WWC, reflecting inadequacies in the studies relied on or subjective decisions by WWC evaluators to include programs where sufficient evidence does not appear to exist. I indicate where this is the case.

Breaking the reading code is the most critical, and for many children the most difficult, step, and I will devote great attention to that. Then I will focus on how to ensure that a child becomes a skilled reader—that is, a reader who reads accurately and rapidly and understands what he reads. I will also discuss what to do for the child who continues to experience difficulties.

17

HELPING YOUR CHILD BREAK
THE READING CODE

One of the most exciting things to happen to your child between the time he enters kindergarten and the time he leaves the next spring is that he breaks the reading code. To do so, he must solve the two parts of the reading puzzle: one involving spoken language, the other written language. For the past five thousand or so years, since man has been reading, each potential reader has faced the same challenge.

To solve the first part of the puzzle, each would-be reader must begin to understand that spoken words come apart and that they are made up of very small bits of language. As you read earlier, these tiny particles of language are called phonemes, and the development of this insight about words is called phonemic awareness. Once a child appreciates that spoken words can be pulled apart into distinct sounds, he is well on his way to solving the spoken-language part of the reading code. Then he is ready to take the next big step: figuring out how printed letters link to these sounds—for example, knowing that the first sound of the spoken word *sat, ssss,* is represented by the letter *s,* and the final sound, *t,* by the letter *t."* And then one day he solves the puzzle: He makes the critical insight that the written word *sat* has the same number and sequence of sounds, *ssss, aah, t,* as the spoken word *sat,* and that the letters represent these sounds. He has broken the reading code! The child has mastered the alphabetic principle. He is ready to read.

This step is terribly important, because once a child knows the code, printed words are no longer a mystery. He has a *strategy*—he knows how to link the letters to the sounds they represent and then to blend the sounds together to read the word. He applies his knowledge of how letters relate to sounds to analyze and to read more and more unfamiliar words. This is referred to as *decoding*. The better and better a child gets at decoding words, the more accurate his reading becomes. A child who knows how to sound out printed words is freed from having to memorize every word he wants to be able to read. Sounding out words allows him to unlock the mystery of reading and to read words he has never before seen.

Recent brain research allows us to link our understanding of the sequence of steps a child goes through in learning to read both to the work that must go on within the brain to solve the reading code and to the teaching of reading. Within the brain, the child is literally building the neural circuitry that links the *sounds* of spoken words, the phonemes, to the print code, the *letters* that represent these sounds. A major role of kindergarten is to provide a child with the kinds of experiences that allow him to build those brain connections as accurately, as strongly, and as quickly as possible.

PART ONE OF THE CODE: SOUND ADVICE

At the very beginning stage of reading, the initial goal is to draw the child's attention to the *sounds* of language. It is important to keep in mind that children will vary in their progress along this road. Some will need to go through each step slowly, while others progress much more rapidly, not requiring instruction in every step of the process. At this stage, a major focus is to help a child to develop *phonemic awareness*—that is, the ability to notice, to count, to identify, and to manipulate the tiniest particles of language making up words.

Phonemic awareness, the most important and sometimes most difficult task in learning to read, is the foundation of all subsequent reading and spelling instruction. Teaching it requires only a small part of the school day, but children in kindergarten and often even in preschool benefit from learning phonemic awareness. Progressively, a child's attention is drawn to the sounds in words, first through noticing rhymes, then by comparing sounds in different words, and finally by learning how to "work on words," pulling them apart, pushing them together, and moving the parts around within the word. While phoneme awareness refers to the sounds of spoken words, it often helps to use letters to emphasize the different sounds and to facilitate transferring this skill to reading. *Always keep in mind that we are teaching phonemic awareness not as an end in itself but because of its central importance in helping a child understand the relation of letters to sounds and, ultimately, to becoming a reader.*

STEP ONE: DEVELOPING AN AWARENESS OF RHYME

The very first step for a child is to develop a growing awareness that words can rhyme. Tuning into rhymes sensitizes very young children to the fact that words come apart. For example, to know that the words *fig* and *jig* rhyme, a child must attend to just a part of each word—the ending, *ig*. The child begins to appreciate that words have parts. His natural joy in playing with rhymes (words ending with the same sound) and alliteration (a series of two or more words beginning with the same initial sound, such as *click, clack, clock*) helps to lay the groundwork for teaching phonemic awareness.

You can help sensitize a child to rhyme by reading stories and poems to him aloud. One richly illustrated rhyming book he will enjoy is *Chugga-Chugga Choo-Choo* by Kevin Lewis and Daniel Kirk. Listen:

> Chugga-chugga
> choo-choo,
> whistle blowing,
> whoooooooo! whoooooooo!

And

> Into tunnels, underground.
> See the darkness. Hear the sound.
> Chugga-chugga choo-choo, echo calling,
> whoooooooo! whoooooooo!
> whoooooooo! whoooooooo!

Children often love to pump their fist up and down as they make the sounds of the train whistle.

There are many books you can read with your child. Knowing you are building a foundation provides added incentive for making reading aloud a routine part of your special time together. In the Appendix you will find a list of children's books that joyfully play with the sounds of language.

It is great fun to read children's books aloud with your child. As he listens to them, or even joins in singing these little ditties, he is taking a step toward becoming a reader. Professor Hallie Yopp Slowik, codirector of the Center for the Advancement of Reading and Writing at California State University, Fullerton, suggests some activities that children can enjoy as they listen to the stories. First, she says, "read the story aloud several times simply for the pure joy of reading and sharing." Then draw the child's attention to the sounds. (I can hear children giggle as they listen to C. P. Ochs's *Moose on the Loose,* about the zookeeper who runs through the town of Zown asking anyone if they've seen a "moose on the loose in a chartreuse caboose.") As you read these stories, simple comments such as "Isn't it funny that so many of the words rhyme?," along with specific examples, help to stimulate rhyme awareness. You can also ask your child to predict the next word in a rhyming story or ask if she notices that all the words in a sentence begin with the same sound: "Listen to the beginning sounds in these words: *Pink pigs picked pretty petals.* What sound do you hear? Yes, the *p* sound. What are some other words that begin with that sound?"

Here are some more selections you can practice reading to your child. I've chosen one of my children's favorites, *One Fish Two Fish Red Fish Blue Fish* by Dr. Seuss:

> Bump!
> Bump!
> Bump!
> Did you ever ride a Wump?
> We have a Wump
> with just one hump.

Or some children may prefer *Madeline* by Ludwig Bemelmans, also written in rhyme. Here he describes little Madeline:

> She was not afraid of mice—
> she loved winter, snow, and ice.
> To the tiger in the zoo
> Madeline just said, "Pooh-pooh,"
> and nobody knew so well
> how to frighten Miss Clavel.

As you read these rhyming stories aloud, it helps if you exaggerate the sound of the rhyming words.

STEP TWO: WORKING ON WORDS

After getting to know rhymes and developing the sense that words have parts, children are ready for the next big step on the road to reading—taking words apart and putting them back together.

→ Pulling apart a word into its sounds is often referred to as *segmenting.*

→ Pushing sounds together to form a word is called *blending.*

These are the two key processes involved in learning to spell and to read. In spelling, a child segments the spoken word into its sounds and transforms each sound into a letter. In reading, the letters are converted into sounds which are then *blended* together to form the word. Working on words forms the central focus of reading programs for young children. Just as a surgeon must learn the smallest details of human anatomy, budding readers must develop a keen awareness of the internal anatomy of spoken and printed words.

In practice, children are introduced to the sounds of language gradually. Most reading programs begin by showing the child how to dissect words into their biggest chunks of sounds, syllables, and then introduce the smaller pieces, phonemes.

Separating Words into Syllables. There are good reasons to begin with syllables. They are the largest units of sound that make up a word, and they are relatively easy for children to identify and manipulate within words. They readily come apart. One activity that helps a child pull apart the sounds in a word is to count (actually clap) the number of syllables in his own name: *A . . . ri* (clap, clap). *Jon . . . a . . . than* (clap, clap, clap). *Ev . . . a* (clap, clap). *Kay . . . la* (clap, clap).

Alternatively, children can take turns clapping out the syllables for each day of the week (*Tues . . . day*), each month of the year (*Sep . . . tem . . . ber*), recognizable objects (*cow . . . boy, scare . . . crow*), and fun words such as *cucumber* (*cu . . . cum . . . ber*) or Cinderella (*Cin . . . der . . . el . . . la*) or *Frozen* (*Fro . . . zen*).

Children can also practice pushing syllables together to form words: "Can you tell me what word *rain . . . bow* makes?" (*rainbow*). "What about *a . . . corn*?" (*acorn*). A child can be asked to blend the sounds of her own name: "What word does *Mi . . . mi* make? (*Mimi*). If you have any questions about the number of syllables in a word, you can always consult the dictionary.

Separating Syllables into Phonemes. As I noted above, separating words into syllables is relatively easy—the words are already perforated and readily come apart. The next major step is appreciating that spoken words can be pulled apart into even finer divisions—phonemes. This is difficult for most beginning readers, and particularly so for dyslexic children.

In the most effective programs, children first practice comparing or matching sounds in different words. The aim is to start a child thinking about how the sounds in these words compare to one another. Always begin by asking him to match the very first sounds in words, and then the final sounds. It is often helpful to have a set of cards with pictures of everyday objects (such as a map, man, knee, cat, dog, mom, mitt, can, cup, tag, top). You can cut the pictures out of a magazine and paste each one on a card or you can use those from children's board games.

You can begin by asking your child to identify the first sound in a word such as *map*. Show him a picture of a map, clearly saying, "This is a *mmmap*." Ask him to name the picture. Next say, "The first sound in *mmmap* is *mmmm*. Can you say the first sound?" Then lay out four or five picture cards, some beginning with the *mmmm* sound. Ask him to say the names of each of the pictures and then to group together all the objects beginning with the *mmmm* sound.

You may also show him three picture cards with, for example, a fan, a dog, and a cake, and ask him to name each picture. Then ask, "Can you show me which one of these pictures begins with the *d* sound?" You can point to one of the pictures (dog), name it, and then ask, "Can you find something else in the room that also begins with the same *d* sound?" Or you can simply ask him to produce a word that begins with the same sound as a target word: "Can you tell me a word that begins with the same sound as *ssssnake*?" (*sit*, *soup*, and *smile*). Sound-matching games are relatively easy, because they do not require the child to manipulate phonemes—ideal for the budding reader in kindergarten.

Once a child is successful at matching sounds, he is ready for the more difficult task of pulling apart words: first the beginnings, then the endings, and then the middle sounds. Children first carry out these kinds of activities with spoken words, and then they progress to games involving the letters and sounds making up written words.

There are all sorts of ways you can pull words apart. You can ask your child to clap for the number of sounds he hears in a word. For example, you can say, "*Ssss . . . ee, see.* Clap for the number of sounds you hear—*ssss . . . ee, see.*" She should clap twice for two sounds, *ssss* and *ee.* Most kindergartners seem to be successful in separating the sounds in two-phoneme words such as *eat, it, say,* and *zoo,* but they are stymied when asked to pull apart even relatively simple three-phoneme words such as *fan* and *mop.* So go slowly. Start with the most basic two-phoneme words and then work your way up to three or more sounds.

You could also add a sound: "What word do you get if you add the sound *t* to the word *see?* (*seat*). You could take away a sound, first a beginning sound: "What word do you get if you take the *ssss* sound from the word *seat?*" (*eat*). Then take away an ending sound: "Can you say *seat* without the *t* sound?" (*see*).

These tasks require segmentation. The converse, phoneme blending, is one of the most important and surprisingly difficult jobs for the beginning reader to master. (Recall that in reading, a child must transform the letters to sounds and then blend the sounds together to form the word.) So you may want to practice blending activities at home with your child. One method is to line up three pictures, such as a man, a map, and a fan. Choose a word represented by one of the pictures, for example, *man,* and slowly pronounce each of its sounds: *mmmm . . . aah . . . n.* Ask your child to point to the picture that shows the word made by pushing these sounds together.

For a slightly more difficult variant, rather than asking your child to point, you can ask him to tell you, to say, the word made by pushing these sounds together.

For both segmenting and blending activities, you can use the following two- and three-phoneme words.

TWO-PHONEME WORDS

is	tie	zoo	fly
to	see	chew	knee
do	toe	row	my
sew	be	mow	it
shoe			

THREE-PHONEME WORDS

cat	bat	cab
sheep	fudge	map
right	jeep	cub
mice	fish	book
feet	man	dog
nap	jet	tag
zap	nip	top
sun	net	mom
cup	bag	tap
bed	bug	top
can	pan	

HINTS FOR BUILDING SUCCESS

The goal of these activities is to draw your child's attention to the smallest parts of words. Here are some of the things you might want to keep in mind as you do so:

→ You want your child's active involvement. When you have it, he is paying attention and learning is going on. The activities should be short and enjoyable. If he is not interested, let it go; little is accomplished if he sits there passively with you talking at him. Try to do these activities when both you and he are alert and in a good mood.

→ You want him to notice each word or word part you say, so speak slowly and clearly and pronounce each sound very carefully.

→ Exaggerate sounds—for example, *mmmman*—and have him do the same when he repeats it back to you.

→ Use concrete objects (blocks or coins) to represent the sounds in words. Your child should indicate how many sounds he hears in a word by the number of coins (or blocks) he places on the table. For example, for the two-phoneme word *zoo*, he would say each sound (*zzzz, oo*) as he lays out one and then a second coin (see Figure 42).

→ Make up your own jingles, rhymes, or silly stories to highlight a particular sound, or even sing a song together. Funny and visually absurd rhymes and alliterations often work best in making a sound more salient to the child. For example, to highlight the *ssss* sound, sing with him, "*Silly Sammy seal slept soundly.*"

Overall, it helps to keep in mind that there is a logic to these activities. They should reinforce what is happening in your child's classroom.

→ Always check with his teacher to ensure that you and the teacher and your child are working in synchrony.

→ Ask about what specific activities he is doing in class and how you can help at home to further his reading skills.

Teachers and parents preferring a ready-made program for teaching phonemic awareness can purchase one of several excellent commercial products. The ones listed here are geared to children as early as preschool or kindergarten and are focused on children at-risk for dyslexia.

→ Headsprout (Learning A–Z, Headsprout.com) is an Internet-based supplemental pre-K–grade six curriculum.

Figure 42. Helping Your Child Count Phonemes in Words
As your child pronounces each sound in a word, such as *zoo,* he puts a coin in the center of the table.

→ Ooka Island (Scholastic.com) was developed by Dr. Kay MacPhee for grades kindergarten through grade two based on the proven concepts she used in creating SpellRead (an evidence-based program).

→ Lexia Core5 Reading (Lexialearning.com), a computerized reading program, provides phonics instruction and gives students in pre-K through grade five independent practice in basic reading skills.

In addition to these, teachers especially may find helpful *Phonemic Awareness in Young Children: A Classroom Curriculum* (Brookes), a program meant for instructing entire preschool or kindergarten classes. Parents may also find *Teach Your Child to Read in 100 Easy Lessons* (Touchstone) a helpful, organized way to teach the basics of reading to their prekindergarten or kindergarten child. I've known parents who have successfully used this program, working with the child for about thirty minutes one to three times a week. Both the children and the parents seemed to enjoy the program and profit from it. They found that entering kindergarten already on the road to reading, or already reading, is a major factor in making school a highly positive experience.

PART TWO OF THE CODE: PUT IT IN WRITING

Once a child has learned about the segmental nature of spoken words and is becoming familiar with individual sounds, she is ready for letters. Most children have already learned to sing the ABC song as preschoolers and enter kindergarten knowing the names of most of the letters. In kindergarten, knowledge of the letters is reinforced—not only their shapes and names but their sounds. By the second half of the kindergarten year, most children can identify and print each of the letters. Now they can begin to use them for the purpose of reading.

Learning about and using different sound and letter combinations to decode words is called *phonics*. In English there are more sounds than letters: about forty-four phonemes and twenty-six letters. These forty-four sounds provide the means by which the developing reader can attach letters and create print, or "visible speech." Many children are ready for linking letters to sounds (phonics instruction) as early as kindergarten, and nearly all children by first grade.

When Barbara Wilson, creator of the Wilson Reading System, was speaking at a Connecticut conference, I thought she made an important observation on teaching phonics when she said to teach the sounds in two directions. In the first one, have the child listen to the sound and ask her what letter represents that sound. For example, say "tuh," and the child should respond saying or pointing to the letter *t*. In the other direction, the child looks at the letter *t* and then says "tuh." Learning a sound and its letter concurrently reinforces the association and helps to ensure that the connection is firm.

THE A TO Z OF TEACHING BEGINNING READING

Although awareness of sounds and letters is necessary, children also need practice in reading stories. They need to apply their newly acquired skills to sounding out and decoding familiar and less familiar words, to reading words in sentences and in books, and to focusing on understanding the meaning of the word and the sentence.

Practice. Brain imaging technology shows the powerful positive effect of practice in creating neural circuits related to the development of what scientists call expertise or skill. Basically, the brain learns by practice. The old dictum about the importance of practice, practice, practice turns out to be correct. Whether it is for learning how to pitch, to play a musical instrument, or to read, it is critical for the development and refinement of the components of the neural systems for reading. Once a child gains an awareness of how letters relate to sounds, he benefits greatly from putting into practice what he has just learned.

After an introduction to the specific letter-sound relationships, the next critical step is for a child to practice the words, both in isolation and then in reading simple sentences and books. To improve his accuracy, a child needs to practice reading, both silently and especially out loud to others. Each time he stumbles on a specific word while reading aloud and with his teacher's or parents' guidance makes corrections and refinements, he is establishing and integrating increasingly more accurate representations of that word's components within his brain. His internal representation of the word reflects the integration of those neural systems serving pronunciation, spelling, and meaning. Writing the word and learning to spell it also contribute to establishing increasingly accurate representations of that word in his neural circuitry.

Learning spelling-sound patterns, practicing them in different words and in reading books, learning how to form letters and to spell the words—all contribute to forging and then reinforcing the connections that will eventually form the neural code for that word. Beginning readers are trying to solve the mystery of the unknown word. The more clues there are, the greater the chances of solving the mystery—and the more likely the child is to recognize the word when he comes across it again.

Even kindergartners, on the cusp of literacy, can practice their new skills. Children love to read even when they know only a few word patterns. Simple booklets, about twelve to twenty-four pages—so-called decodable texts (containing the words with letter-sound patterns that a child has already been taught)—can help him apply his newfound skills by actually reading words in a book. The Bob Books series by Bobby Lynn Maslen is a good example. Many children enjoy the series, not only in print but also on their tablets in the digital version. Other decodable texts are part of reading series, such as S.P.I.R.E. Decodable Readers, available from EPS Publishers; Scholastic Decodable Readers, from Scholastic.com; and Ready Reader and Reading Street, both available from Pearson. All of these provide increasingly complex reading material, both fiction and nonfiction, in a very user-friendly format.

A new approach, Readables, takes text for beginning readers to a new level. A wonderful example is Geodes, a new reading series providing books that build knowledge while developing a child's skills for success and love of reading. This series presents an unusual array of topics for a beginning reader's text, including science, history, visual art, and literature (found in the Great Minds Wit & Wisdom® curriculum), while at the same time aligning with the scope and sequence of Wilson's Fundations® program focused on decoding and word-recognition skills.

Sight Words. Once a child learns about letter-sound relationships in specific words, he eventually generalizes and uses these principles to read other, new words that have the same letter pattern and are pronounced in the same way. So if a child can read *cat* and is able to divide it into *c-at*, he will also be able to decode other words in the same *at* family—*mat, sat, bat*. With time, he will go on to acquire knowledge of more and more complex word families.

However, difficulty sometimes comes where you least expect it. I am thinking of words such as *a, is, are, one, two, said, again, been, could, the,* and *once,* which pop up frequently in books for young children but don't seem to follow the rules. These words do not follow a pattern and cannot be sounded out. Consequently, they must be committed to memory and recognized on sight; not surprisingly, they are often referred to as *sight words.** (Because of the inconsistencies in their letter-sound linkages, they are sometimes called "irregular" words.) Since sight words are sure to be found in children's books, it is important that they become part of a child's reading vocabulary at a very early stage. Making flash cards (actual hard copies) or digital (such as Quizlet, quizlet.com) and reviewing them regularly helps in learning these common words. Having a child print his own card and say the word as he writes it often helps to reinforce its pronunciation. (A list of the most common sight words appears on page 245.)

Writing Letters. Recognizing and writing letters represents a major milestone in learning to read. Emphasis is directed toward writing lowercase letters, because they are more common. Once children can write letters, no matter how poorly, they can engage in a variety of writing exercises that further encourage awareness of the sounds making up words and of how letters represent these sounds. A child's writing out his own name and then common words such as *dog, mom,* and *dad* reinforces that awareness. Early on in kindergarten, children can use alphabet blocks or letter cards to "write" the words. By the time the children are ready to leave kindergarten, they are using letters to spell words phonetically, although not necessarily correctly. These earliest attempts at such spellings are referred to as "invented" or temporary spelling.

Spelling. Spelling is intimately linked to reading, not only because sounds are being linked to letters but because words are being encoded—literally put into a code instead of merely being deciphered or decoded. As I said earlier, in spelling, the spoken word is being pulled apart into its sounds and then each sound is encoded, transformed into the letters representing that sound. Invented spelling functions as a transitional step as kindergartners are trying their hand at matching letters to sounds. When children are try-

---

* *Sight word* can have two meanings: one as I have described, referring to words that don't follow rules for pronunciation, and the other meaning high-frequency words that repeat so often that they should be read "on sight," that is, automatically, without having to go through a decoding process. Here I am employing the first usage.

ing to form printed words based on the sounds they hear in spoken words, it is a good indication that they are on the road to reading. This is why in kindergarten and in the beginning of first grade children are encouraged to practice "invented spelling" and to write a word as they think it sounds—for example, *knd* for *candy, hrs* for *horse,* or *kt* for *cat.* As you can see in these examples, beginning readers most commonly omit vowel sounds. So a child's invented spelling for the word *house* might well progress from *hs* to *hws* to *house.* Notice that the pronunciation of the invented spelling is very close to that of the intended word. A child may have the sounds down, but he has not quite mastered the link between letters and sounds. In kindergarten, that's okay.

Listening, Playing, and Imagining. A kindergarten child's reading program is rounded out by a range of activities that enhance her language skills and her enjoyment of literature. Whether at school or at home, being surrounded by books, listening to stories read aloud, talking about the characters and events in the story, and playing with blocks or puppets all help a child to develop her thinking skills and her imagination, to build her vocabulary, and to become aware of the world around her.

Reading and word knowledge are mutually reinforcing. Reading builds a child's vocabulary, and knowledge of word meanings helps in decoding words and improves reading comprehension. A primary goal of reading is to get to the printed word's meaning. To do so requires a child first to decode the word and then to have the word in his vocabulary. Initially, the words a child is learning to read are quite simple and invariably in his oral vocabulary, but before long the words begin to get more complicated and unfamiliar. At this stage, knowing the meaning of a word helps to facilitate its decoding. As a child encounters an unfamiliar printed word, he tries out different pronunciations. For example, is the *i* in *ink* pronounced like the *i* in *ice* or like the *i* in *it?* (If he knows the meaning of the word *ink,* the chances are far more likely that he will pronounce it correctly.) So, obviously, the larger a child's vocabulary, the more words—and the more difficult and complex words—he is going to be able to decipher and to read. It is never too early to introduce a child to new words and their meaning, which strengthens this key component of his neural model for each of the new words he encounters. Parents can help by making it a point to talk about each new word (or concept) you come across in a story, to define it in simple words, and then to have your child use the word in a sentence. Learning about the meanings of new words reinforces a child's worldly as well as his word knowledge, both key ingredients for strengthening his reading comprehension.

In addition to improving a child's vocabulary, listening to stories has many other positive benefits for reading. At home, you can enhance these positive effects by having your daughter sit next to you as you read, watching as you point to each word as you pronounce it. She will soon notice that words that take longer to say look longer on the page. *Caterpillar* sounds longer than *cat* and looks longer when you see it printed on a page. Such observations help the child learn that the spoken word and the printed word are related. At a later stage she begins to associate the printed word with the sound and meaning of the spoken word. Eventually she will learn to use her finger to point to each of the words as you read them, and then she will want to read and point out the words herself. To get started, I suggest reading aloud the books listed in the Appendix under "Just Right for a Child on the Cusp of Reading."

Kindergarten children also love to make up stories and dictate them to their teacher or their parent, who magically transcribes each word onto paper. As a child dictates his little story and then watches as you point to and read each word back to him, he is learning to associate sound, letters, and meaning. He is building the integrated neural circuits necessary for reading. He is also building his imagination.

Self-confidence. This is probably the most important ingredient in ensuring that a child is setting out on a good path. The most significant development for your child as he leaves kindergarten is how he feels about himself. A critical role of the kindergarten experience is to ensure that every child achieves some degree of success in what he does, that he receives positive comments from his teacher, and that he is encouraged. Without motivation and the sense that he can succeed, a child will have little reason to struggle as he tries to pull apart words that seem to be inseparable.

IS YOUR CHILD MOVING FORWARD IN KINDERGARTEN?

It is important to pinpoint where your child is along the reading pathway and to keep in mind that each child is different. Rather than having a student try to keep up with (or slow down to) a pace that is uncomfortable for him, instruction must be fitted to the child.

Reading instruction often begins in kindergarten, and once instruction begins, there are inevitable questions: Is the reading instruction approach working for the child? Is he making progress or is he falling behind? If he continues at this pace, what does this mean for his future as a reader?

Even children whom one might assume do not need early assessment and monitoring should have it. Some especially bright children may appear to learn to read early and seem to leap frog over learning phonologic skills. These children simply memorize lots of words with seeming ease and quickly build a large reading vocabulary without learning how to analyze these words and pull them apart, let alone figure out how to pronounce a new or unfamiliar word. Invariably there comes a time when such children cannot decipher the insides of these new, relatively long words—especially technical names, as in the sciences (*bicarbonate, polynomial, stegosaurus*), or names of people or places in history or around the world (*Picasso, Lafayette, Laramie, Timbuktu, Kathmandu*). They lack a strategy to deal with them. So it is best to ensure that progress is carefully monitored.

PUTTING IT ALL TOGETHER: YOUR KINDERGARTEN STUDENT

In summary, you must first determine if your child or student is at-risk for dyslexia, making sure that his school uses an effective screener. It is important to make sure that his progress is being carefully monitored. At the end of kindergarten, you can ask yourself or your child's kindergarten teacher, based on her own observations, if your child

→ knows that spoken words come apart and that letters represent these sounds.

→ easily names the letters of the alphabet, in both their upper- and lowercase forms.

→ writes the letters of the alphabet (initially lowercase).

→ is beginning to learn about letter-sound matches.

→ is beginning to decode simple words.

→ is beginning to recognize as many as fifty or more common sight words.

→ uses invented spellings.

→ is beginning to spell some very simple words accurately.

→ knows about print conventions—reading from left to right, from the top of the page to the bottom.

→ has a growing vocabulary.

→ looks forward to reading.

You should feel reassured if your child demonstrates these skills. It is even more solidly reassuring if she is found not to be at-risk following a kindergarten screening—a more reliable indicator that she is indeed off to a running start in reading. Of course, if your child is found to be at-risk, she should undergo diagnostic assessment or begin an evidence-based intervention.

HELPING YOUR CHILD BECOME A READER

First grade is one of the most important for a child on the road to reading, building on the foundation provided in kindergarten. He has already taken the first and most profound step by breaking the reading code. Now he must learn the mechanics of reading, focusing on identifying the words on a page and understanding what those words mean.

Most children enter first grade *ready to read*; they should leave as *readers* with a reading vocabulary of about four hundred words. They should be able to read easy books and even simple directions, spell many short words accurately, and self-correct their reading errors when the identified word doesn't fit with cues provided by the letters in the word or with the surrounding context. These major accomplishments don't occur by chance; they are made possible by well-structured and systematic activities that draw the child ever closer to understanding how words work. The most successful reading programs follow the model I told you about for teaching children how to break the reading code and include the following activities:

→ Learning to read words by
  · sounding out small simple words (first grade)
  · taking apart bigger words (second grade and later)

→ Memorizing sight words
→ Practicing oral and silent reading
→ Writing, including spelling
→ Building word and worldly knowledge
→ Learning comprehension strategies.

Continuing efforts to build self-esteem strengthen the program and hold the ultimate key to its success.

In this chapter I will focus on the basic skills necessary for reading regular words (both big and small), on memorizing sight words, and on practicing reading.

SOUNDING OUT SMALL WORDS

An early step in becoming an accurate reader, as we've seen, is learning to sound out words, beginning with simple two- and three-phoneme words like *zoo* and *dog* and going on to longer clusters of letters and sounds making up more complicated words such as *play, chick,* and *snail.* Children in first grade more rigorously begin to pull apart words and learn how to analyze and to decode progressively more challenging ones. They accomplish this by learning more about the relation of letters to sounds and by learning how to apply this knowledge to reading. As you know by now, this kind of knowledge is technically referred to as *phonics.*

Because phonics has been associated with a particularly boring and repetitive approach to teaching reading, some have questioned its value as part of a reading curriculum. But the criticism of phonics was never a reflection of a lack of effectiveness in helping a child learn to read; rather, phonics was disparaged because of the way it was taught. Phonics has now had a makeover, retaining its core educational value but redesigned to draw children into the reading process. This is good news, since phonics provides the foundation for all subsequent reading. For your child to obtain maximum benefit from phonics, it is important that you are aware of some of the less visible and often overlooked aspects of its instruction.

THE ESSENTIALS OF PHONICS
There are many phonics programs, of varying degrees of effectiveness. The National Reading Panel found that programs that teach phonics *systemati-*

cally and *explicitly* (I will discuss this shortly) are the most effective. Systematic phonics is an organized method of teaching children about how letters relate to sounds. Children learn how to convert letters to specific sounds and then how to blend the sounds together to read a word. They learn how different patterns of letters represent different sounds. They learn rules, and then they learn exceptions to these rules. After a while a child has the knowledge required to analyze and to identify just about any word he encounters. This is the goal of systematic phonics instruction. What is so critical and so unique about learning phonics in this way is that it allows the reader to apply his accumulating knowledge to deciphering and reading words he has never seen before. No other method of teaching reading can make this claim.

In general, systematic phonics programs progressively introduce the child to different letter-sound pairings, beginning with the simplest, most consistent, and most frequent combinations and then gradually expanding to cover more complex and unusual ones. A typical program might begin by teaching:

Simple One-to-One, Letter-Sound Relationships. Here the focus is on those consonants where there is a predictable one-to-one relationship between a letter and its sound (typically in the letters *b, d, f, h, j, k, l, m, p, r, t, v, w,* and *z*). With rare exceptions, there is only one way to pronounce each of these letters so that they are consistently linked to the same sound. *Dear, sledding,* or *pad*—the sound *d* is always the same.

Vowel Sounds. In order to read words, children need to learn about vowel sounds, too. Vowels tend to be more difficult to pronounce than consonants. They can be *long* or *short.* Children are taught that long vowels "say their names" (the *a* sound in *cake,* the *e* sound in *key,* and the *i* sound in *time*). Short vowels do not—for example, *cap, help,* and *tip.*

In systematic programs, specific sets of letters, generally six to eight consonants and two vowels, are taught as a unit. Once they are mastered, children go on to another set. Initial groupings typically contain several of the consonants noted above plus the vowels *a* and *i,* because the sounds of these two vowels are the easiest to distinguish from one another. Once a child knows the sounds of her first group of consonants and vowels, she is excited because she can now begin to link these letters to their sounds and then blend these sounds together and read the word on the page. These simple consonant and vowel combinations are used to make the words in a child's first books and are the focus of reading instruction in first grade. By the middle of first grade, a child should be able to sound out a growing number of these words.

Complex Letter-Sound Patterns. Having mastered one sound–one letter linkages, the child is ready for links where letters and sounds do not invariably have a one-to-one relationship. Here the focus is on patterns where, for example, two letters may represent one speech sound; these combinations are referred to as *digraphs*. Although they sound quite technical, digraphs are very common—*sh-* as in *ship*, *ch-* as in *chip*, *-ng* as in *sing*, *th-* as in *thing*, *wh-* as in *when*. These (and other) more complicated letter-sound relationships are generally taught during first grade and continue into second. As children progress, they are surprised to learn that not only are there digraphs, there are also *trigraphs* (*-dge* as in *wedge*, *-tch* as in *itch*) and even *quadrigraphs* (*-eigh* as in *weigh*, *-ough* as in *tough*). To give you a sense of how helpful it is to learn these patterns, imagine trying to pronounce the letters *-dge (ledge)* or *eigh (sleigh)* if you haven't been taught that they group together to make one pronounceable sound.

Rules. Children are taught useful rules that help them figure out the correct pronunciation of different letter patterns. A child learns that *the pronunciation of a letter sometimes depends on what letters come after it.* In first grade children are taught the *silent e rule:* a vowel followed by a consonant and then an *e*, as in t-*ake*, d-*ime*, or h-*ome* is typically pronounced as a long vowel. They also learn that the letter *c* may be pronounced as *k* when it is followed by *a, o,* or *u* (*ca*ke, *co*me, *cu*p) or followed by any consonant (*cl*ap or *cr*ack), or as *sss* if followed by *e, i,* or *y* (*ce*nt, *ci*nema, or *cy*clone). Such rules help to encourage the rising reader to look at all the letters in a word.

Spelling. Although technically not part of phonics instruction, spelling is intimately related to reading and to the relation of letters to sounds. As children learn to read words, they also begin to learn how to spell these words. In fact, a child should never be asked to spell a word she can't read. Serious spelling instruction begins in the middle of first grade, although children are generally not held to accurate spellings until they are in second grade.

Effective spelling instruction is more than rote memorization of word lists. Spelling (*encoding*, going from sound to letter) strongly reinforces reading (*decoding*, going from letter to sound), and its instruction should be linked to a child's reading lesson. Like reading, spelling instruction follows a logical sequence that begins with phonemic awareness and then learning which letters represent which specific sounds. As a child reads more and learns more words, he begins to appreciate that the same sound can have different spellings; for example, he learns that the *a* sound can be spelled in different ways, such as *mate, weight,* and *straight.* Through spelling lessons as well as his own reading, a child learns the most frequent letter patterns for different sounds. Initially it is best to stress the most frequent spellings and try not to overwhelm a child with every possible spelling of a sound. In this case, *mate* would be the most common spelling for the *a* sound.

As children progress they are introduced to spelling strategies they can apply to help spell new words. They also learn about so-called irregular (sight) words, whose sounds and spellings don't match the rules they have been taught (for example, *should* and *colonel*); the spelling of these words must be memorized. Children vary widely in their spelling abilities, although children who are dyslexic experience by far the most severe and longest-lasting difficulties producing accurate spellings.

TWO APPROACHES TO IDENTIFYING WORDS

In teaching reading, *what* is taught is as important as *how* it is taught, especially for children who may be dyslexic. It is crucial for children to develop a strong foundation of understanding letter-sound relationships. This requires an early focus on phonics, knowledge that children can then apply to decode words they come across in their reading. *Dyslexic children require much more intense, frequent, and extended phonics instruction.* As I said, effective reading programs teach children about phonics not only systematically but *explicitly* as well. There is no subtle or subliminal teaching here; children are not left to their own devices. In general, systematic phonics programs also directly teach letter-sound relationships. "Synthetic phonics" is a prime example of one such program type. Children are systematically taught *code-based instruction*—that is, to transform letters into sounds and then to blend (synthesize) the sounds together to form a pronounceable word. This is in contrast to what are often described as *whole-language* reading programs, which may lightly touch on teaching phonics but not with the early, consistent emphasis on code-based instruction. Typically, within such whole-language programs there is not such a strong emphasis on having children systematically and explicitly work on analyzing and pulling apart words. The focus is not primarily on the sounds of language but on meaning. Early on, children are given books to "read" or to have read to them. Decoding is lightly touched on but is not a major or primary focus of instruction. It is assumed that reading is acquired "naturally," just as speech is (as we know, an incorrect assumption). In this view, letter-sound relationships will be learned naturally, seemingly by osmosis, as children are surrounded by literature and exposed to printed materials. According to this view, there is no reason for a major focus on teaching phonics systematically or explicitly; children will figure it out on their own, or at least with small word-study minilessons rather than a full-blown emphasis.

The two approaches differ markedly in how children are instructed to approach unfamiliar printed words. Phonics instructs a child to try to analyze the word and to sound it out, while whole-language emphasizes guessing the word from the context of the story or from pictures accompanying the story. In phonics, the clues to identifying the word lie *within* the word itself, and children are encouraged to attend to the finer points of the word's structure. In whole-language, the clues are *external* to the word and are to be derived from the meaning of the story. The National Reading Panel found that children who are taught phonics systematically and explicitly make greater progress in reading than with any other type of instruction. Good evidence indicates that beginning teaching phonics in kindergarten or first grade produces the best results. Once a child demonstrates that she has some understanding of how spoken language works, it is time to learn how letters link to these sounds. In general, phonics instruction extends over two school years. All children benefit from it.

First-grade instruction in phonics not only helps children learn how to sound out and pronounce words, improving reading accuracy, it also helps reading comprehension. As children reach higher grades, reading more complicated texts, phonics tends to influence reading comprehension far less. This is reasonable, since reading comprehension in lower grades involves reading very simple words that are usually quite familiar to a child. If he can decode the word, he generally knows what it means. At higher grades, reading involves more complicated and unfamiliar words. To comprehend what he is reading, the reader must not only be able to pronounce the word, he must also know what the word means, which is not taught as a major part of phonics instruction.

Phonics also helps spelling, and the earlier it is taught, the more effective it is. This is not surprising, since phonics concentrates on how letters represent sounds in words, which, after all, is what spelling is about. As a child progresses in school, he learns new words that are more likely to contain irregular spellings; that is, their spelling is not fully predictable by phonics rules. So with time, as discussed below, children learn strategies to figure out larger and more complicated words, and as a result, phonics has less of an effect on a child's spelling.

The quality and quantity of materials available to most effectively teach reading are better than ever. As noted, teachers and parents have available many choices for teaching phonics and phonemic awareness, including programs such as Lexia, Ooka Island, and Headsprout.

To repeat, a typical course for phonics instruction begins in first grade, or today increasingly in kindergarten, and continues through second grade. Once a child is comfortably able to read most one-syllable words, the emphasis switches from phonics to learning strategies for figuring out increasingly longer and more complex words. However, children will vary in their progress, so that some may experience difficulty and need phonics instruction to continue into later grades. The classroom instructional time devoted to phonics per se need not be long—perhaps fifteen to twenty minutes a day. This is reinforced many times a day through all the other activities in a child's reading program and by encouraging the child to read and to write independently.

After first grade, reading instruction systematically moves along. Instruction in the second and third grades reinforces the basic principles of teaching reading with the goal of encouraging the child to read more and to read increasingly more complex books independently. By the end of second grade, a child should be selecting, reading, and enjoying chapter books, as well as spelling words correctly and writing his own little stories.

TAKING APART BIGGER WORDS

In third grade and especially fourth grade and higher, reading instruction adds a new dimension by focusing on big words. In the same way that a beginning reader learns how to pull apart simple words, a maturing reader requires a strategy to analyze the longer and more complicated words he encounters. A child *learns to read* in grades one through three but *reads to learn* (acquire new information) in grades four through eight. Certainly by fourth grade a child begins to study specific subject areas and to do so in increasing depth, bringing about a sharp increase in the number and complexity of printed words he will confront in his daily schoolwork. He is now expected to read multisyllabic words, and the strategies he adopts to tackle them are straightforward, revolving around the basic idea of systematically breaking a multisyllabic word into smaller chunks. The goal is to teach a child how to figure out where there are natural breaks in a word. Once the printed word is pulled apart, the child uses his knowledge of letter-sound relationships to further analyze each smaller part.

Over time children progressively learn how to tackle these longer words. This instruction often begins during first grade, when a child is introduced to compound words such as *inside* and *goldfish* and taught that they can be separated into two parts *(in* and *side, gold* and *fish)*. During second grade children are taught explicit strategies for figuring out how to analyze more complicated long written words, and this is emphasized even more in third grade. The most useful approach is first to divide a word according to where its syllables naturally break. Syllables contain one vowel *sound* (although they may have more than one vowel letter, as in *toe* or *bait)*. Vowel sounds are made by opening the mouth. (Because of this anatomical relationship, you can often tell the number of syllables by simply counting the number of times your jaw moves as you say a word. You can also put your fingers on your jaw and feel its movement as you articulate different words. Say the word *jaw* and you'll note one movement; say the word *scarecrow* and you'll sense two.)

Using these strategies, a child can break apart a surprising number of words. Once a student is able to pull apart a word into its syllables, she can then analyze each syllable into its phonemes, blend the sounds together, and pronounce the word. *Practice is critical.* Children do best when they feel confident enough to take a stab at pronouncing these strange words and are able to laugh at their own errors, and at the same time work to self-correct their mistakes. There is no substitute for practice, guidance, and feedback; practice, self-correction, and more practice.

There are several helpful strategies for children learning to divide multisyllabic words. Here are the most useful.

Dividing a Long Word Between the Two Consonants. A word like *kitten* becomes *kit-ten*. The child sounds out each syllable, *kit* → *k i t,* and *ten* → *t e n,* and then blends the sounds together, *k i t, t e n,* and the word *kitten* is pronounced perfectly.

Knowing Word Derivations. If a child knows how words are built, how they are put together, he will also know how to take them apart. Children progressively learn about prefixes, suffixes, and word origins—first Anglo-Saxon and then Latin roots, and still later Greek origins of English words. These forms are referred to as *morphemes,* parts of words in the English language that, although small, convey meaning. The prefix begins a word (*im* means not; *im*pure means *not* pure). The suffix ends it (*-er* means one who does; a photograph*er* is one who photographs). If your child recognizes a prefix or a suffix, and in some cases both, he is better prepared to unhitch the parts of a potentially unwieldy word. For example, the word *unsinkable* has two common affixes*: the prefix *un-*, meaning not, and the suffix *-able*, meaning having the ability to. Once your child knows frequently used affixes such as *un-* and *-able*, he can pull apart a long word like *unsinkable (un-sink-able)*, pronounce it correctly, and know that it refers to something that cannot be sunk.

Knowing the etymology or the roots of a word is a very powerful aid to reading, shedding light on a word's pronunciation, its spelling, and its meaning. For example, knowing the Latin root *script*, meaning to write, can be helpful in subdividing words like *manuscript* and *transcript* and in knowing that each has to do with writing. In the same way, knowledge of Greek stems is indispensable to those interested in science or technology. It is very useful for a student to know that *ology* (*ol-ogy*) means the study of; now he can decipher and know the meaning of a whole range of potentially difficult and academically important words, including *zoology* (the study of animals), *biology* (the study of living organisms and life processes), and *meteorology* (the study of the atmosphere, especially the weather).

Even the most obscure and complicated-appearing words can be broken down into more manageable units and deciphered if the reader is aware of their derivation or roots. Without this knowledge, it is difficult for students to read any but the simplest of texts. Children who are not able to unhitch long words are forced to skip over these words or to give them made-up names, like *asty* or *Amy* for *astronomy*. Doing so allows the child to read through the text, but since he guesses a different word each time he encounters the word *astronomy*, it is as if he is seeing the word for the first time and hasn't reinforced the neural components—representing the sounds, the

---

* An affix is a letter or a group of letters attached to the beginning or ending of a word or word root that changes its meaning. Prefixes and suffixes are affixes.

meaning, and the orthographic elements—that must come together to form the word. In essence, he does not have the neural components of the word in a well-elaborated, mature, ready position to come together so that he can recognize the word when he comes across it. As a result, he will not recognize the word when his teacher talks about astronomy the next day in class or when he comes across the word *astronomy* again in a book.

Children learn how to read words in a systematic, logical progression. Those who can easily read most small words and have a strategy for approaching longer, more complex ones are mature readers. They are ready to apply their skills to technical or specialized words in textbooks and to uncommon words in literature. They are also adept at using the dictionary to confirm spellings or to look up exact meanings.

WHAT YOU CAN DO TO HELP YOUR CHILD

Below is the most important information you will need to help your child become a reader. I've divided the information into a section on school—the specific questions you want to ask about your child's reading program—and one on the home, focusing on the kinds of activities you as a parent can do with your child to encourage his development as a reader.

AT SCHOOL

One of the questions I am asked most frequently concerns how to choose a reading program that works. The publication of the findings of the National Reading Panel in 2000 represented a milestone in national efforts to translate scientific knowledge of reading into practice. For example, the report includes each of the elements the panel found to be effective, including the amount of time needed to teach reading and language arts (as much as 150 minutes a day in grades one and two), and stipulates that extra time (thirty to forty-five minutes a day) should be built in to permit practice and reinforcement for those children who need extra help or enrichment for those who are ahead of the curve. Since the publication of the report's findings, publishers have worked very hard to update their reading materials to incorporate the most effective, scientifically supported, and, most critically, evidence-based teaching methods.

A BRIEF CHECKLIST CONCERNING YOUR CHILD'S BEGINNING READING PROGRAM

Most importantly, you want to know

→ Is there scientific evidence that the program is effective in improving children's reading—that is, is the program evidence-based? You want specifics. A good place to begin to examine this issue is the U.S. Department of Education's What Works Clearinghouse (ies.ed.gov/ncee/wwc). Keep in mind that testimonials, no matter how moving, are not scientific evidence; nor are articles in newsletters and magazines, nor mentions in the media or online, acceptable evidence.

CAUTION: The term research-based *is not sufficient; it simply means that the program is based on theory, and theory is not evidence or proof that a program actually works.*

WHAT METHOD IS USED TO TEACH READING?

The key questions to ask are these:

→ Is the focus of the initial teaching of reading code-based instruction—that is, are children first taught the alphabetic principle (for example, phonemic awareness and phonics)?
→ In teaching beginning reading, are phonemic awareness and phonics a major focus and taught systematically and explicitly?
→ How are children taught to approach an unfamiliar word? They should feel empowered to first try to analyze and sound out an unknown word rather than solely guessing from the pictures or context. Illustrations and context can be used as a second step to verify if the pronunciation seems to make sense.
→ Does the program include many opportunities to practice reading, to write, and to listen to and talk about stories?

HOW IS INSTRUCTION MATCHED TO A CHILD'S INDIVIDUAL NEEDS?

The critical elements to look for are

→ *Individualization.* The size and flexibility of instructional groups are important. Some components of reading, such as vocabulary, may be taught as a class activity. However, given the variability in reading skills within a class, components such as basic phonologic skills and oral reading are best addressed in smaller groups. You want to ask how large instructional groups are. Children who are dyslexic require *intense* instruction in reading. As the National Reading Panel noted, this translates into small groups, anywhere from four to five students to individualized one-to-one or one-to-two instruction. It is clear that if a student is in a larger group, she is rarely called on, perhaps once during a forty-five-minute period, if at all. When not called upon, the struggling reader begins to disconnect from the group's goings-on because they are too difficult to follow, and she spends much of the time staring out the window or getting into trouble. The grouping should be flexible for each component of reading, so that a child can move in and out of a group depending on his progress and level of skill in that particular area. Children who are progressing more slowly will benefit from smaller groups and more intense attention to that skill. Overall, dyslexic children require a lot of individualized attention, along with repetition and practice, in order to truly benefit from reading instruction.

→ *Feedback and guidance.* Learning should be *active,* with many teacher-student interactions. Ideally, the teacher models reading for the child and then provides feedback and guidance as the student rereads the selection aloud. Of course, how much feedback and guidance is given and its nature are best determined by direct observation while visiting your child's classroom.

→ *Ongoing assessment.* The child's reading skills should be assessed both by informal teacher observation and by more formal measures such as aimsweb or DIBELS (see pages 291–292). As I will discuss later, measuring your dyslexic child's progress in reading needs to be an ongoing process in order to reflect his changing needs. *In the primary grades (one to three), his reading should be assessed at least three times during the school year* to monitor growth, more often if there are indications of failure to make progress.

AT HOME

While a parent should not become her child's primary teacher, she can become her child's biggest helper. With a light hand, good humor, and the suggestions I make here, you can help accelerate your child's progress. In most instances I strongly caution parents against setting out to teach their child all the phonics rules or a complete reading curriculum. Teaching reading is a complex task, one that should be left to a professional. Keep in mind that your child is in class for perhaps six hours a day. You will see him after school, when he is tired and less receptive to learning and when you, too, are not your most energetic or patient self. I recommend that you work with your child fifteen or twenty minutes a few evenings a week; it should remain fun and not a chore for either one of you. For the most part, weekends should be left for enjoyment and not to play catch-up.

Focus on reinforcement. School is where *new* learning should take place; home is ideal for *practice* and *reinforcement.* School helps him build the necessary neural components for reading; home practice strengthens and solidifies these components. Make an effort to coordinate any home reading activity with what your child is learning in school. Children do best when they can focus on one procedure, one approach at a time, so work in harmony with your child's teacher.

Reinforce selected basic skills that will make reading more understandable and ultimately more enjoyable for your child. I will give you a list of children's books that are right for him, that he will enjoy reading, and that will stimulate him to want to read more. I will recommend activities that are easy to do and require little more than a stack of index cards, magnetic letters, a felt marking pen, and a highlighter pen. For some of the activities I recommend colored as well as white index cards. Alternatively, many students have found using electronic index cards—for example, the app Quizlet—is not only helpful but fun as well. With Quizlet you can create your own personal cards or choose one from the extremely large selection of cards other users have created.

Sounding Out Smaller Words. Like most other behaviors, how we read will reflect the habits we develop through instruction and practice. You can play an important role in ensuring that your child develops good reading habits by encouraging certain behaviors. One of the most important is to learn to sound out words and to do so early. Whenever your child comes across a word he is unsure of, encourage him to try to sound it out. You can begin by asking him about the first sound. For example, if the word is *sat,* you can say with some exaggeration, "The first letter is *sss.* What is the sound of *sss?*" Repeat the process with the last sound, *t,* and then the middle sound, *aah.* Once he is able to articulate *s a t,* ask him to blend the sounds together rapidly and say *sat.* Ask him if this sounds right to him. Does it make sense in the story? He will be practicing good reading habits: first decoding an unknown word and then verifying that his pronunciation is accurate. By teaching him to ask himself these questions automatically, you are also fostering his independence as a reader and building his self-confidence.

Speak to your child's teacher and ask what sounds and strategies your child is working on and what you can do to help her practice. If the teacher isn't able to provide you with any suggestions, you can ask her about some of the activities I have listed below. I have selected several of the simplest and most useful strategies. These are relatively easy for you to practice with your child and will help her to pull apart literally hundreds of words that she might otherwise give up on. These are also helpful when you are reading with your child. If she stumbles on a word dependent on one of these strategies for its pronunciation, after you finish a page or story you can use the troublesome word as an opportunity to review the rule. These strategies will allow your child, for example, to correctly pronounce words following the *silent e rule,* so that she knows the difference between words like *mate* and *mat,* and words containing the letter *c,* to determine when a *c* is said softly, as in *cereal* and *cinder,* or makes a hard sound, as in *camel* and *clock* (I refer to this as the *saying c's rule*).

Taking Apart Bigger Words. Once your child is comfortably reading short, one-syllable words, you can help him develop strategies for longer, more complex words. You can help him read

→ compound words such as *newspaper* and *backpack* which are composed of two smaller words.

→ other longer words by learning how to divide those with specific letter patterns—here, having two consonants surrounded by two vowels, such as *goblet* and *mitten.*

→ the beginning, middle, and end of words by calling on prefixes, suffixes, and word roots.

Below are specific details about each of the activities described above. Begin with the ones focused on short words and then, after checking with your child's teacher, introduce your child to the strategies meant for older children with stronger reading skills in place. The most useful rule and the one taught earliest is:

The Secret of the Silent E Rule. Focus on and reinforce the *silent e rule* more than the others. Since it affects many words, it is extremely useful to a novice reader. It is also simple to get across: (1) tell your child the rule, (2) show him several examples, and (3) practice applying the rule to different words. You will need a stack of index cards with pairs of words printed on them, words with and without the silent *e* at the end. Again, electronic index cards using Quizlet may be helpful.

For step one, tell your child, "Some words have a secret code that tells you how to say them. The *silent e* code helps you to read many, many words. Listen, here is the code." Look for a word where there is a vowel followed by a consonant and then a final *e*—which makes the vowel say its own name (vowel, consonant, *silent e*)—and write down some examples, like *rate, fine, same.*

→ Sequentially, point to the vowel, the consonant, and the *silent e* at the end and then say each word.

→ Play a game in which one form of a word is magically transformed into another. You will need the index cards with the pairs of words (for example, *pan/pane*) printed on them.

→ Practice reading one form of the word and then the other. For example, first read the word *hid*, then point to its cousin with the silent *e* at the end, *hide*, and read it, *hide.*

Repeat this for each of the words, then ask your child to do the same. Once he seems to have grasped the idea, alternate, sometimes beginning with the word with the silent *e* and then reading the other word (*rate,* then *rat*). At other times, start with the short form (*rat*) followed by the word with the silent *e* at the end (*rate*).

You can then apply the rule to real-life reading. Ask your child to get a book that he is familiar with and enjoys. With you by his side, have him take a highlighter and mark and read each word following the *silent e* code. Your child will master the exercise and enjoy his ability to read the many words he can now decipher with his secret code.

Here are forty word pairs to begin with. Copy each pair onto an index card (or use Quizlet).

bit, bite	cub, cube	cut, cute
can, cane	cap, cape	cod, code
con, cone	Dan, Dane	dim, dime
fad, fade	fat, fate	fin, fine
fir, fire	hat, hate	hid, hide
hop, hope	kit, kite	Jan, Jane
man, mane	mad, made	mat, mate
not, note	pal, pale	pan, pane
pin, pine	rat, rate	rid, ride
rip, ripe	rob, robe	rod, rode
Sam, same	Sid, side	sit, site
tam, tame	tap, tape	Tim, time
Tom, tome	tub, tube	van, vane
win, wine		

You can also use magnetic letters to make word pairs with and without the silent *e*. Spell out *pal,* for example, and ask your child to spell out *pale*. Conversely, begin with *pale*, ask him to pronounce the word, and then ask if he can transform *pale* into *pal* by removing the magnetic *e*.

Saying C's. You can help your child understand why the *c* in *cent* is pronounced one way and the *c* in *can* another. Share the rule with him, give him some examples, and then practice applying the rules governing how to say the letter *c* whenever it is encountered in words. The rule is

→ When the letter *c* is followed by the letters *e, i,* or *y*, it is pronounced softly, like an *ssss*, as in *cent* and *cinnamon.*

→ When the letter *c* is followed by a consonant or the vowels *a, o,* or *u*, it makes a hard *c* or *k* sound, as in *car* and *cup.*

You can help your child practice the different sounds by writing different *c* words on your index cards, some followed by the letters *e, i,* or *y,* which produce the soft *c,* and others followed by either a consonant or the vowels *a, o,* or *u,* which make the *k* sound. Here are some words to help you get started:

→ Soft *c*'s: *cent, center, cell, cellar, cinder, cider, cite, cycle*

→ Hard *c*'s: *cab, cake, call, can, cap, car, cat, clam, clap, Coke, cop, cup*

Once you have made up about five or six cards in each of the two pronunciation groups, ask your child to group the words pronounced like *cent* in one pile and the ones pronounced like *cat* in another. Next, have him read aloud the words in each group. If he misses a word, put that card under the original pile so he will have another chance to read it correctly.

Compound Words. Compound words serve as a wonderful introduction to longer words. Because they are made up of two real words, it is easy for children to see how the compound word can be divided into smaller pieces.

→ One fun activity that helps make your child become aware of compound words is to go through your house or the pictures in a book or magazine, pointing to and saying the name of common household objects such as *toothpaste, toothbrush, hairbrush, washcloth, dishwasher, staircase, doorway, doorknob, bedroom, baseball,* and *driveway.* In each case, indicate how the spoken word can come apart, such as saying, "*Toothpaste* is really made up of two words, *tooth* and *paste.*"

→ Then say and write each half of the word on a separate index card and show him how they come together to form the word and how they can also be pulled apart. This will help make your child aware that longer words are made of parts that can be pulled apart and pushed back together again.

Two Consonants Surrounded by Two Vowels. This strategy is also relatively easy to master and can make a big difference in the progress your child makes as he comes to slightly longer words.

→ This tool tells a child that in words where there is a vowel followed by two consonants followed by another vowel, you can divide the two consonants down the middle. Although this rule is a mouthful, don't let it put you off. It's really not hard for children to learn.

→ Sometimes I find it helps to refer to this sequence as two consonants surrounded by a vowel on either end (*v, c, c, v*). As shown in Figure 43, you can tell your child to picture the vowels trying to pull the consonants apart. Examples always help, such as *attic (at-tic)* and *magnet (mag-net).*

After telling your child the rule, use a few of the words below as examples, showing her how you would divide the words. Now give her an opportunity

Figure 43. Pulling Apart Longer Words

When there are two consonants surrounded by a vowel on each side, imagine the vowels pulling the consonants apart. Shown here is this strategy applied to the word *magnet.*

to practice. Write each of these words on a separate index card and ask her to show exactly where she would separate the word (the first several should show the division; ask your child to split the remainder). If she hesitates, repeat the strategy and point to the middle consonants as you pronounce each part of the word, such as *pic-nic*.

ab-sent	com-mon	hap-pen	pen-cil	suc-cess
attic	cotton	helmet	picnic	sudden
basket	custom	hidden	plastic	tablet
blanket	dentist	himself	possum	tennis
blossom	fallen	insect	problem	traffic
bottom	funnel	kidnap	public	triplet
button	gallop	kitten	puppet	trumpet
cactus	goblet	lesson	rabbit	tunnel
cannot	gossip	magnet	ribbon	upset
cobweb	gotten	mitten	signal	velvet
		napkin		

Prefixes, Suffixes, and Word Roots. Teaching these meaningful parts of words is my preferred strategy for breaking up multisyllabic words. Learning these small but meaningful bits of language often captures even the most blasé child's interest. He will be delighted that he can literally deconstruct a word and know its pronunciation and its meaning. Introduce your child to a prefix (or suffix), explain its meaning, and then show him the examples below. The prefixes below account for almost two-thirds of all prefixed words (for more on prefixes, see pages 263–264).

For each of the words listed, have him practice as I describe below for the word *imperfect*.

→ For the prefix *im*, you can first write, pronounce, and define it: "*Im* means *not*." Then ask your child to pronounce it.

→ Select a word with the prefix and write the word, underlining the prefix (*im*perfect). Ask her to pronounce the underlined prefix (*im*) and then to read aloud the entire word (*im*perfect).

→ Use the word in a sentence to ensure that your child has a true sense of its meaning. Many of these words are most appropriate for late third-graders or fourth-graders.

PREFIXES

dis-	*im-* or *in-*	*mis-*	*pre-*	*re-*	*un-*
(not)	(not)	(bad)	(before)	(again)	(not)
disagree	impatient	misbehave	prearrange	reappear	unable
disappear	imperfect	mislay	precook	rearm	unbend
disarm	impolite	mislead	prepaid	rearrange	uncover
disconnect	impossible	misplace	preschool	recount	undo
dishonest	impure	mispronounce	pretest	redo	unequal
dislike	inactive	misread	preview	reenter	unfair
disloyal	incomplete	misspell		refill	unhappy
dismount	independent	mistreat		refresh	unkind
disobey	invisible	misuse		reheat	unlock
disorder				renumber	unwrap
disown				replay	
distrust				retell	

SUFFIXES

-able	*-ful*	*-less*	*-ly*	*-ment*	*-ness*
(is, can be)	(full of)	(without)	(resembling)	(action/ process)	(state/ quality of)
admirable	careful	ageless	bravely	agreement	cleverness
agreeable	cheerful	beardless	brotherly	amazement	darkness
breakable	colorful	bottomless	cleverly	arrangement	fairness
curable	delightful	careless	fatherly	development	fullness
desirable	fearful	fearless	foolishly	entertainment	goodness
enjoyable	forgetful	jobless	freely	experiment	happiness
excitable	helpful	painless	honestly	government	kindness
laughable	joyful	sleepless	loudly	payment	loudness
movable	painful	worthless	neatly	punishment	sadness

Learning to sound out smaller words and to unhitch longer ones is only part of a total reading program. Children must also be taught how to read words that don't follow the rules and to apply this knowledge to reading and writing. Below are some suggestions.

Memorizing Sight Words. Introduce your child to and have her commit to memory the most common sight words found in books. Having a core of words that she can read makes it so much easier when a child first attempts to read books on her own. As I said earlier, there is no way to learn sight words other than rote memory. The process is simple.

→ Introduce the word, use it in a sentence, write it on a card, and ask your child to say it.

→ Have her write the word on a card as well. The act of writing a word as she says it aloud helps to reinforce its pronunciation in her memory.

→ Write the word in a sentence, too, since it is important to practice reading words when they are in a sentence as well as when they are alone.

→ Printing each word on a colored index card of its own also helps your child remember a word. Linking a bright color with a particular word helps in establishing memory links for these words.

→ Once you have made the word cards, take a large shoebox, decorate it together with your daughter, label it WORD BANK, and keep the growing number of sight words here. Each week you can reach into the bank, take two or three cards initially, and later on four or more, and review the words on the cards for a few minutes throughout the week.

HELPFUL HINT: Resist the temptation to choose too many words at once. It is much more effective to focus on fewer words, but to do so intensively.

Let your child's progress guide you. If your child seems to have trouble with a word, add it to the next week's words as well. Try to point out and highlight the week's special words whenever you come across them in books, magazines, or newspapers. These words can be put on Quizlet cards and placed in your child's Word Bank.

Since sight words are so common and so well represented among the most frequently used words in print, the best long-term approach to mastering them is to gradually help your child memorize these words, a few at a time. Here is a list of the 150 most common words found in printed English according to the *Word Frequency Book*. The words are listed in order of their frequency. As you skim them, you will immediately notice that the vast majority are sight words.

the	not	into	water	think
of	but	has	long	also
and	what	more	little	around
a	all	her	very	another
to	were	two	after	came
in	when	like	words	come
is	we	him	called	work
you	there	see	just	three
that	can	time	where	word
it	an	could	most	must
he	your	no	know	because
for	which	make	get	does
was	their	than	through	part
on	said	first	back	even
are	if	been	much	place
as	do	its	before	well
with	will	who	go	such
his	each	now	good	here
they	about	people	new	take
at	how	my	write	why
be	up	made	our	things
this	out	over	used	help
from	them	did	me	put
I	then	down	man	years
have	she	only	too	different
or	many	way	any	away
by	some	find	day	again
one	so	use	same	off
had	these	may	right	went
would	other	look	old	number

Because these words are so common, it is necessary for a would-be reader to be able to decipher these little words on sight. Reviewing a few each week should give your child a head start in reading most beginning books. More extensive word lists may be found listed as Fry Sight Words (https://sightwords.com) or in the word lists on VocabularySpellingCity (www.spellingcity.com).

HELPFUL HINT: Begin with the first dozen or so words; they make up a staggering 25 percent of all words appearing in print.

Practicing Reading Connected Text. One of the most important components of a complete reading program is for the child to begin to read connected text, not just words in isolation, and to do so early. Reading connected text should be considered an essential part of a lesson, not just an add-on if there is time remaining. Reading connected text provides a child an opportunity not only to increase her vocabulary and background knowledge but to improve her ability to read individual words by allowing her to see words in different contexts, which may influence both how the words are pronounced and their meaning. For example, the letters *gh* when occurring at the *beginning* of a word are pronounced *g*, as in *ghetto*. However, when *gh* occurs at the *end* of a word it can be pronounced *f*, as in *enough*, or can be *silent*, as in *through*.

Again, practice is the key to becoming a skilled reader. Without it, a child may see a word and then have it float out of his mind forever. With it, he will be able to etch the word components—phonology, the way the word sounds; semantics, what it means; orthography, the way the word looks—into his memory circuits in ready reserve until they are needed again.

Very simple principles guide the choice of reading materials your child should practice with. Beginning readers need to be exposed to books that are easy and fun to read. With these books the child is in familiar territory, since most of the words they contain are already registered in the reading circuits of his brain. These books help to consolidate reading skills already learned. Repeated practice with them is the key to reinforcing and strengthening the neural wiring for the specific components of individual words and word families. Children also benefit from reading somewhat more difficult books, especially books they choose, which reflect their interest or curiosity. With a teacher (or parent) available to provide guidance and to answer questions, reading such books allows the child to expand his knowledge of letter-sound linkages and of exceptions to the rules, and to learn new vocabulary words and concepts.

In practice, a child starts with the simplest books, and as he learns more letters and sounds and more sight words, he gradually works his way up to longer and more interesting books. But at the very beginning, for the sheer pleasure of being able to read on his own—and for the confidence it builds—nothing works as well as the small decodable booklets and the more recently available Geode readers I mentioned earlier.

Once you read one to your child, chances are that he will want to reread it on his own over and over again. Children like to read these simple books because they *can*. In the previous chapter I mentioned several good series to start with, in particular Bob Books by Bobby Lynn Maslen. At the beginning, I would make it a habit first to read a book to your child and then to sit with him as he reads it back to you. This ensures that he can read the book and that you will be there to offer assistance if he comes across an unexpectedly difficult word. When you read with him, you can support him by

→ reminding him that part of the fun of reading is figuring out the word and that it often takes more than one try.

→ sharing with him your own missteps in reading.

→ encouraging him to first sound out the word and to listen to how the word sounds—to ask himself, *Does it fit in? Does it sound right?*

→ encouraging him to trust his instincts and, if a word doesn't seem right, to apply the strategies he knows and to check his Word Bank to see if it is a tricky word that must be memorized. If so, he can check its pronunciation with you.

Figure 44. Reading with Your Child

To read a book with ease, a child should be able to read about nineteen out of twenty words on a page correctly. This is especially important for children in kindergarten or first grade, who are just beginning to learn to read. Views about how best to select reading materials for boys and girls in second grade and above are undergoing change from the previous belief that children in all grades should be matched to books at their instructional level—that is, that can be read with 90 percent to 95 percent accuracy—to more recent thoughts that it might be best for them to be exposed to and read texts at varying levels, giving the students' choice and interest more weight. For their independent reading, students in fact tend to choose texts that are above their instructional level. A proponent of having students read books of varying text difficulty, from easy to difficult, Professor Tim Shanahan, Distinguished Professor Emeritus at the University of Illinois at Chicago, argues that in selecting reading materials, students' "curiosity about the content of the harder materials outweighs their fear of failure." Studies have not supported the widely held belief that limiting reading materials to texts at the student's instructional level leads to improved reading or interest in reading. However, if a student selects a book at a higher reading level, it is important for an adult to be close by to help him if he cannot pronounce a word, or to define it if he does not know its meaning.

During a recent summer, when my seven-and-a-half-year-old granddaughter, about to enter second grade, visited, I learned how important a child's interest in a book's topic can be. She had chosen a story of sisters and their relationship to one another. I read the first few chapters with her, the two of us alternating reading a few pages. She became engrossed in the story. Soon she left for home, clutching the book in her hand. I wondered if she would continue reading it after she left. Several hours later her father sent me a wonderful photo of her sitting on the couch with a big smile on her face and the book held up in front of her. She had finished the book and was now asking her dad to order another book on the same topic. Earlier I had turned the book over and had seen that it was recommended for children in the fifth grade.

An unusual group of texts, Wit & Wisdom®, available from the nonprofit Great Minds® organization, offer the early reader knowledge while appealing to his interest in an unusually wide range of topics. Wit & Wisdom® has now joined with the Wilson Reading System Fundations® program to produce the aforementioned Readables, specifically the reading series Geodes, which provides interesting content while being aligned with the Fundations® program for teaching beginning reading.

For parents of beginning readers, it is particularly helpful to know the reading level of a book. There are aids to help you do exactly that. One, the Lexile Framework, matches students to books by using a common measure, the Lexile unit. Lexile levels extend from zero to about 1,700, with second- to third-grade students typically reading in the Lexile range between 420 and 820. In this system *The Cat in the Hat* is scored at 260 Lexile units, *Charlotte's Web* at 680, *Ethan Frome* at 1,160, and *Silent Spring* at 1,340. Parents and others can access the Lexile Framework at www.lexile.com. Another system, the Degrees of Reading Power (DRP) (www.questarai.com/drp), lists children's books according to their reading level. The information is available in the form of "Readability Reports" in two broad categories, "Readability of Literature and Popular Titles" and "Readability of Textbooks."

For discovering a book's level of difficulty when you are at a library, you can turn to *Children's Books in Print*, which sorts books by title, author, and subject matter. Libraries subscribe and are able to access this resource.

In the Appendix you will find lists of books for children, including picture books, poetry, easy readers, picture books with rhymes/pattern repetition, and transitional chapter books. Consider that selection a starting point for further exploration. Your local library has a treasury of good books, and you might also want to consult the lists of recommended works in such guides as

> *A to Zoo: Subject Access to Children's Picture Books,* tenth edition by Rebecca L. Thomas (Santa Barbara: Libraries Unlimited, 2018).
>
> *Books to Build On: A Grade-by-Grade Resource Guide for Parents and Teachers,* edited by John Holdren and E. D. Hirsch, Jr. (New York: Dell, 1996).
>
> *Boys and Literacy: Practical Strategies for Librarians, Teachers, and Parents,* by Elizabeth Knowles and Martha Smith (Westport, CT: Libraries Unlimited, 2005).
>
> *How to Get Your Child to Love Reading: Inspiration and Ideas from a Teacher Who Really Knows,* by Esmé Raji Codell (Chapel Hill, NC: Algonquin, 2003).
>
> *100 Best Books for Children* by Anita Silvey (Boston: Houghton Mifflin, 2004).
>
> *The New York Times Parent's Guide to the Best Books for Children* by Eden Ross Lipson (New York: Three Rivers, 2000).

Magazines provide inviting bite-sized reading materials for every age and every interest. Providing materials of genuine interest helps to make reading a habit, encouraging practice that builds automatic reading systems. Beginning readers eagerly anticipate the latest arrival of their very own magazine full of simple stories that they can read, as well as fun activities and puzzles. For children three to six, *Ladybug* magazine introduces poems, rhymes, and songs. *Spider* is geared for children six to nine who are becoming independent readers. *ASK* targets the same age range but is aimed at what the publisher refers to as "curious" children who like to ask questions. *ASK* particularly focuses on the arts and science. *MUSE* has a somewhat similar focus on science and the arts but also includes technology and many jokes for children nine to fourteen. *Cricket*'s audience is nine- to fourteen-year-olds; it has wonderful stories, puzzles, and contests that are appealing to children who are now reading to learn and to enjoy themselves. CricketMedia also publishes a number of more specialized magazines for children, such as *Dig into History*, a history and archaeology magazine for children nine to fourteen. You can subscribe to any of these by contacting www.cricketmagazine .com. *Ranger Rick*, devoted to nature, science, and the outdoors, is filled with beautiful photographs and is ideal for curious children seven to twelve, while *Ranger Rick Jr.* is geared for children four to seven. Subscriptions are available from the National Wildlife Federation at www.nwf.org. To introduce children to current events, I suggest *TIME for Kids*, a newsmagazine geared for students in grades K–6. All editions contain feature news stories, many contributed by children. Suggested activities and subscriber information are available at www.timeforkids.com. For children ages six through twelve who crave sports-related stories, *Sports Illustrated Kids* (www.sikids) covers all major organized sports teams (NBA, NFL, NHL, and MLB) as well as extreme sports.

Even if you choose a book or magazine that turns out to be too difficult for your child, you can always do shared reading, in which you read most of the publication but pause when you come to a word your child can read, sound out, or predict successfully. Once your child learns to read words accurately, she is ready for the next step. In that phase of her reading development, she will not only read words correctly, she will read them smoothly and rapidly—without giving it a second thought. How she becomes that fluent reader and what you can do to help her achieve that is the subject to which I now turn.

19

HELPING YOUR CHILD BECOME
A SKILLED READER

FLUENCY

Fluency, the ability to read a text accurately, quickly, and with good intonation (prosody), is the hallmark of a skilled reader. Children who are fluent readers love to read. Since reading requires little energy for them, it's relaxing and easy for them to get lost in a book. They are the ones who come to the dinner table with book in hand or want to keep their bedroom light on at night to finish "just one more chapter." In contrast, for a nonfluent reader, opening a book is hardly a welcome event.

To appreciate fluency, let me introduce you to a pair of young readers, Amelia and Christy, who are classmates in the fourth grade. Both girls are extremely bright and attend their school's Talented and Gifted (TAG) program.

Recently Christy has developed headaches that often keep her at home. She was never considered a problem reader, although her reading skills seemed to lag behind her capacity to understand stories read to her. Christy will do anything to avoid reading. When she does read, she has to devote all her attention to deciphering the words—often making several attempts before correctly reading a word aloud—and therefore has little energy left to consider the meaning of the passage she is reading. She reads in an expressionless monotone that is difficult to follow. Amelia, on the other hand, glides across the lines of text, adjusting her tempo or modulating her voice to reflect its meaning. Amelia is fluent, Christy is not.

If parents and teachers understood the essential role of fluency in skilled reading and how effectively it can be facilitated, fluency would no longer be the most neglected reading skill. The most recent National Assessment of Educational Progress (NAEP) data examining fluency reported that 45 percent of fourth-grade boys and girls were not yet fluent readers. That is shocking, given the fact that children are expected to develop into fluent readers by the end of second grade. Perhaps this is because many educators believe that once a child can read words accurately, the work of becoming a reader—and of teaching reading—is complete. In fact, reading words accurately only brings a child to the jumping-off point for skilled reading, and this is where a parent (and teacher) can play a significant role in helping her land there.

Children learn to read a word accurately, and then, after sufficient practice, to read the print fluently. Listen to a skilled reader as he reads aloud and you notice that he is not simply decoding words but rather has progressed to comprehending the meaning of the words before him. Fluency describes how a skilled reader reads. I reserve the term *automatic* for the brain processes that make fluent reading possible. Remember, we believe that the neural circuitry within the brain has reinforced all the features of a word so that a mere glance at that word instantly integrates the orthography, phonology, and meaning of the word in the word form area located in the back of the left side of the brain. That is why we believe fluent readers are able to read not only accurately and rapidly but, most important, with good intonation reflecting good understanding.

Fluency forms the bridge between decoding, that is, reading a word accurately, and comprehension, understanding what you read (see Figure 45). Children acquire it word by word, or word group by word group, reflecting the words they have read and fully mastered by repeated exposure—*if* they begin pronouncing the word correctly. It isn't as if suddenly a child reads all words fluently. For some people as few as four correct readings are necessary for automatic recognition of a word, while as many as twelve correct readings may be necessary for others. And while accuracy is a necessary precursor to fluency, accuracy does not necessarily evolve into fluency. This is especially true of a dyslexic child, who is not able to call upon the specialized word

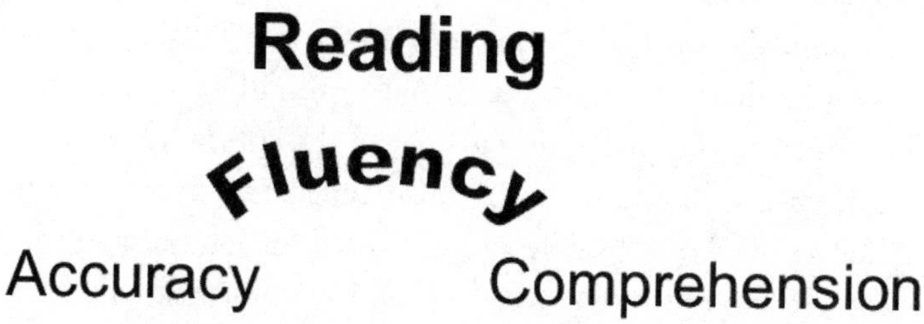

Reading

Fluency

Accuracy   Comprehension

Figure 45. Fluency Forms the Bridge Between Decoding and Comprehension

form area and must instead rely on slower, ancillary neural pathways for reading. These pathways are effective for accurate reading but not for fluency.

One of the most important components of a complete reading program is for the child to begin to read connected text, not just words in isolation, early on. Reading connected text provides the all-important pathway to meaning, to understanding that the printed words on the page are connected and sending the reader a message. Fluent reading frees up the reader's attention so that it can be devoted more fully to comprehending what the message says. Reading connected text also provides the child a vital opportunity not only to increase her vocabulary and background knowledge but to improve her ability to read individual words by allowing her to see words in different contexts, which may influence how the words are pronounced and their meaning. That is why children need to begin to read connected text (first decodable, then authentic text) just as soon as they begin to learn how to decode individual words.

TEACHING FLUENCY: WHAT WORKS BEST

When the National Reading Panel reviewed almost a hundred research articles on teaching fluency, they made the important discovery that the programs that work most effectively and produce the largest gains share three key features:

→ (1) a focus on a child's *oral* reading
→ (2) opportunities for practice, allowing a child to read and to reread words *aloud* in connected text
→ (3) ongoing feedback as a child reads.

More recent studies replicate and support the findings of the NRP.

Reading aloud makes feedback possible. Silent reading does not. The feedback is essential, because it enables a child to modify his pronunciation of a specific word and simultaneously to correct the components of the neural model of that word so that it continuously reflects the exact pronunciation and spelling of the word. Needless to say, the feedback must be carried out in a constructive and positive manner. You must not risk having the child lose self-confidence or develop a fear of reading aloud in front of others. The person who is listening to the child read has to be skilled enough to see where the breakdowns in the reading process are occurring and intervene at that time and then say, "Okay, read that sentence again." When the person listening realizes that the child is beginning to become frustrated with her poor reading fluency, the experienced listener will say, "Why don't I take a turn? Let me read the next paragraph or two."

This overall technique of having the child read aloud to an experienced adult is referred to as *guided oral reading.* The guidance can come from a teacher, a tutor, or a parent. Once a child has received feedback from an adult and is fairly fluent, he can practice with a peer or even by listening to what he has read on a recorder or smartphone. Approaches using this method include echo reading, paired reading, repeated reading, shared reading, and assisted reading. In echo reading, the child's teacher first models by reading a paragraph aloud, then the student reads the same paragraph aloud under the teacher's guidance. The child reads the same selection at home that evening with a parent listening and gently providing corrective feedback. Then students pair up in class the next day (paired reading) and take turns reading each page (or paragraph) to one another. Paired reading not only helps reading fluency but also plays an important role in the development of self-esteem, as well as improving a child's phonics skills. With this method, a teacher can keep track of and monitor a large number of children as they read in pairs. There is evidence that *paired reading, where meaning and discussion of content is an integral part of the practice, is effective in improving reading comprehension.* Recent studies, however, question whether guided oral reading should focus on the same material read multiple times (*repeated oral reading*) or on new materials (referred to as *continuous reading*). Results indicate that while both methods will increase word recognition and improve fluency, children who read new materials show greater growth in reading comprehension.

As part of these studies, researchers had the opportunity to examine children who only listened to the teacher reading the stories instead of reading the stories themselves. These children did not make any gains in word recognition, fluency, or comprehension. Professor Melanie Kuhn of Purdue University College of Education notes that "while reading aloud to students is important in fostering a love of reading, learners must actively engage in the reading of connected text if they are to become skilled readers."

Guided oral reading helps poor and good readers in different ways. For someone struggling, reading a passage out loud over and over while receiving feedback is one more means of obtaining input to master a specific word. It helps the student build a more accurate neural representation of that word's components within her brain and pronounce the word correctly. Rereading a passage aloud allows a good reader to integrate the word's meaning with the surrounding text, giving him a better sense of the prosody or the rhythm of the passage.

From its extensive review, the NRP concluded that guided oral reading programs "help improve children's reading ability, at least through grade 5, and they help improve the reading of students with learning problems much later than this." Nonetheless, the proven effectiveness of guided oral reading to increase fluency is too often ignored, which is unacceptable. In fact, *the evidence is so strong that I urge the adoption of these reading programs as an integral part of every school reading curriculum throughout primary school.* Guided oral reading may even be helpful to some students in middle or high school. This approach might also help readers master the special language peculiar to different academic disciplines, like biology and history. Remember, if a child has trouble pronouncing a word, he is unlikely to have integrated the features necessary to build an accurate representation of that word in his brain, and it will be difficult for him to store or to retrieve any information associated with that word. If troublesome words are pre-identified and practiced aloud with guidance, the child will more easily build the neural model for that word and connect relevant information to it.

Fluency training, one component of a comprehensive reading lesson, takes perhaps fifteen minutes or less per lesson. Keep in mind that a child must first read a word accurately before she can read it fluently. So in carrying out guided repeated oral reading with a child, it is important to have her practice reading passages at her comfort level. As I said earlier, children are comfortable with a book's text when they can read about nineteen out of twenty words correctly; such a book is at the child's *independent reading level.* A teacher can recommend books that she feels are appropriate. If your school participates in either the Lexile or the DRP reading programs, your child's teacher will be able to give you a list of books matched to your child's reading level and interest.

For older elementary- and middle-schoolers, a good source of high-interest, low-readability (indicating relatively easy text) books can be found at www .pearsonschool.com, under "Pearson School Library Collection." For example, Pearson's version of *Moby-Dick* is written at a third- to fourth-grade level; *The Red Badge of Courage* is written at a fourth- to eighth-grade level. Other good sources of this kind of reading material are available from Gander publishers and the Oxford School Shakespeare series, available from Amazon. This series includes not only the complete Shakespeare plays but also clear explanations of difficult words and passages, plot synopses, summaries of individual scenes, and notes on the main characters.

PARENTS CAN HELP BUILD FLUENCY

As an informed parent you can play a strong and active role in helping your child become a fluent reader. Typical readers generally begin to read fluently by the middle of second grade, dyslexic readers much later. You can gauge your child's fluency by listening to how smoothly he reads and how well he is able to read with expression.

Listening to him read aloud often is the single most meaningful activity you can do in this process. Put aside time for it each evening, or at least three or four times a week. Consistency is far more important than the amount of time devoted to each practice session. In these sessions, you can have him read to you or you can take turns reading aloud to one another—a paragraph, a page, a short chapter at a time. For a child who is struggling, it is often useful for you to read a little and then have him reread aloud the same material that you have just modeled reading. If you can, get a duplicate set of books so that each of you can follow along as the other reads. It is helpful to give your child a ruler or other level marker to place under each line as he reads or follows along so he keeps his place.

The best texts for fluency-building are those that encourage reading with feeling, such as poetry and plays, which are meant to be performed orally, so that the child's reading these out loud sounds natural. The objective is to have the student select reading materials that he can eventually practice enough to be able to read before an audience. Doing this emphasizes that rather than increased speed when practicing fluency, the goal is to read to extract *meaning*. This is why reading connected text is a must and preferable to simply reading words in isolation to gain speed.

Too often educators focus their time and energy on having their students practice reading word lists and even connected texts quickly. While speed is important, extracting meaning from the text is more so. Educators must understand that the definition of fluency is the ability to read not only accurately and rapidly but also, and most important, with prosody. *Prosody is and must be a significant component of fluency.* Once a student has begun to master decoding, more attention and time need to be spent practicing reading connected text and on emphasizing the need to extract meaning from print—what the specific words tell the reader about the meaning of the phrase or sentence or paragraph. The automaticity that is integral to fluency allows the reader to shift his attention to meaning and comprehension. *Accurate and fast reading does not necessarily translate into increased comprehension, which is why fluency practice where meaning and comprehension are stressed is critical. It is not only practice and more practice that is necessary, but practice that stresses comprehension of what the print is telling the reader.* This is where the importance of prosody (reading with expression) becomes evident. Prosody, and with it comprehension, are revealed by oral reading pauses, speed, phrasing, loudness, and intonation.

Evenings spent reading together build a lifelong pattern of enjoyment. I believe that the pleasures associated with this experience become built into the reading networks within a child's brain. As a child matures into a teenager and then a young adult and no longer reads aloud with her parent, each time she picks up a book and begins to read she resurrects this happy memory of days when she snuggled next to her father or mother and learned about *The Cat in the Hat, Goodnight Moon, Danny and the Dinosaur, Robert the Rose Horse,* and, later on, *Pumpkin Smasher Witch* and *The Westing Game.*

The methods I have described here are meant for the typical reader: A dyslexic reader may require more (see pages 285–303).

To summarize, here is what parents and teachers should know about reading fluency:

→ It includes accuracy, automaticity, and prosody (reading with expression).

→ It is improved by a child reading aloud to an adult for fifteen minutes four or five times a week. The reading

material should be at the child's reading level and should
be read at 85 to 90 percent accuracy. As the child's fluency
improves, he can be exposed to texts at a more difficult
reading level.

→ It is improved by *repeated* oral reading of the *same passage;*
however, *continuous* oral reading of *different passages at a
similar reading level* provides an even more powerful boost
to increasing fluency as well as accuracy and vocabulary.

Once a child begins to read more and more words accurately and rapidly,
he can turn his attention to more complex texts. At this stage, his higher
thinking and reasoning skills come into play and, together with his vocabulary and knowledge of the world around him, help him derive meaning
from his reading.

BUILDING READING COMPREHENSION: VOCABULARY AND BACKGROUND KNOWLEDGE

A mature reader must have the vocabulary and the background knowledge
to understand and to place in context what he has just read. For example, if
he comes across the words *Eiffel Tower,* he needs to know how to pronounce
each word. But to give the words meaning, he also needs to know what the
Eiffel Tower is. This calls on his word knowledge—his vocabulary. It is even
better when he can place the word more fully in context—when he knows
that the Eiffel Tower is located on the Champ de Mars in Paris, that it was
designed by Alexandre Gustave Eiffel for the exposition of 1889, and that it
is a major tourist attraction. This sort of information is the reader's worldly
knowledge that helps him gain meaning from what he is reading. Word
knowledge is found in the dictionary, worldly knowledge more likely in an
encyclopedia, or today, in Wikipedia or on Google.

A child does not simply open a book, decode the words, and magically
understand the content. The extent and depth of his vocabulary and background knowledge play a major role in determining how well equipped
he is to gain meaning from the text. Important, too, are the active thinking processes the reader uses to integrate his existing knowledge with the
new information he encounters. In a general sense, these thinking processes
can be regarded as reasoning. Based on this model, approaches to teaching

comprehension focus on strengthening and extending the child's word and worldly knowledge and providing him with a series of strategies to organize and direct his thinking as he reads.

BUILDING VOCABULARY: THE MEANING OF WORDS

The size of a child's vocabulary is one of the best predictors of his reading comprehension; children with the largest vocabularies tend to be the strongest readers. A growing body of evidence suggests that increases in the quality and quantity of exposure to language and vocabulary in early childhood and verbal interaction with parents and caregivers are associated with improved language and vocabulary later in childhood.

A significant factor in a child's vocabulary in the early school years relates to the sophistication of the parents' vocabulary during conversations with him and the degree to which the parents speak directly to him. In fact, parent-child interactions at a young age are far more important for language development than, for example, incidental exposure to language through television. Furthermore, the quality of teacher-child verbal interactions for preschoolers is positively related to children's vocabulary and comprehension gains in the early school years. Language measures, including *receptive vocabulary*,* in the early years were predictive of reading comprehension in elementary and middle school for both children at-risk of dyslexia and typically developing children.

Strategies focusing on the early introduction of reading books to young children offer a potential avenue to improving language, particularly vocabulary. It has long been known that books stimulate vocabulary because they contain a wider vocabulary than occurs in ordinary conversations, and even simple stories for toddlers include more complex words and phrases than the content of typical parent-child conversation.

In a recent report we used data from the Connecticut Longitudinal Study to examine growth in reading comprehension and verbal ability from grades one through nine. We found that in the early grades vocabulary was not strongly related to reading comprehension. However, for grade five and beyond, vocabulary had a much stronger influence. These data, along with what we know about the importance of early language development for read-

* *Receptive vocabulary* is vocabulary understood by the child though not used in spoken language.

ing comprehension, point to a combined approach to reducing the achievement gap in reading between typical and dyslexic readers:

→ Encouraging oral language in general, and building vocabulary in particular, throughout early childhood and continuing through the school years

→ Encouraging evidence-based instruction that systematically develops the skills of phonologic awareness, decoding, word reading, fluency, and reading comprehension in school

BECK'S FRAMEWORK: BUILDING VOCABULARY

Isabel Beck, professor emerita in the School of Education at the University of Pittsburgh, and her University of Pittsburgh coauthors, Margaret McKeown and Linda Kucan, provide a framework for systematically building vocabulary. They speak of three tiers of vocabulary. Tier 1 consists of common words found in oral language. Tier 2 consists of words found in written language which are of high utility. Tier 3 consists of words that are limited to specialized domains, such as law, medicine, and archaeology. Since Tier 1 words are already known and Tier 3 words are so specialized, the focus should be on learning Tier 2 words.

At the beginning of school, most of the words a child encounters in print will be in his spoken vocabulary—Tier 1 words. Then, sometime around third grade, the words he comes across in print are less likely to be familiar to him—Tier 2 words. This marks a pivotal point in the child's growth as a skilled reader. He must therefore build up his vocabulary so he can recognize these less familiar words when he reads them. Beck and her colleagues provide examples of Tier 2 words, including *comment, occurrence, tended, mention, emerging, admit,* just to name a few.

The basic principles of vocabulary instruction are for the children to have frequent and varied experience with the target words and robust teaching. Vocabulary is best learned by multiple exposures to the word over time, usually about twelve encounters. While it is possible for teachers to introduce as many as ten new words each week, a more reasonable approach is to focus on five to seven new words.

Vocabulary instruction that works best involves the child in actively getting to the word from every possible perspective. The more connections he can make between this new word and the words and world he knows, the more salient the word is and the more likely he is to make it a part of his usable vocabulary. Once a new word is introduced, an essential part of the learning process is to encourage the child to use the word as much as possible—initially just to make sentences with it, then to incorporate it into his everyday life. Having repeated encounters with the same word also helps to ensure that its meaning is tightly woven into the automatic reading circuits within the child's brain and that subsequently it can be retrieved quickly.

In teaching vocabulary, the intent is for the child to regard any new word as a fully formed idea, more than just a label. Rather than reciting a word's definition by rote or rattling off a synonym, the goal is for the child to make that word an active part of his thinking process. A word is most likely to spring to life if a child can integrate it with familiar ideas or with his own past experiences. For example, if the new word is *transmission* and the student is familiar with cars and how they work, he will quickly integrate *transmission* into his knowledge base and use the word as part of his active vocabulary. It is important to keep in mind that it is the ideas that take precedence and that the lion's share of the time spent teaching should be devoted to discussions of the ideas represented by specific words. In fact, it is often better to discuss the word's meaning and relevance before naming it. For the word *emerging,* for example, think of referring to a technology that is just about to come out. In effect, this creates a cradle to catch and hold the word once it is mentioned.

A child's vocabulary grows by about three thousand new words a year. These words can be directly and explicitly taught as part of a vocabulary program, or the words can be acquired implicitly, as most of them are in middle-class children, through everyday life experiences—hearing his parents' conversations, those of his peers and teachers, on television and in the movies, on trips to the museum and to the supermarket, or through reading. For the middle-class child, only a small proportion of his vocabulary will come from direct instruction. In contrast, for disadvantaged, often minority students, there may be a paucity of vocabulary from the adults around him, which means that if vocabulary is to improve, it will need to be taught explicitly in school—most importantly early on, in kindergarten and first grade. For children in the earliest grades, the best way to teach vocabulary is for the teacher to read books aloud and then, after the story has been read, to select words for review.

For school-age children overall, a useful guide is *The Educator's Word Fre-*

quency Guide, available from Questar in hardcover as well as on a CD (the CD is somewhat expensive). This guide, created from more than sixty thousand samples of text from more than six thousand textbooks, works of literature, and popular works of fiction and nonfiction, presents a quantitative summary of the printed vocabulary encountered by students in American schools and is helpful in designing instructional materials tailored to the age and educational progress of students. Other useful resources include vocabulary.com, VocabularySpellingCity (Spellingcity.com), and textproject.org/teachers/word-lists. This latter program has a number of very helpful word lists, including the thousand most frequent words in texts for grades three to nine. Word lists available from Oxton House (www.oxtonhouse.com) are also helpful.

THE PARENTS' ROLE

A parent can easily follow the teacher's lead. For example, as your child delves into a book, you can ask her to pick out one or two new words in each chapter for later study and discussion. She can highlight the words, place a Post-it on the page, or jot the word down. Later the two of you can look at how the word is used in the text. Talk about what it might mean. Look for clues in the surrounding words. Does it elicit any images or connections to your child's life experiences? Does she know any similar words? Talk about what the words have in common and how they differ. What other words can you both think of that bring to mind the same idea or concept? Write the words down. Take a pad and pencil and see how many words the two of you can generate as part of your personal brainstorming. Categorize the words together. Have your daughter use each word in a sentence. Have her play word detective and see how many times she sees or hears the word in the next twenty-four hours, and then revisit the word and continue your discussion.

Creative teachers and parents have developed other useful aids for vocabulary instruction that go beyond books and beyond the classroom. In one highly successful program, children competed to become a "Word Wizard," receiving points for bringing to class a newly learned vocabulary word they had used, seen, or heard in a different context outside of school.

Dictionaries are useful as a starting point but often provide sparse definitions of words. All children in third grade and above benefit from knowing how to look up words in a dictionary, but learning about words should not be limited to this one source. Dictionary.com is a helpful online resource for looking up word definitions, synonyms, and antonyms, with an additional "word of the day" feature that produces a constant stream of new words.

Dictionary definitions, concise as they usually are, are sometimes hard for children to understand. One very helpful dictionary for young children is the *Longman Dictionary of Contemporary English* (www.ldoceonline.com), which presents clear, child-friendly definitions using only the two thousand most common words in English. The utility of this dictionary is enhanced because the definition is usually accompanied by an example as well. Another useful dictionary website is Onelook.com, which offers access to many dictionaries.

The nature of instruction should be *flexible*, depending on the word itself. Some words, such as *democracy,* require a full discussion along with examples; other words, such as *double helix,* are best learned through images as well as explanations. Everything is fair game, even using English muffins or a Tennyson poem, "Flower in the Crannied Wall," to learn about *nooks and crannies.* Learning new words also benefits from *repetition* and *reinforcement.* Conceptually challenging words need to be introduced and discussed over and over again; a child does not assimilate and use a word after meeting it once, twice, or even three times. It takes many repetitions.

Children can be encouraged to use their knowledge of word parts, specifically prefixes, suffixes, Latin roots, and Greek stems, to pull an unknown word apart and determine its meaning. Just as you don't try to teach your child every unfamiliar word in a story, it is helpful to focus on learning only the most important of the hundreds of affixes and roots. If your child comes to know these really well, he will be able to substantially add to his reading vocabulary. Fortunately, only a relative few of these word parts occur with any real frequency. Just twenty prefixes account for 97 percent of all words with prefixes found in English schoolbooks, and nine of these prefixes account for 75 percent of all prefixed words. Here are the most common ones:

RANK	PREFIX	PERCENTAGE OF ALL PREFIXED WORDS
1	Un- *(not)*	26
2	Re- *(again)*	14
3	In-, im-, il-, ir- *(not)*	11
4	Dis- *(not)*	7
5	En-, em- *(put into)*	4
6	Non- *(not)*	4
7	In-, im- *(in)*	3
8	Over- *(excessive)*	3

9	Mis- *(bad)*	3
10	Sub- *(below)*	3
11	Pre- *(before)*	3
12	Inter- *(between)*	3
13	Fore- *(earlier)*	3
14	De- *(reverse)*	2
15	Trans- *(across)*	2
16	Super- *(above)*	1
17	Semi- *(half)*	1
18	Anti- *(opposite)*	1
19	Mid- *(middle)*	1
20	Under- *(too little)*	1
All others		4

Since a child gains much of her vocabulary incidentally, it is important to maximize her ability to learn new words as a result of her everyday experiences. The guiding principle here is that the richer and denser her network of *worldly knowledge,* the greater the possibility that a new word will find a relevant hook to latch on to, activate her existing knowledge, and be integrated within this network. Similarly, the larger a child's *vocabulary* is, the more likely she is to understand and to relate to a new idea or bit of information. The more she is exposed to a range of meaningful experiences and *the more she reads, the greater her vocabulary and knowledge network.*

Parents and teachers can encourage children to pay attention to and learn new words. One father I know, Hayden, used his hobby of keeping an aquarium as a learning experience for his son, Jeremy. As a child, Jeremy accompanied his father to the pet store to buy supplies and to select new fish. Before setting out, father and son looked through their many books on fish, read about the different types of fish and their care, selected the ones they were most interested in, and wrote the names on a list. At first Hayden did all the reading, with Jeremy looking at the pictures and following along. Soon Jeremy was able to recognize the words in print and pointed to the words *zebra fish, shark, angelfish, parrot fish,* or *clownfish* whenever that name appeared. After a while father and son began to read together aloud. As Jeremy entered the fourth grade, he read more and more of the selections to his father. Discussions between father and son soon centered on not only fish but algae, corals, anemones, chlorophyll, predators, infections fish are prone to, evolution, and ecology. What began as an interest in aquarium fish expanded into a much broader interest in the sea and creatures of the sea. Other parents have used a child's interest in baseball, soccer, dogs, cars, skateboarding, dollhouses, politics, space, or horses as the impetus for the development of a larger vocabulary and a broader, enriched view of the world.

There is no one formula for teaching a child new words. Rather, it is important to keep in mind that each and every experience is an opportunity for learning—going fishing, visiting Grandma, playing baseball, putting groceries away after shopping, taking a walk in the woods, hearing news on television, and reading (books, magazines, newspapers, and even labels and signs). The key is to expose a child to as many different kinds of experiences as possible and to talk to him about these experiences and the words they generate in an open-ended way so that there are no wrong answers—and then to encourage him to use the new words as often as possible.

For adolescents thinking of college and young adults planning on professional or graduate school, an important group of words to learn are those in the Academic Word List (https://www.victoria.ac.nz/lals/resources /academicwordlist/most-frequent), developed in New Zealand and generated from academic texts in a variety of subject areas. Examples include *distribute, conclusion, proceed, logical, obtain, acquire, retain, exclude, attribute, assume, capacity, enable, perspective, relevant, perceive, component, restrict, generate, distinct, assess, alter, amend,* and *contrast.*

HOW PARENTS CAN NURTURE READING COMPREHENSION

Just as parents can have a positive effect on fluency and vocabulary, they can have a similar effect on reading comprehension. A study carried out with six- and seven-year-olds determined that home reading habits are strong predictors of a child's later performance in reading. Skilled readers who were read to more often by their parents were more likely to read with their parents and to talk about books and stories with them.

As the child goes from early elementary school to later elementary and middle school, reading comprehension takes on an even more important role. As you know, reading comprehension is the result of a complex interplay of the child's decoding skills and reading fluency in combination with her vocabulary, background knowledge, and reasoning ability. Comprehension is often thought of as the process of making connections between new information in the text and the known information the reader brings. We often talk of three levels of comprehension:

→ literal or factual comprehension
→ inferential comprehension
→ critical or evaluative comprehension

While in the past literal comprehension was the primary focus of instruction, with the introduction of Common Core State Standards (CCSS)* it is the inferential and critical levels of comprehension that are emphasized. What this means is that parents (and teachers) need to focus comprehension strategies on building background knowledge by *providing vivid examples, fostering imagery, teaching the child how to compare and contrast, and fostering hypothesis generation and prediction.*

Comprehension strategies should begin to emphasize getting away from low-level factual questions to asking and answering higher-level questions that require the child to use not only the information presented in the text but also her background knowledge and reasoning skills. In addition to reading, the child should also be taught to combine her writing skills to summarize and synthesize what she has read.

I like to divide reading comprehension activities into three parts: those things you can do before opening the book, others that are most helpful as the child reads, and others that help him to organize his thoughts and sum up the events of the story after he finishes reading. These activities are fun but intensive. *By no means should you feel obligated to go through each one of these for each book you and your child read together.* I am laying out a range of possibilities. You must decide which of these best suit you and your child. There is no one right way or one set formula to teach comprehension skills to your child. The idea is to transfer a way of actively thinking about reading to your child. Once he has assimilated this, he is an independent reader.

Before the Book Is Opened. Right from the start, it is important to establish a purpose for reading—one of which is to find out what is going to happen to the main character or how the mystery will be solved. There are a number of

* Common Core State Standards refers to the educational initiative developed in 2010 and adopted by many states that sets standards for what K–12 students should know in English language arts and mathematics at the conclusion of each school grade.

ways you can start a child thinking about purpose. Point out the book's title and author. If there is an illustration on the cover, you might talk about what it says about the book and its purpose. Scanning through a book prior to reading helps to familiarize the reader with its content and to give him a sense of where the book (or chapter) is going. Consider *The Story About Ping* by Marjorie Flack and Kurt Wiese. The cover drawing is of a scampering duck and a background of Chinese junk boats resting in a river as the sun is going down. Turning the pages reinforces images of Ping, the scampering duck; other ducks; boats; water; a little boy holding Ping; and Ping surrounded by a host of ducks amid lots of hay at the book's end. You can ask your child what he thinks this book seems to be about and what he is anticipating will happen in the story.

The cover picture of Ping can spawn a discussion of ducks, visits to a farm, or what you know about ducks. Do ducks like water? Do they swim? What do you think Ping is doing in the water? What about the boat? Looking at the cover, you can see a setting sun visible in the distance. Ask your child what time of day he thinks the cover portrays and why. Ask him if he thinks this is going to be a happy or sad story and why. You are modeling for your child how a good reader proceeds. *A skilled reader is an active reader, not someone adrift; he reads purposefully, making predictions and wanting to know if his predictions are validated or if he needs to alter them.* Making such predictions draws your child into the story and makes him eager to go on.

It is always a good idea to anticipate any potentially difficult or unfamiliar words or concepts. If you have the time, you should preread or scan through the chapter or book to identify the worrisome words. In this story, the word *Yangtze* is likely to be a new one for your child. You could show him on a globe where the Yangtze River is located, or you could talk about rivers, including any that are in your local area or that you have visited. In this way you are helping to activate your child's background knowledge by relating this story to things that he already knows about or that are meaningful to him. Finally, you might try to relate it to other stories, such as those about ducks. These strategies are quite flexible. You can apply them with your child while you read to him or later on, as he reads to you.

During Reading. Start out by commenting on the first few sentences or first paragraphs of the story. This ensures that your child is actively involved from the very beginning. Opening sentences are critical, for they often set the stage for the entire story and introduce you to the place and characters. In *The Story About Ping*, we immediately meet Ping and his "mother and father and two sisters and three brothers and eleven aunts and uncles and forty-two cousins." We find out that "their home was a boat with two wise eyes on the Yangtze River." Just a few short paragraphs later, we learn that "Ping was always careful, very careful not to be last, because the last duck to cross over the bridge always got a spank on the back." At this point you can pause to reflect on what has been learned so far and what the author might be preparing the reader for. You can ask, "Do you think there's a chance that Ping might be last on the boat when we turn the page? Why? And if he is, what do you think he might do? What would you do?" By talking about the story and how your child might relate to it, you are drawing him into the narrative and raising his interest level. This is important to do right from the start, because once a child loses interest, it is hard to bring him back into the flow of the story. On the other hand, once he is part of the unfolding events, he is hooked and will want to know what happens next.

It is also important to establish the *who, what, where,* and *when* right from the beginning. This information represents the structure of the story; it provides reference points for discussion and anchors the narrative. Knowing these story facts can also help your child to visualize what is happening, encouraging him to bring up images that he associates with the place or people. Holding such an image in mind is one more connection to the action in the book and to better comprehension. Just as you did earlier on, you can identify and talk to him about difficult words or concepts; in *Ping* these could be *wise, scurrying, running away,* or *reunion.* Ask your child to point out what he sees as difficulties affecting the characters in the story and then to suggest some possible solutions. You can talk about the ramifications of each. As you read on, ask him if his predictions need to be changed or modified, and ask him to explain his thinking.

From time to time you can ask your child to summarize the plot, to recount the sequence of events that has occurred, or to tell you the main idea in a paragraph you have just read. The last of these helps your child differentiate important from trivial or redundant information, a key factor in getting the gist of a story. To help enrich the story and to bring your child closer to it, ask him more about the characters. In this example, you can talk about Ping or the Master of the Boat who calls out "La-la-la-la-lei!" each evening, or even the little boy who plucks Ping from the water. Does he like each of them? Why? Do they remind him of anyone he knows or someone he has read about before? How does the story make him feel: happy, sad, lucky, frightened, eager, or excited? Why? Follow the same pattern as you read other stories, always integrating your child and his experiences into the fabric of the story.

After Finishing the Book. Ask your child to summarize its plot, to tell you what happened and his feelings about it. See if he can retell the events in order. If not, this is the time to chart out aspects of the story visually, such as cause and effect or a particular sequence of events, using software such as *Inspiration* (see page 450). Ask your child if he wants to read the story again. Ask if he wants to read more stories about the same topics: in this case, ducks, boats, rivers, or China.

Because these activities involve basic reasoning skills, you can introduce them to your child even before he learns to read, and certainly before decoding instruction begins. Reading aloud to your child is an excellent way to get him started. One of the very first things you can teach your child about reading is that its purpose is to gain *meaning* from the printed word. You can show him what to look for and listen for in a story. You and your child can even act out characters. These kinds of activities bring the child into the story, encourage him to be an active listener, and help him acquire the kinds of skills that build meaning. They are developing his oral language skills, his vocabulary, his background knowledge, and his understanding of how books work.

Books are the center of a child's reading development. There are so many wonderful children's books available, for every age, interest, and reading level (see pages 527–532 in the Appendix). Introduce your child to books and reading very early on. Read to him as he gets older. Make sure he reads books (at his reading level or slightly above) aloud to you and that he reads books to himself as well. Initially he should be reading at least twenty minutes daily, and later, thirty minutes or more. Reading *consistently*—at least four to five days a week, even if for shorter periods of time—is more important than the length of time spent reading each day. Reading more frequently, especially aloud, will improve his reading skills, and that, more than anything else, will encourage him to want to read for longer periods of time. The basic skills he will require in order to comprehend and the strategies he will need to call on remain relatively stable over time. Once he begins reading instruction, fluency, vocabulary, and comprehension skills should be taught alongside those focused more on decoding.

CAUTION: It is important for you to learn about the Common Core State Standards (CCSS), which are discussed in the Appendix. Keep in mind that CCSS *do not* focus on teaching decoding skills in the early grades. Such an omission is not in the best interests of dyslexic students.

Take your child to the library early, even before she is a reader. As soon as she is eligible, get your child her own library card. Going to the library frequently, attending story hour, and getting to know the children's librarian at the neighborhood public library (and later on at school, too) are smart things to do for your child. The children's librarian is an expert in children's books and can be a wonderful resource for you and your child. As she gets to know your child and develops a sense of her interests, of her favorite books and authors, and of the kind of child she is, the librarian can alert her to books of interest, to upcoming events, and to newly arrived additions.

WHERE DOES WRITING COME IN?

While educators have made considerable progress in developing effective programs for reading and mathematics, writing is a topic that is often considered the orphan member of teaching the three *r*'s. Within the past few years writing has begun to receive attention from educators and researchers, especially in the work of Steven Graham and Karen Harris, professors at Arizona State University, who have studied and written extensively about writing, primarily for typical readers but with particular relevance for dyslexic students. Graham and Harris note that "the most critical element in creating an environment where students can prosper and grow as writers is for them to write." They and others advise educators to make sure their students write frequently across the curriculum, write for real audiences, and make personal choices about what they write. Teachers should encourage the development of unique interpretations of assigned writing topics. Two additional scientifically based best practices include (1) teaching students about specific genres or characteristics of good writing, including teaching students that stories have a setting, a starting event, characters, actions, and a resolution, and (2) providing students with good models.

Graham and Harris emphasize the importance of teaching writing strategies, especially encouraging students to view writing as a process. Particularly important strategies include those for planning, drafting, revising, and editing text. *The Educator's Practice Guide: Teaching Elementary School Students to Be Effective Writers,* published by What Works Clearinghouse in 2018, uses helpful mnemonics for these strategies. For example, for planning, the student can remember POW, standing for

→ *Pick* your ideas—that is, choose what to write about
→ *Organize* notes—that is, brainstorm and organize ideas into a writing plan
→ *Write*, and continue to modify as you write.

COPS is another helpful mnemonic for revising and editing, standing for Did I:

→ Capitalize each word at the beginning of sentences and the first letter of proper names?
→ Check the overall appearance of the paper?
→ Use the correct punctuation at the end of sentences and use commas correctly?
→ Spell words correctly?

To reiterate Graham and Harris's view, the most important component of writing is *frequent writing*.

These general pieces of advice, such as assigning lots of writing activities and providing exposure to good writing, will usually not be sufficient in teaching a dyslexic child to develop competent writing skills. A program that is highly successful with many children, including boys and girls who are dyslexic, and one that has become a particular favorite of mine, is Judith Hochman's *Teaching Basic Writing Skills* (see pages 301–302).

Graham and Harris also stress the need for children to learn to type and use word processing in composing written essays. This allows the child to focus on the creative writing process rather than on physically forming letters and sentences.

THE TROUBLE SIGNS

One of the added benefits of sitting alongside your child and reading with him is that it gives you insight into his ability to read and to comprehend what he is reading. Below are some of the signs of a child who is having trouble reading.

→ He doesn't seem to get much from his reading.

→ He has trouble answering the question "What was the book about?"

→ He doesn't enjoy reading.

→ He spends the same amount of time on easy passages as he does on difficult ones.

→ He doesn't finish what he begins to read.

→ He doesn't seem to be able to relate his reading to things he knows.

→ He has trouble drawing inferences from his reading; his interpretations are always extremely literal.

→ He can't quite come up with main ideas or summarize what he has read.

→ He can't distinguish important ideas from lesser ones in the text.

→ He has trouble making predictions.

→ He rarely looks back to earlier pages to check his reading.

→ He says reading is boring or tiring.

→ He avoids reading.

If this seems to describe your child, you need to confirm your impressions. Speak to his teacher and perhaps have your child tested at school. It is important to determine not only if he has a problem reading but the nature of the problem. Most often if a child is experiencing reading problems, it is a reflection of difficulty in decoding words accurately and reading fluently. He may be struggling with trying to get to the word itself.

Knowledge is empowering, and now that you understand reading, you should have the confidence to help your child. Don't worry about each and every detail. You know what to do and can act accordingly.

PART IV

TURNING STRUGGLING READERS INTO PROFICIENT READERS

20

MICHELE'S PROGRAM: A MODEL THAT WORKS

Michele has not had an easy time at school. The daughter of a journalist and his wife, an engineer, Michele was having difficulty noticing and appreciating rhymes, and it seemed that she couldn't stop mispronouncing words—prolonged baby talk, her pediatrician called it.

Michele's difficulties persisted, clearly affecting her grasp of print and reading. When she struggled to read in first grade, her mother was told to "give her time." By third grade her parents were very concerned. She rarely picked up a book on her own. When her mother tried to read with her, Michele found every excuse to avoid it. Her mom felt that Michele should be tested, but her teacher said it wasn't necessary; after her own informal testing, her teacher had concluded that Michele did not have a reading problem. Michele's parents were fearful of antagonizing her teacher and school officials, especially given that Michele was not progressing and was dependent on her school's support.

The situation deteriorated. At home Michele sulked and was starting to avoid schoolwork of any kind. At school she was assigned a peer tutor and given extra time on the school computer. Nothing seemed to make a difference. Desperate, her family came to the Yale Center for Dyslexia & Creativity, where Michele was tested and found to have dyslexia. With that evaluation in hand, Michele's parents returned to their school, where a Pupil Placement Team (PPT) meeting was held and Michele was identified as a special needs child with a reading disability, specifically dyslexia. It was now February. She started on a reading program quite popular with teachers in the New England area. By June there was little progress. One moment she was a self-assured, graceful ballerina or a deeply immersed Lego designer, the next an insecure, flustered, stumbling reader soon to enter fourth grade. Then I met with all parties concerned.

The school personnel were genuinely interested and caring, but the reading program was misguided. It included a hodgepodge of phonics activities that were not taught systematically or explicitly. As far as I could ascertain, there was no scientific evidence to support the general effectiveness of the reading instruction being employed. Moreover, this nine-year-old was being taught with two different reading programs (one with her special education teacher, one in her regular classroom) reflecting two very different reading strategies. Fluency, vocabulary growth, and reading comprehension were virtually being ignored.

I was especially concerned because, as you now know, in fourth grade the emphasis is no longer on learning to read but on reading to learn. Michele continued to have difficulty with isolated words and struggled to read the required textbook for social studies. She was sure to fail subjects she had the intellect and the motivation to master. That to me was, and is, unacceptable, because effective, evidence-based reading programs for dyslexic children exist and can help almost any child.

The program we designed for Michele focused on three major goals:

(1) To provide a proven type of reading intervention (that is, evidence-based) delivered with sufficient intensity by a knowledgeable teacher

(2) To integrate Michele's reading intervention with the rest of her regular classroom and schoolwork

(3) To ensure that her reading status was steadily monitored

From the beginning it was critical that a responsible, knowledgeable educator take charge of Michele's educational plan. This role seemed most natural for Ms. Griffin, the school's new reading specialist. She and I reviewed a list of evidence-based interventions and chose the Wilson Reading System. We were impressed by the program's evidence base and comprehensive nature, both in the content of the program itself and in the inclusion of supporting materials for the teachers carrying out the program. Reading is taught as part of an enriched total language experience that includes each of the major components of reading, including decoding, sight-word recognition, fluency, vocabulary, and comprehension, as well as spelling. Ms. Griffin made sure to devote substantial time both to reading connected text aloud and to reading independently. Since Michele had much catching up to do, we wanted to ensure that the intervention was delivered with sufficient punch. She needed to receive instruction for a large block of time and in a small group, with maximal student-teacher interaction. We were also pleased that the Wilson Reading System offered an intensive professional development program for teachers, helping to ensure that the program would be taught effectively. Ms. Griffin had earned a Wilson Level II certification.

In place of Michele's regular classroom instruction, she is receiving at least ninety minutes of reading and language arts daily in a special small group of four students. Fluency practice—with single words, with phrases, and, importantly, with connected text—is provided daily, not as an extra if there is time but rather as an essential part of her program. The competitive part of Michele particularly enjoys the regular graphing of her fluency rates, where she can see the steady increase in the number of words she can read correctly per minute. (See pages 296–297 for details on graphing fluency progress.)

Michele is receiving another forty-five minutes of resource room instruction four days a week. The resource room functions as a bridge both to her regular classroom and to her home. Much of this time is devoted to integrating Michele's reading needs with her instruction in other subject areas. Here the focus is on *prereading* and *rereading*. Working alongside her teacher, Michele is able to preread her reading assignments (for social studies, literature, and science) before she tackles them alone. She practices pronouncing difficult words by copying each word onto an index card or by using Quizlet to make index cards. Time permitting, Michele can look up the word's meaning, use it in a sentence, read it as quickly as she can several times, and draw a relevant cartoon to remind herself of the meaning. All classroom handouts are also reviewed. Michele takes her written tests here; her teacher helps her read the questions. She receives additional time on her class tests as well as on the statewide proficiency tests.

Rigorous efforts are under way to help Michele in her regular class. As I've said, of all in-class activities, reading aloud in front of classmates is the one feared most by students with dyslexia. In Michele's case, if she is to read aloud in class, she is told the day before and practices the passage in question with her resource room teacher and/or with a parent at home. (A child often feels more confident when she has recorded herself reading the passage aloud several times and is able to listen to how she sounds.) If she feels ready to read aloud in class, she simply raises her hand.

Spelling and writing activities are integrated between her regular classroom and the resource room. Michele is expected to spell accurately the words she has learned as part of her reading program. Since spelling is more difficult than reading, if she cannot read a word, she should not be expected to spell it.

Writing (handwriting and composition) is challenging for dyslexic readers. Given Michele's penmanship difficulties, time limitations, and her facility on the computer, I recommended that she learn and become skilled at keyboarding. A good program is Typing Instructor for Kids, available from Amazon, which her parents obtained and then started using to teach her touch typing. In addition, more than other children, dyslexic readers require and greatly benefit from being directly taught writing strategies and a structure to follow in constructing sentences and in developing paragraphs. A program that is highly successful with many children, including boys and girls who are dyslexic, is Judith Hochman's *Teaching Basic Writing Skills* (see pages 301–302). Michele is gradually writing more of her compositions in her regular classroom. The next step is beginning the assignment in her regular classroom and reviewing it in the resource room.

Efforts in the resource room are also focused on ensuring that Michele's afternoons and evenings at home are productive. In the resource room, Michele reviews all her homework assignments to ensure that she has read each correctly and knows what to do. I also recommended that she keep an extra set of textbooks at home. This allows a parent (or tutor) to follow along with Michele as she reads a passage aloud.

The keystone in this entire process is a system of constant, seamless communication between her special small-group sessions, the resource room, and the regular classroom, as well as between school and home. A formal plan for such communication has been set in place. As I have emphasized, home is a place for reinforcement of skills and for pleasure reading. Michele should not have to depend on tutors or her parents to teach her *new* skills. At the same time, since dyslexic students benefit from practice, I view partnerships between school and home as a positive means of extending learning. I encouraged such a collaboration for Michele's family. Her home reading reflects three specific goals:

→ First, she reinforces what she is learning in class by practicing reading school-based materials *aloud* to her parents (or tutor).

→ Second, Michele is encouraged to read silently for pleasure. The Lexile Find a Book option, which matches a student's reading level and her interests with the Lexile-determined text complexity of a book, helps her find just the right book to read for pleasure—one aligned both with what she enjoys reading and with what she is able to read without too much difficulty. Since Michele has a wonderful imagination and loves anything to do with fantasy, a list of books with a fantasy theme—at and a bit above her reading level—was easily generated, encouraging her to enjoy reading and want to read even more.

→ Third, time permitting, her parents read books to Michele that are above her current reading level, or she watches short videos on YouTube as introductory material to school assignments or to her own areas of interest. In this way (in essence reflecting the unexpected nature of dyslexia), Michele maintains access to information that is at her high intellectual level but above her lower reading level and that increases her vocabulary and expands her knowledge of the world around her that would otherwise be unavailable to her.

Together these home activities support her school instruction and help her develop fluency skills as well as the word and worldly knowledge critical for skilled reading. (Of course, these activities are not all carried out in the same evening. Reading school-based materials aloud should receive the highest priority.)

Michele's reading is continuously monitored. Progress monitoring is built into the Wilson Reading System. This ensures that the teacher is absolutely up-to-date and can use these data to build upon and plan and pace the next lesson. At the end of the school year Michele is given standardized reading tests to see how she is doing compared to others her age. For this purpose I recommend the Woodcock-Johnson IV (WJ-IV) test of achievement, the Test of Word Reading Efficiency, second edition (TOWRE-2), for single-word fluency, and aimsweb to measure fluency when reading connected text. (I will discuss approaches to teaching and measuring fluency in dyslexic children shortly.)

The major work of learning should take place during school hours, when the child is most alert. At the same time:

→ A tutor can perform a valuable function in helping to reinforce a child's reading skills. A tutor can help the student practice reading aloud and apply her growing reading and writing skills to her homework assignments.

→ Tutors can introduce children to study skills, especially ways to organize their time and their notebooks, which have long-term benefits.

Michele's family wanted to have the additional support of an after-school tutor. I urged that her efforts and that of the school's reading specialist be closely coordinated; it is essential that they work together as a unified team. Under no circumstance should the tutor operate as a separate agent.

Ten months after Michele began the program, she had increased her reading scores by almost two years. Most important, she had come to think of herself as a reader. She no longer guessed wildly at unfamiliar words. If she stumbled over a word, she eventually pronounced the word correctly—an unheard-of event just a few months before. When she could not fully decode a new word, she made a reasonable attempt to do so; she used her knowledge of how letters represent sounds and her excellent vocabulary to figure it out. She increasingly raised her hand in class and was for the first time even eager to be called on. She was focusing on reading words more fluently, learning how to pull apart long words, and learning about Latin-based word origins so that she could read the new words that were constantly appearing in her social studies and science texts. Michele was benefiting greatly from previewing her textbook reading assignments, identifying and then practicing difficult words. She was coming home from school eager to read a passage aloud to her parents. Spelling was still difficult, but she was improving. She had mastered keyboarding skills and was now writing her essays and book reports on a computer. This had relieved her of so much pressure that she could now concentrate on content rather than on letter formation. Having been doubtful that anything would or could help her become a reader, she was first relieved and then truly delighted by her progress, as were her parents and her teachers. The teachers started adapting Michele's program for their other students with reading problems, who are now being evaluated for dyslexia.

Successful programs like Michele's should be the goal for just about any dyslexic child. Keep in mind that teaching a dyslexic child to read is founded on the same basic principles used to teach any child to read. However, since the neural systems responsible for transforming print into language may not be as responsive as in other children, *the instruction must be relentless and amplified in every way possible so that it penetrates and takes hold.* Earlier I told you that an effective reading program is predicated on a three-pronged approach involving reading intervention, integration of the intervention with regular schoolwork, and frequent monitoring of reading progress. Of course the key lies with the intervention, and this deserves special mention.

ESSENTIALS OF A SUCCESSFUL READING INTERVENTION

Several essentials make up a successful reading intervention. The first is program content, which I will discuss in detail later. Here I want to focus on issues of implementation—when, how, by whom, and for how long the intervention is presented to the student—and the setting or environment in which the intervention takes place. These factors, often overlooked, will determine the ultimate success or failure of even the very best interventions. They reflect the components that were built into Michele's program. You must ensure that each is made a part of your child's reading program.

Early Screening, Early Diagnosis, Early Intervention. The first two items are the essential first steps in successfully teaching a dyslexic child to read. The earlier they take place, the better. Because I have witnessed so many parents lose precious time by wanting to "wait a little longer" or to "give it time," I want to emphasize that no amount of denial or rationalizing will change the situation; it only puts your child further behind and damages her self-esteem. A child needs help *before* she fails. Earlier I noted that since the publication of the first edition of this book, our group has published important data showing that the *achievement gap between dyslexic and typical readers is already present—and is large and persistent—in first grade.* These data provide validation for the need for early screening, early diagnosis, and early intervention to get ahead of dyslexia and not have to struggle so hard to catch up. So, to repeat, a major step to make early identification a reality is *to ensure that all young children are screened during kindergarten or first grade.*

The Shaywitz DyslexiaScreen™ is now being used to help identify those young children who are most at risk for dyslexia. Children who receive help early may become readers who develop accuracy and substantially improve their fluency so that they look forward to reading. A dyslexic child who is not identified until he is in third grade or later is already thousands of unlearned words behind, a gap that must be addressed if he is ever to catch up with his peers. Not being able to do what your peers seem to be able to do so effortlessly profoundly assaults a child's sense of himself and his dreams for the future. The best approach is early screening leading to diagnosis and early intervention.

Intense Instruction. This is essential for the dyslexic reader. It reflects the *dyslexic child's requirement for more instruction, which is more finely calibrated and more direct.* The dyslexic student is behind his classmates and must make more progress than they do if he is even to begin to narrow the achievement gap. He must make a leap; if not, he will remain a long way behind.

Effective reading instruction is responsive to the child's unique needs, to her actions and to her behaviors. Her teacher must know to slow down, to repeat, to speed up, to change the pace, to find an alternative explanation, to stop. This means that the teacher must interact with her often enough to be able to detect change and adjust instruction accordingly.

Optimally, a child who is struggling to read should be in a group of no larger than four to five students with an instructor. And she should receive this specialized reading instruction four to five days a week. A larger group will greatly undermine the possibility of success, since she will have far fewer opportunities to interact with her teacher and therefore will often lose interest or attention during the long pauses between being called on. With larger groups, a dyslexic student may be called on only once or even not at all during a reading session. Sadly, I have witnessed this situation far too often.

High-Quality Instruction. This is provided by a highly qualified teacher. To be a successful reader, the dyslexic child must become an active learner, one engaged by her teacher. Learning to read is a highly interactive process, rapidly resonating back and forth between teacher and child. Gaining the child's attention requires constant effort on the part of the teacher. She must work diligently to involve her student, asking her questions or asking her to justify a response. ("Michele, I wonder why you would say that. Can you tell me? Where in the text is the information telling you this?") Reading is extremely hard work for a dyslexic student, consuming and draining away huge amounts of her attention. The teacher must prevent her student from drifting away and daydreaming. A teacher is constantly delivering the necessary knowledge while at the same time working hard to ensure that it is accompanied by a "hook" that she thinks will be meaningful to the child. She is constantly thinking about how to convey this information to the child.

Good evidence highlights the difference a teacher can make in the overall success (or failure) of a reading program. In one instance the same instructional method was used in two studies but with two different outcomes. According to the researcher who carried out both investigations, the study in which the program was most effective "employed highly skilled teachers who all had a number of years' experience teaching children with reading disabilities," while the other study "employed inexperienced teachers." This is a powerful argument for ensuring that anyone who takes on such a responsibility must be a reading teacher or a teacher who has had recent training and experience in scientifically based—optimally evidence-based—methods for teaching reading. The *primary* job of teaching a dyslexic child to read should not be left to classroom aides, peer tutors, or teachers who do not possess the necessary knowledge or experience.

In another study computers were used to teach dyslexic children comprehension strategies. The children seemed to learn the specific approaches very well but did not necessarily use what they had learned. When their teacher was seated alongside the student at the computer, the children applied the mastered strategy. When the children were alone at the computer, they did not appear to apply the strategies they had been taught.

> *CAUTION:* Learning a strategy and using a strategy are not necessarily the same. Computers are not a substitute for a good teacher.

Sufficient (Long) Duration. One of the most common errors in teaching a dyslexic child to read is to withdraw prematurely instruction that seems to be working. Dyslexia is a lifelong condition and requires ongoing intervention and support to ensure that not only the child's accuracy but also her fluency improves enough that she is reading accurately as well as rapidly *and with good prosody.*

CAUTION: A child who is reading accurately but not fluently at grade level still requires intensive reading instruction.

A child with dyslexia who is not identified early may require as much as *150 to 300 hours of intensive instruction* (at least sixty minutes, optimally ninety minutes or more, for most school days over a one- to three-year period) if she is going to begin to close the reading gap between herself and her peers. The longer identification and effective reading instruction are delayed, the longer the child will require to catch up.

You can help ensure that your dyslexic child will learn to read by

→ getting help early, ensuring that your child is screened for dyslexia in kindergarten or first grade.
→ maximizing instructional time with frequent sessions of long duration in small groups.
→ insisting on qualified teachers.
→ insisting on evidence-based reading programs.

The key factors critical to successfully implementing an intervention program for dyslexic students—early screening, early diagnosis, early intervention, and intensive instruction of high quality and sufficiently long duration—were available to support Michele at her local school, but unfortunately that remains somewhat unusual. With increasing awareness and understanding of dyslexia, progress is being made, although we are not nearly where we need to be.

That is why the environment or climate of a school becomes so important, affecting the impact of any implementation program for a dyslexic student. If you sense that the school your child is attending is not addressing his or her needs and doesn't make dyslexic children feel welcome, you should consider alternatives if at all possible. A specialized school for dyslexic boys and girls would make a positive difference. At such schools all the necessities are typically present, with the addition of having *all* the teachers (rather than only the special education, resource room, and reading teachers) understand dyslexia. Moreover, the attitude toward the dyslexic student is one reflecting understanding and twenty-first-century scientific knowledge of dyslexia. The dyslexic student is part of a community of fellow dyslexic students and educators who both understand and embrace students who are dyslexic.

21

TEACHING THE DYSLEXIC CHILD TO READ

Above all else, parents and teachers of a dyslexic child must know that he can be taught to read, though it is not easy and may take some time. But it is happening more and more. Your child's becoming a reader should be your expectation, and your child must be made to know this, too. At the same time, most often for a dyslexic child learning to read is not a sprint but a marathon. Like a marathon, the reward at the finish line is profound relief matched with a strong, newfound sense of empowerment, a feeling for the child that he can conquer any obstacle. I have witnessed scores of hardworking, bright, dyslexic young men and women transform into readers who turn to books and magazines for pleasure, who are not put off by reading materials that are part of their work, and who continue on to the highest levels of education and succeed in their careers and professions—and in life.

The approach I recommend for teaching a child with dyslexia to read follows naturally from what I have already discussed. The approach is consistent with the template I described earlier for teaching children to break the reading code, so please forgive the repetition. All children must master the same elements of reading.

The first two elements, phonemic awareness and phonics, are often grouped together as alphabetics, and there is extensive scientific support for the critical role they play in early reading. The process of mastering each of these steps is much more difficult for the dyslexic reader. The principles are the same. For the dyslexic child, the approaches must be

→ more intensive.
→ more frequent.
→ persistent and continued for a substantial period of time.

TEACHING THE VERY YOUNG CHILD AT-RISK FOR DYSLEXIA

By now you don't need me to tell you that as a parent you cannot and should not wait to identify dyslexia in your son or daughter. Simple, efficient, reliable screening measures (for example, the Shaywitz DyslexiaScreen™) are now available that will identify at-risk children in kindergarten and first, second, and third grades.

Prereading to Early Reading Programs. For children in preschool or those in kindergarten who are screened and identified as at-risk, parents can consider several commercial programs, including Headsprout, Ooka Island, and Lexia Core5 Reading (which I described previously; see page 216).

Connected Text—Reading Decodable Texts. It is not enough to teach a child the elements of phonemic awareness and phonics. What is important for typical readers is even more essential for those readers who are dyslexic. It is critical that *simultaneously with being taught these elements children begin to read connected text.* For the child at-risk for dyslexia, it is best for the connected text initially to be in the form of controlled vocabulary words that the child has already mastered. As you now know, these texts are referred to as "decodable texts" (containing the words with letter-sound patterns that a child has already been taught) and now also as Readables (see page 218). These books can help him apply his newfound skills by actually reading words in a story. The Bob Books series and Geodes are good examples (see page 218). Reflecting their importance in encouraging reading in dyslexic boys and girls, I want to draw your attention to these and other decodable texts, among them S.P.I.R.E. Decodable Readers (EPS Publishers) and Scholastic Decodable Readers (Scholastic). These reading series provide increasingly complex reading material in a very user-friendly format, incorporating both fiction and nonfiction. Alternatively, parents and children can have fun in creating their own personal minibooks. It takes little more than keeping a list of the words your son or daughter has learned and combining them into very simple sentences. ("Pat the cat. The hat is on the cat.") As the child learns new words, you can keep adding to the series, thereby showing your child how much he has learned. Then he has his very own library to call upon when he wants to "read" a book. Generally these little books are most helpful in kindergarten and early first grade. Other strategies I discussed earlier are also used, including teaching sight words (see page 245 for a list of the most common sight words), writing letters, and spelling.

TEACHING THE SCHOOL-AGE CHILD WITH DYSLEXIA

For the older child diagnosed with dyslexia, programs referred to by the WWC as "Phonological Awareness Plus Letter Knowledge Training" are effective. Common threads run through all these programs. They follow the same template I described earlier for teaching children to break the reading code.

ALPHABETICS

As I noted earlier, phonemic awareness and phonics are often grouped together as alphabetics, and there is extensive scientific support for the critical role they play in early reading. While many commercially available programs contain elements of these components, few provide enough scientific evidence to meet the rigorous standards developed by What Works Clearinghouse or the *strong or moderate* evidence in the guidelines in the Every Student Succeeds Act. Below I have reviewed some examples of programs that meet What Works Clearinghouse standards for effectiveness. I want to emphasize that these are just some examples. What Works Clearinghouse lists other programs that are effective and is constantly updating its roster of programs. There may be other programs that will prove effective, but at the time of this writing there is not enough evidence to declare them unequivocally effective.

You may have heard reading programs referred to generically as Orton-Gillingham, a tutorial approach developed for struggling readers over seven decades ago by Samuel Orton, M.D., a neuropsychiatrist, and three teachers with whom he worked, Anna Gillingham, Bessie Stillman, and Beth Slingerland. It is typically taught one-to-one or in small groups. Orton-Gillingham focuses on phonemic awareness and phonics embedded within a highly structured and systematic framework that tries to engage all the senses in learning about letters and sounds. This multisensory approach embodies auditory, visual, and kinesthetic elements. For example, a child taps each finger to his thumb as he sounds out a word. Orton-Gillingham is conceptualized as an *approach* with many features or principles, and many commercially available programs, including the ones listed below, use components of the Orton-Gillingham approach. At this time there is no concrete evidence that multisensory strategies significantly improve reading, though in the future evidence may emerge.

The Wilson Reading System is a structured literacy program based on phonological-coding research and Orton-Gillingham principles designed for children in grades two and above. Children meet with a teacher for sixty to ninety minutes two to five times a week, individually or in a small group (up to six students) for forty-five to ninety minutes daily. A new version of the Wilson Reading System (WRS4) released in early 2018 represents a major step forward by having students begin reading connected text at the very first lesson. This represents a breakthrough for dyslexic students and places them on the path to becoming readers who not only decode well but who read fluently and with good comprehension. The Wilson program provides students with many rules and may be especially appealing to brighter students who are highly verbal. In addition to the structure and content of the program itself, what is particularly outstanding about Wilson is the range and degree of support provided to educators teaching the program. Training for teachers begins with a ninety-hour online certification course, which includes just about everything a teacher would want to know about implementing the program, including information about the structure of language, how to use curriculum materials with students, how to plan lessons, and how to make sure that each of the components of teaching reading is covered. In addition, teachers are required to carry out a supervised practicum for at least sixty lessons with a student. Each lesson is planned individually—there are no preplanned lessons, but there is a standard lesson outline that the teacher fills in for each student. Progress monitoring and assessment are built into every lesson. Students need to meet criteria for automaticity and fluency before they can move on to the next subset. Certified Wilson instructors are required to participate in an extensive professional development program.

According to What Works Clearinghouse, this system produces significant positive effects on alphabetics.

Fundations, a version of Wilson available for students in kindergarten through third grade, is designed to be used with a core literature-based language arts program for an integrated and comprehensive approach to reading and spelling. Fundations provides all students in kindergarten through grade three classrooms with a systematic program in the foundational skills for reading and spelling, emphasizing the basics of reading, including phonemic awareness, phonics-word study, high-frequency word study, fluency, vocabulary, handwriting, and spelling. Fundations is used both as a supplemental whole-class thirty-minute daily teaching program and for at-risk students in a small group for an additional thirty minutes three to five times per week.

I have focused particularly on the Wilson Reading System because it contains many of the elements, in terms of both content and presentation, that along with teaching-related materials, provide an exemplar of a well-thought-out evidence-based instructional program for the dyslexic reader.

Another program based on Orton-Gillingham principles is SpellRead, a small-group program for dyslexic readers in grades two to twelve. The program focuses on developing the child's ability to automatically identify the sounds of increasingly complex letter combinations and then to effortlessly apply this knowledge to reading and spelling words. The program has positive effects on alphabetics and potentially positive effects on fluency and comprehension.

READ 180 was found by What Works Clearinghouse to have potentially positive effects on comprehension and general literacy achievement in children in grades four through nine. The typical program is ninety minutes long and consists of whole-group and small-group direct instruction. The goal of READ 180 is to address gaps in students' skills through the use of a computer program, literature, and direct instruction in reading skills.

The Lindamood Phoneme Sequencing (LiPS) program also uses Orton-Gillingham principles, but in contrast to programs that teach children to focus on how the word *looks*, it encourages children to learn about how the specific sounds are articulated and how each sound *feels*—that is, to learn the *oral motor* characteristics of individual speech sounds. Subsequent activities then focus on direct instruction in letter patterns, sight words, and context clues in reading. What Works Clearinghouse found LiPS to be effective in teaching alphabetics and reading fluency.

Sound Partners, a phonics-based tutoring program that provides supplemental reading instruction for kindergarten through grade three, is credited by What Works Clearinghouse as having positive effects on alphabetics, fluency, and comprehension.

Success for All (SFA), developed by Robert Slavin and Nancy Madden at Johns Hopkins University, was rated by What Works Clearinghouse to be an effective ninety-minute-a-day, whole-school reform model for teaching phonemic awareness, phonics, comprehension, and general reading achievement.

In the computer interactive category there is Read Naturally, a program that aims to improve reading fluency and uses a combination of texts, audio, and computer software. What Works Clearinghouse says that there are "potentially positive effects with Read Naturally on general reading achievement, mixed effects in reading fluency, and no discernable effects on alphabetics and comprehension for beginning readers." Newer studies may provide additional information.

Read, Write & Type! is a software program developed for students six to nine years of age. What Works Clearinghouse found it to be effective in teaching beginning readers phonemic awareness and phonics through typing the words.

You will hear many claims of the superiority of one program over another, but in fact all the evidence-based programs have been shown to be effective. Other variables to be considered include teacher training, availability of materials, and, most importantly, *time included for teacher training* and *time provided by the child's school to fully implement the program.* The Wilson Reading System is designed to be completed in two to three years according to the minutes per day designated for both one-on-one and small-group lessons. This program, and indeed all other evidence-based reading programs, must be implemented as they were designed to be and should not be abbreviated, either for the time spent each day or for the duration of the program.

A WORD ABOUT RTI

Response to Intervention (RTI) was written into the 2004 reauthorization of the Individuals with Disabilities Education Act (IDEA; see page 468 for a detailed discussion) and has been adopted by the majority of school systems in the United States. RTI provides a range of assessment, instruction, and interventions within a tiered system offering different reading interventions depending on the child's level of reading difficulty. Concerns about the "unappreciated paucity of empirical support for RTI and an overly optimistic view of its practical, problematic issues" were raised as early as 2009, especially on the "potential negative long-term impact on students with disabilities." In fact, a very large-scale study of RTI proved that these concerns were valid, and as a headline in *Education Week* said, "RTI Practice Falls Short of Promise, First Graders Who Were Identified for More Help Fell Further Behind." The article highlighted the results of a large federal study released in 2015 which indicated that first-grade children who received reading interventions via an RTI framework actually read less well than their virtually identical peers who did not participate in RTI. It is not clear whether the many school districts using an RTI framework will try to find more effective approaches or will simply ignore the evidence. The original intent of RTI was to serve students with relatively mild learning problems. However, currently RTI is being used to serve children who are often severely dyslexic and who, after typically six weeks or so of instruction, are funneled back into their regular classrooms, unremediated and, sadly, destined to fail.

OTHER INFLUENTIAL FACTORS

As I mentioned earlier, in addition to the reading program, there are other important factors that can significantly influence your child's eventual success in reading and in life, including the school environment (especially the school's understanding of dyslexia), the intrinsic qualities of the child, and the significant adults who interact with him.

In addition, the presence, severity, and effectiveness of approaches to any associated conditions, including anxiety and ADHD, will affect his future. Especially important is ongoing support and encouragement of a champion, whether it be a parent, a grandparent, or a teacher. As you will read later, through real-life examples of highly successful dyslexic men and women, the magical ingredient is often such a champion, who provides support for both the child's spirit and his reading. Such a champion is vigilant in ensuring that the child is screened early, diagnosed, and receives the necessary instruction, at the same time nurturing the child's self-esteem, pointing out his strengths and encouraging him to pursue his interests. The champion is often the key to giving the child hope so he can envision a bright future.

It is important that the child's progress be monitored and tracked using standardized progress-monitoring tools such as aimsweb and DIBELS (Dynamic Indicators of Basic Early Literacy Skills). The aimsweb Reading Curriculum-Based Measurement (R-CBM) is a brief, individually administered, standardized test of oral reading fluency designed to be used for progress monitoring from the beginning of first grade through eighth grade. DIBELS Next, the most recent iteration of DIBELS, is very similar to aimsweb, providing a measure of oral reading fluency from the middle of first grade through sixth grade. Each provides short passages at the child's reading level which the child reads aloud for one minute as accurately and quickly as he can; scores are reported as correct words read per minute (CWPM). Both aimsweb and DIBELS Next are typically assessed at the beginning, middle, and end of the school year. If a child performs well, three assessments are adequate. If not, each can be administered more frequently to monitor progress and to make any necessary instructional refinements.

I am proud to say that the Shaywitz DyslexiaScreen™ has been a game-changer in making possible the identification of children as early as kindergarten, first grade, second grade, or third grade as at-risk for dyslexia. I am referring again to the screener here because for boys and girls who are dyslexic, the cost of not being identified by schools is alarming and, I believe, unethical. As you know, schools identify about 4 percent as dyslexic, when the actual number is 20 percent. Many of these overlooked boys and girls see others learning to read, and so begins a downward spiral in feeling that they are not bright and that they can't read, which evolves into feeling that school is not for them. This tragedy is preventable. This may well be the most important action that can significantly benefit a dyslexic child's future. Parents should demand that schools screen children early.

I receive a great many requests for information about programs teaching phonemic awareness and phonics, and that is why I am emphasizing throughout this book that teachers and parents have many available choices, including the aforementioned Lexia, Ooka Island, and Headsprout. Additional programs many have found useful in teaching the elements of phonological awareness include *Phonemic Awareness in Young Children* and Siegfried Engelmann's *Teach Your Child to Read in 100 Easy Lessons*.

Good evidence indicates that beginning teaching phonics in kindergarten or first grade produces the best results. It is not only important to teach phonics, it is incredibly important for struggling readers to have opportunities to *listen* to books that spark their interest, books that they will look forward to hearing over and over again. I have enjoyed watching as little Ari and Eva reach for a favorite book, curl up, and smile as each listens to familiar stories and then slowly begins to sound out simple words such as *hop* and *pop*. As the child listens and reads aloud more and more, the words come sliding out more quickly and smoothly. What joy a child experiences as he discovers he can read many of the words on the page and he is encouraged to keep trying to decipher others. It isn't easy, but he can do it, bit by bit.

MOVING FROM ACCURACY TO FLUENCY

As you are aware, fluency is a major, critical factor facilitating reading comprehension. Reflecting its great importance, especially to a dyslexic child, I will emphasize once again the key factors to keep in mind when teaching fluency. These are to practice oral reading, practice reading connected text over and over again, and provide ongoing feedback as the child reads.

Recent research emphasizes the importance of fluent reading of connected text, in contrast to fluent reading of isolated words, as an important element in the development of reading comprehension. Reading connected text should not be delayed, for this is what reading is all about: getting to meaning. By seeing the same word in different contexts, the child realizes that the same word he sees on the page may have several specific meanings, depending on the surrounding words or paragraph. Reading connected text increases a child's vocabulary and background knowledge, two critical factors influencing comprehension.

It is important to use a reading intervention program that rapidly gets the child into decodable texts—an entry into reading for meaning—just as soon as the child starts to learn decoding. It is critical that the child read aloud to a teacher, parent, or tutor who thoughtfully and gently provides feedback.

Much has been learned about how best to implement repeated oral reading programs. Repeated reading of old materials and reading new materials both have a positive influence on word recognition and reading fluency. A child's reading comprehension, however, profits most when he reads new materials. Simply listening to a teacher read a story does not improve word recognition, fluency, or comprehension. "While reading aloud to students is important in fostering a love of reading, learners must actively engage in the reading of connected text if they are to become skilled readers." Dyslexic readers will both enjoy and benefit from reading books referred to as high interest, low readability (see the Appendix).

If your child hates to read and doesn't read very well, regularly reading aloud with him will quickly alert you to a problem if one exists. A child who avoids reading is among those most in need of practice and guidance and is especially helped by your reading aloud with him and your helpful feedback. That will avoid a common downward spiral often observed in dyslexic readers: such children will avoid reading, and as a result they never receive practice or the helpful feedback and correction that will improve their reading. Children who become less involved with reading fall further and further behind their classmates. They never build the fast-paced neural circuits necessary for rapid and automatic word recognition. Even with the most reluctant reader, it is possible to find books that are comfortable—and interesting—to read aloud with you. The most pressing goal is to practice reading the printed word aloud and not to teach literature.

Reading should always be encouraged for pleasure and for knowledge. However, *if a child is a halting or tenuous reader, simply encouraging him to read silently to himself will not make him a better reader.* He may simply be repeating errors—or daydreaming. It is only by reading aloud, with feedback and correction, that real gains in reading are noted.

It is important to find enjoyable ways to encourage reading aloud. Having a child practice reading poetry is an excellent method for improving his fluency. Poems are usually short, they have rhyme, and they are tailor-made for reading quickly and with expression. Some teachers stage a "poetry party." The children select a poem and for the next several days practice reading the poem aloud again and again. The day of the party, lights are turned out and flashlights or lamps provide dim light and atmosphere. After sufficient practice, even the most disabled readers seem to be able to read their poems with accuracy, smoothness, and expression in this environment. (See the Appendix for recommended books of poetry.)

Alternatively, many students enjoy and benefit from staging dramatic readings of a selection from a script. They enjoy the drama and are willing participants. Dramatic readings truly improve fluency. In one recent study students participating in this approach, called Reader's Theatre, for just a ten-week period made an entire year's gain in improving their reading rates. Students can face an audience or simply sit in a circle and read from a script. Readers Theatre is successful with children beginning as early as second grade and continuing through high school.

Students also enjoy reading and rereading (or singing) songs or limericks. Lyrics to a song can be handed out to a class or to a small group and the song can be read or sung for several days in a row. Song lyrics are particularly suitable for choral reading, where the teacher first reads the words and then she and the students join together to read along as a group, rereading the lyrics four or more times.

Older dyslexic students serve as wonderful reading tutors for younger students. Each student benefits from reading aloud; in addition, the dyslexic student gains a rare sense of satisfaction and accomplishment from being able to help someone read.

At Home. Parents of dyslexic readers can also help to improve their child's fluency. In Professor Tim Razinski's reading clinic at Kent State University, parents participate with their child in a method that I mentioned earlier called *paired reading,* a variation of repeated oral reading. It requires only fifteen minutes an evening. Parents read a brief story or passage to their child, both parent and child then read the same passage together a few times, and then the child alone reads the text back to his parent. Comparisons showed that children who read aloud with their parents made substantially larger gains in fluency than those who didn't.

I urge parents of dyslexic children to make fluency training—repeated oral reading—their number-one priority. Because it involves reinforcement rather than teaching a child a new concept, it is ideally suited for the home. Fluency training is something all parents can do for their child—it requires little time and minimal expertise, and it invariably works. You are training your child's brain and helping build the accurate neural models necessary for quick and accurate reading. There are very few activities for a dyslexic reader that provide as much improvement for the amount of time spent as guided repeated oral reading does.

HELPING OLDER CHILDREN BECOME MORE FLUENT

Fortunately, it is not too late for older dyslexic students to benefit from carefully targeted fluency practice. The same principles I described for younger children apply, only adapted to the needs and interests of older students. Reading passages aloud is helpful to students of all ages, but older students have difficulty finding suitable materials to read aloud. Older students can make use of the Lexile and DRP systems for help in choosing interesting books at their reading level. Whatever the material, students should practice reading aloud passages that are relatively short, preferably less than two hundred words.

Older students also benefit from reading single words aloud, especially so-called irregular words. Around the fourth grade there is a tremendous surge in the proportion of words that are irregular—those that do not follow the regular rules for pronunciation and do not lend themselves to sounding out. Unless a child who is dyslexic is provided with specific practice targeted to these words (words such as *bough* and *though*), she will suffer a setback. The sudden onslaught of such hard-to-decipher words may be one factor behind the well-known dip in reading performance that seems to occur around the fourth grade, colloquially referred to as "the fourth-grade slump." Fortunately, more and more reading programs that are effective for older dyslexic readers are incorporating fluency exercises with these irregular words. For example, words such as *canoe, courage, promise,* and *bouquet* are likely to be found in reading materials meant for students in higher grades but must be learned through repeated practice as part of fluency exercises. The word lists obtained from Oxton House Publishers (http://www.oxtonhouse.com) include more than one hundred irregular words.

Some older children may gain the most from practicing relevant words—words they will need in order to pursue their studies. I recommend that they practice reading aloud words taken from their reading materials in different subjects, such as social studies and science. These words can be put on index cards or on the computer (for example, Quizlet) for practice reading aloud. Without such supports, assigning a dyslexic student reading in a textbook may lead only to frustration and disappointment.

Measuring Fluency. Almost as important as teaching and practicing fluency is measuring it. There are useful guidelines to help you judge a child's fluency relative to others his age. For children in lower grades, the expected fluency rates for oral reading of passages are

GRADE LEVEL	CORRECT WORDS READ PER MINUTE (CWPM)
spring first grade	40 to 60
spring second grade	80 to 100
spring third grade	100 to 120
fourth grade and above	120 to 180

These rates are approximations. Both aimsweb and DIBELS Next have taken the next step and set specific thresholds for oral fluency rates in the early grades.

As important as measuring a child's fluency rate is tracking his improvement on a simple graph, at least weekly, as he continues to practice. Both aimsweb and DIBELS Next are very useful in monitoring progress. Keep in mind, too, that a child's rate of reading growth follows a distinct pattern: Growth is greatest in the early school years and diminishes with each succeeding grade. Growth is most often at its maximum at the beginning of each school year and tapers off toward spring.

All children do not respond in exactly the same manner to a program. Most important is the trajectory. A relatively flat line indicating little progress loudly proclaims the need for a change, just as a steep upward incline proudly trumpets progress. Happily, reading programs like those I have recommended and increased scientific understanding join together to make reading fluency more approachable.

> *CAUTION:* In reading about fluency, here and elsewhere, rate and speed of reading are frequently mentioned. These are important parameters, but I cannot emphasize enough that you must remember that the most critical factor in fluency is *understanding—getting to the meaning of what has just been read.* This is what is inferred by the word *prosody* in the definition of fluency. When you listen to a child practice reading aloud, he should be reading with prosody: inflection, speed, and tone reflecting the meaning of what he has read. His reading should not be flat or monotonic, as if he were reading an unknown language. If it is, he is not reading fluently. When a child is practicing reading aloud, listen carefully for intonation and ask him relevant questions to help ensure he is reading fluently—with meaning.

Additional Aspects of Fluency. Lack of fluency in reading is often accompanied by lack of fluency in speaking, that is, difficulty in quickly retrieving words when called upon in class to explain or describe a topic. Many of my students have found it extremely helpful to spend part of their study-hall time working individually with a teacher (or at home with a tutor), reading and then talking through the content of an assignment. Speaking and using the specific vocabulary relevant to a subject often functions to lubricate a student's retrieval processes and enables him to access exactly what he wants to say more readily. Previewing, reviewing, and talking through are critically important for the dyslexic child if he is to participate in the class and discussion periods more fully.

Building Vocabulary. For dyslexic children it is especially critical to build a large reservoir of vocabulary words, so that even if the child's ability to decode words is weak, her extensive vocabulary will enable her to figure out the word. I earlier detailed Isabel Beck's three-tier framework for systematically building vocabulary. You will recall that Tier 1 words reflect very common words used in everyday conversations; Tier 2 words are those likely to be found in books and are extremely useful to all readers, especially the up-and-coming reader; Tier 3 words are specialized words and terms that are more often used in specialized disciplines. It is Tier 2 words that have the greatest utility and that logically form the focus of effective vocabulary instruction. It is important to keep in mind that the major focus in teaching vocabulary should be on the *idea* that the word represents, even to the point of beginning any discussion with the idea rather than the definition of a word.

Both teachers and parents need to be selective about which Tier 2 words a child is expected to learn. While there is no perfect way of selecting which words to prioritize, a useful approach is to focus on those words most likely to stretch the child's mind—those that are conceptually difficult (but not beyond his grasp)—and that he would not likely get on his own. Above all, a word should be relevant and useful. For example, you can preread an assigned book and select specific words for intensive vocabulary instruction. A child can also suggest words that give him a hard time. A word chosen for vocabulary instruction should tie into a central concept of a story or topic the child is learning. It often works best to focus on a group of words unified by a common theme or topic, so that learning each individual word reinforces learning the other words in the group as well. Words that relate to that common theme but have opposite meanings also work well.

For example, in one approach the theme was "fear." The teacher and her students contributed to generating a word list based on fear, writing each of the words on the board. When a large number of words accumulated, the words were subdivided into categories. Students were encouraged to think of words that conveyed not being fearful—words such as *intrepid, heroic,* and *brave.* In this brainstorming method, several other categories were developed as well:

> *Emotions/reactions:* terror, horror, fainting, quaking
> *People/creatures:* monster, ghost, bogeyman, alien, humanoid,
>   poltergeist, Dracula

Places: cemetery, graveyard, haunted house, funeral parlor, unlit street, dark alley.

In the process of generating these words, children engage in far-ranging discussions of their own experiences relating to fear, activating their background knowledge and relating it to the current book the class is reading. An important by-product of this activity is that the children have enriched and extended their understanding of fear. They now possess a new arsenal of words. The very act of forming categories for these words has given structure and direction in how to think about words and their potential meanings.

For dyslexic readers, who benefit from more practice and more reinforcement, computer programs are helpful in supplementing vocabulary instruction. Several Internet resources are very helpful in supporting vocabulary development; see, for example, vocabularyandspellingcity.com and vocabulary.com.

TO SUMMARIZE: TEACHING DYSLEXIC CHILDREN VOCABULARY—WHAT TEACHERS AND PARENTS CAN DO

→ Follow the basic principles: (1) frequent and varied word experiences; (2) intense, interesting instruction.

→ It may be possible to introduce ten new words each week. More usual is three to five words per lesson, reaching ten new words in two weeks of class time.

→ The required number of encounters with a new word for learning can be as little as four, but fluency may require as many as twelve.

→ The best source of words for kindergarten and first grade is trade books read to children by a teacher.

Comprehension: The Goal of Reading. Never lose sight of the major goal of reading: to comprehend what the author of the text is communicating to the reader. To emphasize once again, reading comprehension is the result of a complex interplay of the child's decoding skills and reading fluency in combination with her vocabulary, background knowledge, and reasoning ability. As important as comprehension is to all readers, it is especially important to boys and girls who are dyslexic. Happily, this is an area that dyslexics typically master. As I emphasized early on, though dyslexics have problems decoding and reading slowly, these problems can be thought of as circumscribed and surrounded by a sea of strengths in higher cognitive processes such as reasoning, concept formation, critical thinking, and *comprehension* (see Figure 12, page 56). Comprehension deficits are not and should not be included in the state-of-the-art definition of dyslexia.

Comprehension is an area dyslexics can and often do excel in. To most fully develop the dyslexic student's strong potential in this critical area, it is important to gain an understanding of exactly what comprehension is, along with the known paths leading to high levels of comprehension. Comprehension is often thought of as the process of making connections between new information in the text and the known information the reader brings. Rather than literal comprehension being the major focus of instruction, it is the *inferential* and *critical* levels of comprehension that should be emphasized more and more. Here the child uses her own background knowledge, vocabulary, and reasoning skills to grasp the meaning of what she has just read. The basic idea is

→ to encourage your child to be an active listener, the forerunner of an active reader.
→ to ensure that all the steps you take are directed toward grabbing his attention and pulling him into the reading.
→ to keep in mind that the goal is for the words and the ideas they represent to take on *meaning*.
→ to continually look for ways to connect what is happening in the pages of the book you and your child are reading to what is familiar or meaningful to your child.

DYSLEXIA AND COMMON CORE STATE STANDARDS (CCSS)

CCSS expects second-graders to read and comprehend complex expository, informational literature at the second- and third-grade levels proficiently by the end of the year. As you might expect, such an expectation significantly impacts dyslexic students, the vast majority of whom are not reading proficiently by the end of second grade. What has become abundantly clear is that the CCSS and accompanying tests such as Partnership for Assessment of

Readiness for College and Careers (PARCC), Smarter Balanced Assessment Consortium (SBAC), and tests developed by individual states based on the CCSS (all designed to assess whether students are meeting the CCSS) are totally inappropriate for students with dyslexia. Such tests are based on the mistaken belief that all students, including dyslexic students, will be fluent readers by the end of second grade and that all students should read at grade level and above, an expectation that clearly ignores all that has been learned about the development of reading in dyslexic children. Furthermore, CCSS is predicated on comprehension-focused reading instruction using "complex text" as its basis, ignoring whether dyslexic students can actually read those words. If this were not serious enough, tests such as the PARCC, SBAC, and other tests based on the same presumptions have the pernicious effect of encouraging schools to eliminate or drop all other reading instruction, including the much-needed decoding instruction for dyslexic students. The CCSS is wholly inappropriate for children in very early grades, especially dyslexic students, who are invariably still struggling with and working hard to master decoding. Later on I will discuss how the GED test for adults is now based on the Common Core test and may actually be harming hard-working adult dyslexic readers.

What About Writing? For the dyslexic student, direct, explicit instruction is critical for the development of good writing skills. One very effective writing program for dyslexic students is found in the aforementioned *Teaching Basic Writing Skills* by Judith Hochman, developed at the Windward School in White Plains, New York. For Hochman, "The two primary goals of this program are to raise the linguistic complexity of students' sentences and to improve the organization of their compositions." This is accomplished by systematically presenting goals, strategies, and activities for writing sentences, paragraphs, and compositions. Starting with sentences, the program teaches students to distinguish between sentences and sentence fragments, to sequence words correctly in scrambled sentences, to use coordinating and subordinating conjunctions, to add and move phrases and clauses within sentences, to combine multiple sentences into one, to expand sentences, and to summarize. Once the student has mastered sentences, the program advances to paragraphs and compositions, teaching students to write a topic sentence, then organize several additional sentences into a cohesive paragraph. In this component of the program, students first learn to develop a quick outline and write a draft based on it. During the revising and editing process, students are taught to use transition words and to practice writing introductions and conclusions. Hochman notes that "although spelling, handwriting and related skills are important, students should be encouraged to focus on developing the higher-level skills they need to write."

Professors Steven Graham and Karen Harris at Arizona State University emphasize the importance of students' learning to use word processing in the writing process: "Many schools still use 19th century writing tools such as pencil and paper, even though scientific studies demonstrate that students in grades 1 to 12 show greater improvement in their writing over time when they use word processing to write at school versus writing by hand. . . . We obviously need to move writing instruction more squarely into the 21st century, making it possible for our students to take advantage of word processing and other electronic methods for composing."

Accommodations: The Next Step. As the child progresses in school and reaches middle school and then high school—particularly any child who has the potential to participate in demanding academic programs—the focus expands to ensuring that necessary accommodations are provided. This is important because it places emphasis not only on the child's reading difficulty but on his strengths. It reminds everyone that the isolated phonologic weakness is only one small part of a much larger picture. Far too often the focus is only on the weakness, and the child's strong capabilities (and potential) are overlooked. Whatever those strengths are—the ability to reason, to analyze, to conceptualize, to be creative, to have empathy, to visualize, to imagine, or to think in novel ways—it is imperative that they be identified, nurtured, and allowed to define that child. For many dyslexic children, accommodations represent the difference between academic success and failure, between having a test result reflect their ability versus being penalized for slow reading, between a growing sense of self-confidence and an enduring sense of defeat.

Perhaps the most important accommodation is the provision of extra time on assignments and examinations. Dyslexics' slow reading persists into young adulthood and beyond. One important consequence is that even though dyslexic readers are often very bright and have excellent verbal intelligence, they will appear not to comprehend printed materials unless they are given additional time to carry out the reading. This means that for a dyslexic to demonstrate his true understanding of written material, whether it be a homework assignment or an exam, he must be provided with additional time. Unless a dyslexic is allowed that time, his grade will be a measure of his disability rather than a measure of his ability. That's not fair.

22

PROTECTING AND NOURISHING YOUR CHILD'S SOUL

In addition to providing the loving and nurturing that come naturally with parenting, parents (and teachers, too) of children with dyslexia should make their number-one goal the preservation of their child's self-esteem. This is the area of greatest vulnerability for children who are dyslexic. Teachers and parents often hold high expectations for a child who is harboring a hidden disability and then are surprised, disappointed, or even angry when the child does not perform well in school. If he is accused often of not working hard enough, of not being motivated, or of not really being that smart after all, the child soon begins to doubt himself. The enormous effort and extraordinary perseverance he must expend just to keep up seem to have no payoff. That is why understanding by parents, by teachers, and ultimately by the child of the nature of his reading problem is critical in helping a child to develop a positive sense of himself.

An unwavering commitment to the intrinsic value of a child with dyslexia is essential. Every dyslexic child is invariably going to endure ups and downs in his school experiences. So early on, each child needs to know that no matter what, he can always count on his parents for unconditional support. All dyslexics who have become successful by any account share the unfailing love and support of their parent(s) or a grandparent, teacher, or spouse.

Parents must always be there for the child where and when he needs it: in ensuring early screening by his school, in seeking out an accurate diagnosis and then an effective intervention, in helping to make sure that school is a positive experience, in exposing him to the world, and in reading with him and perhaps for him. Those who support a dyslexic child must always remind that child of his value as a person.

There are a number of specific steps you can take to build your child's self-image. The first is to let him know the nature of his reading problem. Children are greatly relieved to learn why they have so much difficulty reading. Such knowledge brings with it the light of self-awareness and self-advocacy, which in turn build the road to self-esteem and empowerment. Fred's story illustrates that.

Fred, a middle-school student, had just been diagnosed as dyslexic. His parents worried that knowing his diagnosis would discourage Fred and somehow throw him off track and have him lose confidence in himself. They feared that knowing his diagnosis might somehow negatively alter Fred's self-perception and disrupt his relationship with peers.

I sat down alone with Fred and explained his dyslexia, reviewing his history and performance. I spoke about dyslexia in the context of what it is, what it affects, and what it does not affect (intelligence). Fred asked some probing questions, and then, with a great sigh of relief, responded, "I am still the same person. The only difference is that I now know the nature of my difficulty and its name, and most importantly, I understand myself and how I learn and what I can do to help myself."

Fred looked as if the weight of the world had been taken off his shoulders. He is doing very well at home, with peers, and in school, where he is now able to speak to his teachers about any modifications he needs, including test accommodations and not taking two finals in one day. His new self-awareness led to confidence and positive experiences and opened him to new visions of a future far beyond his previous limited dreams.

You can help your child build a positive mindset by:

→ growing and maintaining his self-esteem.
→ ensuring that he develops self-awareness of who he is and how his brain functions.
→ helping him learn to self-advocate.
→ taking the long view—helping him keep in mind that there can be a pot of gold at the end of the rainbow

To accomplish these goals, there are certain facts you should share with a child when she first finds out that she has dyslexia:

→ Tell her that her difficulty has a name, *dyslexia.*

→ Tell her that dyslexia is a problem intelligent people have with reading. You can actually draw out our model of the reading difficulty surrounded by a sea of strengths, appropriate to the child's age and level of understanding.

→ Reassure her that dyslexia has nothing to do with how smart she is. Tell her that in fact some of the brightest people have dyslexia.

→ Explain that her problem is caused by a difficulty in pulling apart the sounds of words; refer to them as *sticky sounds.*

→ Tell her that words have sounds that come apart and that people with dyslexia have trouble pulling the sounds apart, so that instead of hearing three sounds in a word, she may only notice one or two. You can use a set of blocks to make your point.

→ Explain to her the kinds of problems *sticky sounds* cause: in sounding out words; in confusing words that sound alike, like *general* and *gentle;* in reading aloud; in spelling as well as pronouncing words. You can tell her how some people often know the answer to a question but have a problem getting the right words out of their mouth.

→ Reassure her that this is a common problem that many children have and that it can be helped by the right teaching.

→ Tell her that she will learn to read.

→ Tell her about the many famous people who are smart, successful, and also dyslexic. Depending on her age and interests, these could include the Nobel Prize–winning molecular biologist Carol Greider; the brilliant cardiac surgeon Dr. Toby Cosgrove; the actor Orlando Bloom; the children's book authors Patricia Palocco (*Thank You, Mr. Falker*) and Dav Pilkey (*Captain Underpants*); the entrepreneur and founder of Virgin Airlines, Sir Richard Branson; the economist and financial expert Diane Swonk; the entertainment superagent Ari Emanuel; the Academy Award–winning producer Brian Grazer (*A Beautiful Mind*);

and the groundbreaking geneticist George Church. Cite the stories of struggling dyslexics mentioned in this book, people who are now highly successful.

→ Reassure her that her brain is normal, that there are no defects in it.

→ Explain to your child that the express route (or highway) to reading is blocked in dyslexia and that she has to take another route, the secondary, bumpier, and slower back roads. As a result, she can get to her destination, but it just takes much longer.

After one of my patients, Phillip, shared his MRI with his third-grade classmates and told them he had dyslexia, his teacher wrote me to say the children were curious about it and had the following questions (my responses follow):

→ *What is dyslexia?*
Dyslexia means being smart and having a hard time learning to read.

→ *When you have dyslexia, will you always have it? Can it be made better?*
Children who have dyslexia will always have dyslexia, but they can learn how to read. Sometimes it takes awhile for reading to get better, but it always improves a lot.

→ *How do you get dyslexia?*
Children get dyslexia just like they get brown or green eyes or get to be short or tall—they are born with it.

→ *Does dyslexia come from a virus?*
No. Dyslexia is not catching. You cannot get it from another person.

→ *Why is it called such a funny name like dyslexia?*
The word *dyslexia* comes from the foreign (Latin) word *dys,* meaning *difficult,* and the foreign (Greek) word *lexia,* which means *words;* together the two parts add up to a *difficulty reading words.*

Early on it is critical to help your child identify an interest or a hobby, an area in which he can have a positive experience, whether it be pure enjoyment or perhaps the ability to stand out or excel—an interest in fish or in rocks, a talent for baton twirling or juggling, skill in skating or swimming, a talent for acting or drawing, a propensity to understand science or computers, or a love of poetry or music. It doesn't matter what it is. For the writer John Irving it was his involvement and skill in wrestling that helped him endure particularly harsh experiences at school. Having an interest or athletic activity often means a child's life is not always negative or entirely about studying or catching up. It allows the child to see himself as a victor, as a competent individual who has mastery over a topic or an area; he gets to experience the feeling of winning. You must encourage him to explore a range of possibilities—in school and after school, at the Y, in sports, in music, in art, in photography—and support his participation. Not every child falls in love with an activity immediately, so it is important to expose him to different possibilities and to help get him through the initial rough spots when his first response may be to want to give up. Remember that in most areas stars are not born but made after lots of practice and hard work. Praise should not be treated as a rationed item; be giving with your praise as long as it is honestly deserved. Your child will be receiving more than his share of criticism from teachers and others. You should see to it that he knows the soothing sounds of praise.

Encourage your child to view himself as a person who has something to say and whom people should respect. Discuss important decisions with him. These might involve issues that affect the entire family, like where to vacation, or issues that are more personal to him. You might talk about a current events issue, such as a national, local, or school election, or about a movie or television show. Be patient and listen as he expresses himself. Take his comments seriously and respond to them. As he gets older, there will be many occasions when he will need to speak up and advocate for himself. Getting into the habit of speaking out and being heard is invaluable preparation for that. It is important that he fully understand the nature of his reading problem and its implications, so that as he goes on in school he can determine his need for educational modifications or accommodations and feel comfortable in speaking up for himself. Role-playing with a parent or a trusted teacher or tutor can be helpful practice.

It is terribly important how you as a parent view your child. If you feel that a diagnosis of dyslexia means that his future is doomed, he will come to feel the same way. Having a diagnosis of dyslexia should not preclude a child from pursuing his dreams. Given adequate intelligence, interest, persistence, and support, a child who is dyslexic can pursue virtually any area that interests him. Men and women who are dyslexic have distinguished themselves in every area imaginable, including areas the uninformed might believe are not possible for a person who is dyslexic: writing, law, medicine, science, and poetry. Children who are dyslexic should not be reflexively shunted onto a path of nonacademic ambitions unless that is clearly their preference; they should at least be helped to understand that they have a much fuller range of options.

Many parents of children who are dyslexic have experienced reading problems themselves. If you have, tell your child about them and about how you felt when you were growing up. Allow him to see that people whom he admires are not perfect and are able to succeed in life.

Be aware that some of his behaviors, such as procrastinating with regard to reading or studying for a test, may reflect anxiety associated with previous difficulties in reading or taking tests. If these begin to interfere with his daily functioning—that is, if he always seems anxious, sad, or worried—it may be helpful to consult with a child psychiatrist or psychologist. Your pediatrician will provide you with an appropriate referral. A visit to a professional often helps you better manage a problem or prevents a more serious one.

Finally, don't patronize a child or dumb-down his expectations. Always treat him as a person with many dimensions, not simply as a person who has a reading problem. *Let his strengths and not his weaknesses define him as a person.*

PART V

CHOOSING A SCHOOL

23

———

FINDING THE SCHOOL FOR
YOUR DYSLEXIC CHILD

For any parent, choosing the right school for his or her child is a top priority. This is even more so when a son or daughter is dyslexic. One distressed
parent raced up to me at Hartsfield-Jackson Airport in Atlanta, sobbing, "Dr.
Shaywitz, Dr. Shaywitz, please help me. I just learned that my son, Aiden, is
dyslexic. What should I do? I don't even know where to begin."

Having spoken with and problem-solved with scores of parents, I've found
that concerns about schooling are often paramount. Will attending a regular
school, public or private, keep a child in the mainstream, where she will be
offered the same opportunities, classes, and circle of friends as her peers?
Will she be seen as just another student, part of the group? Or will attending
a public or independent school be too much for her? Will she find herself
unable to keep pace, continually overwhelmed, losing confidence in herself?
Will she come to see herself as a failure and become, whether in reality or
in her own perception, an outsider and not part of the mainstream group?
Conversely, if she attends a special school for children who are dyslexic, will
she somehow feel lesser, different, or marginalized? Will she be missing out
on something, either academic or social, and feel coddled or not part of the
group, either way limiting her future opportunities? Will this choice take her
off the mainstream road to success and cost her a fulfilling future? Or will

attending a special school provide her with the reading and other academic skills and self-awareness to facilitate her academic progress and solidify her self-concept, allowing her to move forward to realize her potential? Will being with and interacting with other students who are dyslexic show her that they are no less intelligent than nondyslexics and actually creative and fun to be with? These are difficult questions, with different answers for each child, and at varying times in her development they need to be reassessed. Let's explore the options.

THE MOST CRITICAL ELEMENT: CLIMATE—HIDDEN BUT EVERYWHERE

In the years since *Overcoming Dyslexia* was first published, I have been a dyslexia detective, working constantly to learn more about what really counts, including what makes the most difference in the school a dyslexic child attends. Reading program, overall curriculum, and availability of the arts and sports are important. But having visited endless schools, for the better part of a day sitting in on a range of classes, speaking to parents, meeting with educators and school administrators, and getting to know many, many dyslexic boys and girls, I came to realize that these features, as important as they are, are only parts of a much deeper and I think critical aspect of a child's experience.

Climate. The most critical element in a school a dyslexic child attends is the *climate.* It permeates every classroom, the auditorium, the cafeteria, the playing fields, and, most especially, virtually every interaction between your child and his teachers and peers. Climate is truly at the heart of the school experience for dyslexic children. Other components certainly influence the child's day-to-day experience, but the climate supersedes all else in impacting your child's future, how she views herself as a learner, her self-esteem, and her sense of empowerment. Does she see herself as a worthy person who is bright and can succeed or as someone who always has to be on the lookout for a misstep? If she makes an error, is there someone at the ready to point a finger and pounce on her as if to send the message *See, you're really not up to par—you're a damaged person?*

If a dyslexic child attends a school that does not "believe in dyslexia," or even if the school pays lip service to providing a reasonably good reading program or accommodations and the climate is not truly welcoming, that child will suffer.

As you know by now, when a dyslexic child is called upon to answer a question, she will on occasion (no matter how well she knows the answer) blurt out a word that sounds similar but is incorrect (such as *"confession* stand" rather than *"concession* stand"). At such times, will an ill-prepared teacher who knows frightfully little about dyslexia stare at her in dismay while her classmates smirk or giggle? Or alternatively, will her teacher, who understands dyslexia, smile and provide her with a choice—"oh, Eva, did you mean *concession* stand or *confession* stand?"—thus providing an opportunity for the child to hear, recognize, and verbalize the correct response? In this case Eva will have an opportunity to respond again, choosing the correct word, and then sit down feeling much better about herself. In a school where the climate is right, where knowledge of dyslexia is required, teachers will know that Eva, as a child who is dyslexic, may know the correct response but will often be unable to retrieve the correct word.

A school without a full understanding of dyslexia is not a school for your child. Things are bound to go wrong. For example, Mimi is a wonderful, hardworking, and bright middle-schooler who is dyslexic. Her school decided that she had to ask for extra time before each and every test. Of course, in granting Mimi this accommodation, the school acknowledged that she required that time to be able to show her knowledge. But why should she have to ask for it each time? That does not represent a good climate. Not surprisingly, for one exam Mimi was so nervous that she neglected to make the request. Not only was she not allowed the extra time the school knew was necessary for her, but her school went on to count the grade on this test, which she was predictably unable to finish. That is not the kind of school you want your dyslexic child to attend.

In Connecticut the requirement for consideration in the gifted and talented program in at least one Fairfield County community required a student to be two years above his grade level in reading. Disregarding that dyslexia is defined as an "unexpected difficulty in reading in relation to a person's intelligence," this ignorant rule assured that bright dyslexic students would be ineligible for placement in the program. Similarly, some schools make the false assumption that if you read slowly you must also think slowly and do not allow dyslexic students access to higher-level math and science courses. A student named David was as brilliant and eager a student as I have ever met, and yet his school would not even consider that he should be allowed to take AP math or AP chemistry courses.

One of the most egregious cases of confusion of slow reading with slow thinking occurred at a top-rated college, where a dyslexic student, Leo, had been approved for extra time on exams as an accommodation for his slow reading. He was very pleased with that and told his professor so. In response, the professor began speaking to Leo very, very slowly, as if Leo could not process the spoken words at a normal rate. The professor's assumption was that if Leo was a slow reader, he must also be a slow thinker. Not surprisingly, Leo was devastated by his professor's profound lack of understanding of dyslexia.

Considerations. Most parents I know prefer to have their child attend the local public school, but the typical public school, including its special education program, does little to move a dyslexic child forward. You must always keep in mind that the availability of special education is by itself no guarantee that a child is receiving an appropriate education. In general, public schools are slow or fail to identify dyslexia, provide too little instruction, and often use unproven and incomplete programs taught by teachers who know little about teaching reading. Most egregious and harmful to their students who may be dyslexic is that typically neither the administrators nor the classroom teachers have any real sense of dyslexia—a condition affecting at least 20 percent of their students!

Still, there is something to be said for having a child remain in his local school. Public schools generally offer much more diversity than can be found in a private school setting. This diversity extends to the composition of the student body, to the possibilities for choice within the curriculum, and to the opportunities for participation in important nonacademic activities such as sports, music, and art. Public schools are also free. So it's important to know that public schools can be made receptive to new ideas—they can change. Parents and teachers can use the information in this book, together with the Report of the National Reading Panel and the What Works Clearinghouse to bring evidence-based reading instruction to every child's classroom. The information and data are available and are beginning to be more widespread. Awareness is growing along with the hope that such interventions will soon become a widespread reality for dyslexic readers, that teachers will come to expect success from their dyslexic students and children will benefit enormously. Joining visionary educators in this forward-looking movement are parents who, frustrated by their experiences advocating for their child and fueled by their love for their child, have banded together to form groups such as Decoding Dyslexia, a national confederation of state groups advocating for children who are dyslexic.

Parents must always be vigilant to ensure that their child is receiving the most effective reading instruction, that it is evidence-based and integrated with his other academic subjects, and that their child's strengths are not overlooked.

THINKING ABOUT CHANGING SCHOOLS

In some instances it is in the best interests of a child to change schools. Serious consideration should be given to changing schools if you have tried a public school setting and

- → the school has provided a special reading program and your child is still lagging behind.
- → the school cannot organize an effective program and your child is falling behind.
- → the constant battle to have the school provide promised services is adversely affecting your family.
- → the continual lack of understanding is taking a toll on your child, in terms of his or her self-concept and desire to learn and go to school.
- → your child is demonstrating the onset of negative behaviors.

Selecting that new school represents a critical juncture for parents. Jack and Inez Johnson were at this point when they came to see me. Their son, Eddie, was the spitting image of his large, bearlike, ruddy-faced father. Eddie had begun his schooling in a private day school in a suburb of New York City. As is so often the case, Jack and Inez had complex reasons for sending Eddie to a private school. On the surface it seemed obvious that this was *the* school to go to if you were smart and your parents could afford it. His father longed to give Eddie the kind of early education he never had.

Jack, the managing partner of a large law firm, knew what it was like to have reading problems. As a child and into the present, reading was and is a struggle for him. At first Jack rejoiced in Eddie's advanced vocabulary and quickness at grasping new ideas and concepts—his general brightness in the preschool years—so he enrolled him in this much-desired school. Once there, Eddie was not the academic star his parents imagined he might be. Perhaps it was just a period of adjustment, they thought. He's a boy, and boys may be slow to adjust. Knowing deep down what they could not outwardly admit to themselves, they set about creating a safety net for their son. They hired a tutor. Eddie went to school from eight to four; then after dinner he worked with his tutor or with one of his parents. The days were very long, and it was a losing battle: Eddie was in free fall, heading for the bottom of his class. The school itself was tone deaf to his needs. "He must buckle down and work harder," the headmistress told Eddie's parents. At age ten Eddie was seriously dyslexic. He was daily facing what seemed to him to be an avalanche of new words—long, complicated, multisyllabic words. He could no longer get by just with memorization. The only saving grace was that Eddie was a star soccer player.

Worn down by their nightly homework sessions with Eddie and concerned about his increasingly gloomy persona, Jack and Inez came to the Yale Center for Dyslexia & Creativity. After discussing Eddie's evaluation and his needs for specialized reading instruction, understanding the impact of dyslexia and the importance of support, we talked about school options: special education in a public school, another independent day school, or a school specializing in educating children with dyslexia. We talked about the benefits and costs associated with each type of school.

Many parents turn to independent day schools because they find the local public schools wanting. Private schools offer smaller classes and potentially more opportunity for individualization. Discipline tends to be firmer, and the students often have a greater chance of going on to a competitive college. Of course, like public schools, independent day schools vary in quality. At their very best, such schools can provide a constructive refuge for a child like Eddie. If the administration and faculty are willing, a small private school can provide the bricks and mortar of an effective program for improving reading. Relative smallness can bring with it both flexibility in modifying the rest of the child's academic program and also a sense of intimacy in getting to know the student and in providing him with a supportive network of encouragement that extends throughout the school.

For most children, the options for private schooling are more likely to be country day schools or boarding schools. More often than not, these schools turn out to be a disaster for a child with dyslexia. In contrast to most public school settings, I have found that children in private schools are more likely to be in lockstep with one another. Such schools may often pride themselves on uniformity, which often extends to the curriculum and to reading instruction. For some of these schools, maintaining decades, if not centuries, of tradition trumps the desire for change and indeed of moving forward. The lack of student diversity in this setting most often does not serve the interests of a student with dyslexia; the opportunities for innovation, for providing different courses, and for creating special programs are limited.

However, there are private schools with quality reading programs and understanding of the needs of their students who are dyslexic. It is your job to assess the climate of the school overall and in each of its programs before enrolling your child. Moreover, if your child has already been identified as having dyslexia, or if you suspect dyslexia, proper testing and evaluation should be conducted before you approach any school. A diagnosis of dyslexia pinpoints the specific nature of the problem, which in turn leads to a much deeper understanding of your child and better equips you to ask the right questions and find the best school.

For the parent of a child who is dyslexic, enrolling him at an independent school often represents the belief that he will get more attention and a better education there than is possible in a public setting. But be aware that attendance at an independent school may evade the real issue: acceptance that your child is in any way different from the idealized American boy or girl. Many parents worry that if their son or daughter attends a special school for children with dyslexia, he or she will be removed from the mainstream of society and will not be able to participate in it fully. Society often tends to view people with disabilities not only as different but as lesser individuals, in this case often incorrectly confusing dyslexia with mental slowness or intellectual disability.

Parents like the Johnsons can be torn. Jack, having suffered both in school and in life from poor reading relative to his intellectual strengths, now wanted his son to have a better education, to be a reader. He was in favor of sending Eddie to a nearby specialized school for children with dyslexia. On the other hand, the thought of Eddie attending some special school for dyslexic children frightened Inez. "Let him stay one more year at the school he's in," she pleaded with her husband. "We'll work harder, get a better tutor. It will work out." She worried that the children at a school catering to dyslexics would not be as bright or as sweet as Eddie. Going to a special school might mean giving up soccer, from which he gained so much positive feedback, surrounded by friends. He was so bright, she thought; surely reading would come. Jack, based on his own experience, believed that going to a special school would help Eddie develop a more positive view of himself. Inez believed that if her son were to attend such a school, his sense of self would be diminished; he would view himself as somehow defective. Inez did not take into consideration that most specialized schools only go through the eighth grade, meaning that Eddie would have to go to a regular school for ninth grade and that attending a specialized school might allow him to go forward with stronger academic skills and a higher level of self-esteem.

The Johnsons made many inquiries, followed by visits to several schools in their area. They wavered and vacillated; they wanted so much to do what was best for Eddie. Finally Eddie's parents settled on their priorities. The school should be within commuting distance; have a mixed student population rather than one focused entirely on dyslexia; offer a strong sports program, one in which their son could participate in interscholastic matches; and already have in place, or at least be willing to create, a comprehensive educational program with an evidence-based reading intervention at its core.

The Johnsons chose a setting that would emphasize Eddie's strengths—socialization with a nondisabled peer group and active participation in team sports—rather than his differences. They found a school that was primarily for typical learners but was willing to accept a child who had dyslexia. The school's reading program was earnest and had several components often found in evidence-based interventions. However, it was not clear if these parts would be sufficient to provide Eddie with the explicitness, practice, and intensity he required in order to become a skilled reader. As the school year evolved, it would be important to keep a sharp eye out to see if the school's overall climate was truly supportive. The goal was for Eddie not only to progress academically but to develop a positive self-image, allowing him to leave the school feeling good about himself.

THINKING ABOUT A SPECIALIZED SCHOOL

Like all schools, specialized schools for boys and girls who are dyslexic vary in the quality of their instruction and in the makeup of their student body. Quite a number of highly accomplished college students spent part of their formative years at schools for children with dyslexia (in most instances from two to four years). My impression is that students with dyslexia who attend such schools are more likely to have an overall positive outcome than students who attend a traditional independent school. Schools of this type exist across the country. A list of them can be found on pages 325–326. I have observed classes or met with faculty and students at many of these schools.

Each time I visit, I come away wishing that each dyslexic boy and girl were able to experience the benefits of such a school. Although not every child succeeds totally, each child leaves with his or her dignity intact and a sense of competence.

These schools are so special because all the children there, at each grade level, require some degree of help in reading, *which ensures that educational need serves as the central focus of the school's instructional program rather than being incidental or peripheral to it.* There is no need to negotiate or to argue for the proper reading instruction; students receive highly effective reading instruction provided with sufficient intensity and over the necessary duration. Teachers are highly experienced and share a sense of camaraderie. The expectation is for success. Faculty members are encouraged to keep abreast of new developments in the teaching of reading. Instruction is optimized: Children are taught in small groups, reading and other language-related skills occupy a significant segment of the day, classes are coordinated around a specific theme or a child's needs. Each child's reading needs and his academic curriculum are fully integrated: Reading is related to writing; writing is related to what the child is learning in literature or social studies. The science curriculum is not watered down, but difficult-to-pronounce words are reviewed before pages are assigned for reading. No child needs to fear that when he is called on, the class will laugh when he stumbles over pronouncing words.

Although many parents, like Inez Johnson, shy away from such schools because they are fearful of the effects on their child's self-esteem, the result is often just the reverse. In fact, I believe that, even more than improvements in reading, the most important long-term benefit of being a student at such a school may well be its positive influence on a child's self-esteem. As I've emphasized, building self-esteem is essential for later success. Attending a school where you are not one of the very few who have been identified as dyslexic but instead are in the majority represents a very different kind of experience. You can relax. You no longer need to hide your dyslexia. Such students also meet bright people like themselves. It is akin to looking in the mirror and liking what you see. Students find others who share their experiences and concerns. They also learn that they do not have to like all people who have dyslexia; like everyone else, such individuals run the full gamut of personalities and traits. Because these students are in a supportive environment, they can learn about and practice the self-advocacy skills they will need when they are in other settings.

Without the pressures often found at other schools, children at these schools have the opportunity to grow and to discover who they are. They develop an awareness of the full range of their skills, their strengths as well as their weaknesses. For example, at the Windward School in White Plains, New York, and in New York City, students attend a course that helps them develop insights into their own characteristics as learners. At the Park Century School in Culver City, California, each graduating (eighth-grade) student creates a video document demonstrating his understanding of the basis of his reading difficulties. Contrary to the view that enrollment at a special school is a sign of defeat, attendance at such a school may be the initial step in a student's pathway to success. One of the first dyslexic students I encountered at Yale, Kayla, whom you met earlier, had attended a school specialized for students with dyslexia. After graduating from Yale, Kayla went on to fulfill her long-held dream of a career as a screenwriter. Typically, as I said, children spend two to four years at such a school and then return either to their local public school or to a nonspecialized independent school. Explicit plans are made to help ensure that the child makes a successful transition.

A VERY PERSONAL MESSAGE FROM A PARENT

Dear Dr. Shaywitz,

I'm writing to you to let you know how much you changed my son's life—and my life—for the better.

I went to a talk you gave in California. At that point, I was wrestling with what to do: keep [my dyslexic son, Levi] at our mainstream public school, or send him to a special school to address his reading head-on.

Dr. Shaywitz, you were laserlike in your immediate response: "Time is your son's most precious resource. This is a top priority for your son." You didn't mince words, and I was—and am, to this day—profoundly grateful. Your message was what I needed to hear. That same day, I shared with my husband your call to action. Thankfully, your research, and your honest, highly informed clarion call, made the case for my husband as well. Life presents unexpected challenges, and I am so grateful that you unequivocally pointed us in the right direction, at a time when we—and our son—most needed your expertise and guidance.

Our son went to a specialized school for first through fourth grade, where every day he practiced learning to break down words together with reading these words in connected text. Every day he learned he was not alone in this challenge, a challenge that with meaningful, intensive, and frequent practice he could overcome. He not only learned reading skills, he also learned perseverance. Even at that young age he dug in deep and never gave up, because, thanks in great part to you, he had adults around him who understood his learning needs—and took action. He learned the value of hard work, the intrinsic satisfaction of giving his all to learning.

After four years of step-by-step, explicit reading instruction, Levi successfully returned to our public school for fifth grade. That same fall, his younger sister started first grade. Our daughter was a typical learner, so the mainstream first-grade curriculum was an easy, comfortable fit. Seeing what she and many classmates could do in first grade made me realize how much a typical first-grade classroom would have been nothing less than a total nightmare for our son. He would likely have paid a high emotional price along the way, feeling like a failure surrounded by many kids experiencing reading success. I don't think Levi ever felt like a failure, thanks to your sound advice. Instead he got to use his precious time learning the things he needed to learn, so that he could thrive in the classroom, and in life.

For Levi, the process of overcoming dyslexia has not been a quick fix. Levi continued to have a tutor, twice a week, from fifth grade through ninth grade, and through it all, he has had to put in countless hours of effort. Through it all, Levi has never given less than 110 percent. I am beyond proud of him, and again, so grateful to you.

A couple of months ago, when my cell phone rang, it was my son, Levi. "Mom, I just got my SAT scores. I got a 650 in math, a 750 in writing, and Mom, I got a 780 in reading." I can't put into words the intense, overwhelming emotions of that moment. I started to cry. I was just so proud of him,

and so happy for him. We are more grateful than you'll ever
know!

Take good care,
A grateful mom

SUMMING UP: ADVANTAGES AND DISADVANTAGES ASSOCIATED WITH EACH CHOICE OF SCHOOL

Public Schools. Public schools allow a child to interact with a diverse group of children and to participate in a range of local activities. They are convenient and free. On the downside, there is scant evidence that most current special education reading programs in public schools benefit a child who is dyslexic. Does the school use the word *dyslexia* and understand the impact of dyslexia? If your local school is willing to provide an evidence-based intervention (you can use Michele's program on page 275 as a guide) to integrate your child's reading needs with the remainder of his academic subjects and to regularly monitor his reading pulse, it is certainly worth a try. As school systems and educators become aware of dyslexia and of evidence-based interventions, schools should be—hopefully, will be—increasing their receptivity to utilizing more effective reading programs. Six months (or a semester) represents a reasonable trial period; it is long enough to see positive results if they are going to occur and short enough not to inflict permanent harm on your child. Above all, if you are troubled by the results, don't wait any longer. If you have the resources and your child is floundering, consider an alternative school.

Private Independent Schools. If you have your heart set on sending your child to an independent day school, it is important that you learn about the school's climate—that is, its attitude toward bright children who struggle to read. Explicitly ask the school how it addresses the needs of dyslexic children. Private schools generally do not offer specialized reading programs for such children and may not have the flexibility or knowledge to modify or adapt their courses. Such schools vary in their understanding and in the availability of accommodations. They may offer advanced science or computer courses for those students who have the interest and the ability. Extracurricular activities and social activities are often a strength.

Specialized Schools for Children with Dyslexia. For children struggling with dyslexia whose parents have the financial resources, a specialized school should be seriously considered. Specialized schools care deeply about children and frequently extend themselves to offer scholarships and financial aid packages to help assist families.

I think these schools should be given consideration even for children with less severe dyslexia. These are the ones who fall through the cracks, who often struggle unnoticed and unassisted, and who could benefit from the enrichment of their reading skills and the building of self-esteem that a specialized school offers. Here is a list of recommended schools of this type:

AIM Academy	Conshohocken, Pennsylvania
Assets School	Honolulu, Hawaii
Atlanta Speech School	Atlanta, Georgia
Benchmark School	Media, Pennsylvania
Brehm Preparatory	East Carbondale, Illinois
Carroll School	Lincoln, Massachusetts
Charles Armstrong School	Belmont, California
Churchill School	New York, New York
Churchill School	St. Louis, Missouri
Cove School	Northbrook, Illinois
Delaware Valley Friends School	Paoli, Pennsylvania
DePaul School	Louisville, Kentucky
Eagle Hill School	Greenwich, Connecticut
Eagle Hill-Southport	Southport, Connecticut
Forman School	Litchfield, Connecticut
Gow School	South Wales, New York
Greenwood School	Putney, Vermont
Hamilton School at Wheeler	Providence, Rhode Island
Hamlin Robinson School	Seattle, Washington
Jemicy School	Owings, Maryland
Kildonan School	Amenia, New York
Lab School	Washington, DC
Landmark School	Prides Crossing, Massachusetts
Louisiana Key Academy	Baton Rouge, Louisiana

Marburn Academy	New Albany, Ohio
NewBridge School	Poway, California
Newgrange School	Hamilton, New Jersey
Oakwood School	Annandale, Virginia
Park Century School	Culver City, California
Prentice School	Santa Ana, California
Schenck School	Atlanta, Georgia
Shefta School	New York, New York
Trident Academy	Mount Pleasant, South Carolina
Vanguard School	Lake Wales, Florida
Villa Maria School	Stamford, Connecticut
Westmark School	Encino, California
Windward School	White Plains, New York
Windward School	New York, New York
The Winston School	Short Hills, New Jersey

24

A CLOSER LOOK AT SOME OF THE FINEST SPECIALIZED SCHOOLS

I want to give you a firsthand sense of what happens at a top specialized school, not only in terms of the students' experience but on the part of the people and principles driving the success of these schools. Such schools offer programs and other benefits for dyslexic children that many parents and educators may not be aware of.

THE WINDWARD SCHOOL, JOHN (JAY) RUSSELL, HEAD

I first met John Russell when I was invited to meet with a group of superintendents on the BOCES (Board of Cooperative Educational Services) for New York State to talk about dyslexia. I asked these educational leaders if they had heard of the National Reading Panel, on which I had recently served. Of the dozen or so superintendents, only John Russell raised his hand. As usual, he was ahead of the crowd.

During his tenure as superintendent, Russell was unexpectedly offered the head position at Windward. He had begun as a math and science teacher at a high school, gone on to middle-school level, served as an elementary-school principal, become an assistant superintendent for curriculum instruction, and, last, served as a superintendent. His vast experience in education left Russell curious about some things he had noticed.

> Even when I was a superintendent, I would constantly ask my director of special education, "We keep identifying kids. Why don't we ever exit kids from this program? Where's the benefit of the program if the kids aren't able to return to the mainstream?"

Russell had three major goals when he arrived at Windward.

1. *Emphasizing that our students are smart.* To preserve the school's mission, which is dedicated to clarifying for our parents and, to some extent, the larger general public that our kids are *smart.* Many people continue to harbor the mistaken idea that our kids are not cognitively able to do so many things that they, indeed, can do. That's one of the things that I recognized I had to really pay attention to, and I continue to fight that battle. Part of our mission statement is to share our expertise with the broader community. We want to disabuse independent schools and public schools of their notions about our kids so that when they leave here, they can go into a mainstream setting. And so to me, job one was to set important expectations for what our students are capable of doing. In order to serve our students optimally, the school, too, is very specific about which students we will accept—we know the profile of the students who will respond to our program, the curriculum we offer, and the instructional program that we use. *We are absolutely steadfast in not accepting any student who doesn't meet that profile.* One of our great strengths is knowing exactly who are the kids we serve, not trying to serve all students with all disabilities but really being laser-focused on kids with dyslexia.

2. *Requiring teacher preparation of all Windward teachers.* First there is *professional development,* the input that teaches them *what* they need to know. Next is *implementation,* where the teachers go into the classroom and learn how to apply the *what.* This gives our future teachers a chance to practice that for a year, and sometimes two years, without actually being responsible for a class. Last, there is *evaluation and supervision,*

which gives the teachers feedback on how they're implementing *what* they learned. Nobody teaches the first year they're here.

Everything that happens at Windward is *intentional.* This was formalized by John Russell, and I believe it is one of the major reasons Windward is so successful. For example, the mentoring program includes formal expectations based on mentors' specific responsibilities. "We really codified what we expected in the mentor-mentee relationship, and we built on that," he told me.

3. *Instituting a more rigorous reliance on data and data analysis.* To accomplish this, Russell insisted that the school aggregate and collect standardized data.

> Moving forward, I wanted to aggregate the data so I could tell parents, *You know what? For the 1,539 kids who have left during my tenure, I can tell you for a fact, because I have the data and I've analyzed it, that 98 percent of them are reading in the average to above-average range in reading comprehension on the standardized test, and that's normed against a so-called regular ed population.* The standardized data were analyzed on a yearly basis. Teacher input and the teacher evaluations are continuously examined in a serious way.

Russell's MBA gave him an edge.

> One of the things I realized during my MBA is that virtually every innovative corporation invests considerable amounts of money in research and development, but I knew of no school that was actually doing that in a direct manner. So I created a position here—a teacher who's designated as a research associate. That's our commitment to staying current with the research. We've really focused on research. When we say we use evidence-based practices, we actually mean it and can substantiate it.
>
> At the school there is an internal joke: *At Windward, in order to get in, you have to fail the entrance exam, and then, when you start doing really well, we ask you to leave.* I don't know of any other school that does that as conscientiously as we do. On average, children stay about four and a half years.

When are they ready to leave? I asked.

When they've internalized all the skills that we've taught them. One of the things we pay very close attention to is teaching the kids self-advocacy skills. If a child is able to read and write well—and by that we mean approximating at least grade-level work—then they have to also be able to self-monitor and self-regulate. We're famous for the structure we provide. That external structure is a surrogate for the self-regulatory skills that these kids have yet to develop.

I believe that anything that we call a skill, if it's a skill, then it can be taught. If it can be taught, it can be learned, and if it can be learned, it can be internalized. When you get the academic skills in place, the self-advocacy and self-regulatory skills in place, that's when the kids are ready to go.

Here is yet another way Windward is *intentional* about its program: By the time the kids leave eighth grade, they will have written four annotated research papers that are of significant length. We don't assign that. We teach them that. In the course of teaching that, one of the things they get is a planner. They plan their first paper with the help of the teacher, and then, on the second paper, on their own. They lay out a calendar of benchmarks they have to reach to write a paper in eight weeks. We work backward from that. We lay that out for them very specifically. We teach them something as simple, that often doesn't get taught, as note-taking. For example, students learn how to extract key words and phrases from text so that you have the key words and phrases and then you write your own sentences, as opposed to plagiarizing and just copying.

For another example, the kids are responsible for having a homework notebook. That's not that uncommon. Every school in the world does, but we have a period at the end of the day called *focus*. For the last fifteen minutes of every middle-school kid's day, they return to their homeroom *focus* classroom where the teacher with the child goes through his homework list. Does the student understand what his homework is? Does he have the books that he needs to do the homework with him? Is there any part of this homework that he feels he's not ready to deal with? If that's the case, the teacher in his homeroom will either help start the homework or will send the student back to the teacher who assigned the homework so he can ask clarifying questions. Yes, we're intentional about everything we do. We don't just walk in and wing it.

When the kids are getting their reading instruction, they are in groups varying in size. The average size is ten. There are usually two adults in the classroom. One is the master teacher, and one is an assistant teacher. We group homogeneously for the skills-based instruction. Reading is language arts and we do *three periods* of language arts each day. We do reading in terms of decoding in the younger grades, and then we do connected text, and then we do education and writing. That takes three periods. It's about two hours.

Russell does not believe children have to have one-to-one instruction.

We don't provide one-to-one. In order to get into Windward, you have to have a diagnosis from an evaluation that clearly identifies the child has average to above-average intelligence and dyslexia. That's the first hurdle. Then you're in the pool. Then we have you come in, and you visit with us. You actually attend classes for two half-days when we also assess your decoding skill.

We have to have a group for you. We usually do have a group, but some kids are outliers, and we don't have a group for them. As long as we have a group, you're in. Those are skill-based groups, and they're fluid. Homogeneous grouping for the purpose of instruction and skills-based classes is, I think, absolutely necessary. Here, kids move from group to group. The groups are fluid, and we move them around. At the end of the day, when the children leave here, whether it's in sixth grade, fourth grade, or eighth grade, 98 percent of them are reading in the average to above-average range in reading comprehension, and a similar number, 96 percent, in vocabulary.

The Windward Teacher Training Institute enhances the program. It was first developed to train our own teachers. We want everyone to understand our larger mission, and so we want to share our expertise with the broader educational community. Last year we provided our own faculty with 11,000 hours of professional development, and we had 1,500 teachers go through our teacher training institute.

There are three core courses that everybody has to take. There's a direct instruction in reading course. There's a direct instruction in writing, and then there's a course in language that's taught by a speech and language pathologist called language and literacy. Those three courses are an absolute bare minimum. Then we offer courses in direct instruction and mathematics. Every teacher who teaches here is expected to re-enroll in those courses, minimally, once every five years. Our teacher training never ends.

Windward's program has excellent components, but of course the outcome is what matters—and that's the best part of all. Before students begin at Windward, they take a standardized test in reading and math. We get a baseline of where they are compared to a so-called normal population. Then they take those tests at the end of each year and then I compare their last year's performance with their pre-performance, and that's where I get those comparisons—98 percent in reading comprehension at or above their grade level, and also in reading vocabulary with 96 percent at grade level or above.

That's one measure. I'm kind of suspicious of data, so if you presented that to me, I would say, *Well, that's fine. It's an immediate effect, but is it a persistent effect?* For the past ten years, we have surveyed the schools that our kids go to after they leave Windward, and we wait at least two years. We asked the administrator and guidance counselors in these schools, *How are our kids doing academically compared to their grade-level peers?* Now, 98 percent of that group were in mainstream settings. At the last school where we asked, *How are our kids doing,* 97 percent were at or above their grade-level peers academically.

We've also asked them questions like *How are they doing in terms of their social-emotional adjustment to the school?* Again, we had great results with that. We asked them about participation in extracurricular activities and giving back to the school—fabulous participation in that.

The last thing we do is track the secondary schools that our kids have attended. The list of independent and public schools has continued to grow over the past twelve years. More and more schools that would not have in the past taken a kid with a learning disability (most commonly dyslexia) are now taking our kids. Every time one of our students goes into one of those schools that has not seen the light, they serve as an ambassador for all those kids who are going to come after them, and our kids do unbelievably well, and they—our kids, our alumni—are the ones who are doing the most heavy lifting and getting the best results in disabusing people of their mistaken ideas about our students' capabilities.

We have kids in every competitive university in this country, including Yale, U. Penn, and Brown. The kid who spoke at graduation who was an alum this past year, he was a Phi Beta Kappa junior year at Harvard, and he just graduated from Harvard. So our kids go to every possible school.

Russell believes passionately in early identification.

Mandatory early identification would be my advice. Early, early, early. What I would tell parents is early detection, early identification, is really the first step into a longer process. If you don't take that first step, the longer you wait, the steeper the hill you have to climb to get over that hump. The kids that we had detected early, the academic remediation may be as effective with older kids, but the difference is the kids' self-esteem, which in most cases—not all—has not been damaged. Those kids who have been identified early have not been mistreated by well-intentioned, ill-informed people who have told them to try harder, or you're so good in art, or you're so good in math, surely you can read. Just try harder. That would be my advice. Early, early, early identification.

THE PARK CENTURY SCHOOL, JUDITH FULLER, HEAD

The Park Century School (PCS) in Culver City, California, is a highly regarded school for dyslexic students in grades two through eight. I got to know this school in the best way possible, that is, through consulting with the parents of dyslexic children, Natalie and Peter, who were students there. Since I knew both children before they enrolled at PCS, I was able to contrast their experiences at the two schools, including their abilities and their frequent disappointments and frustrations that led up to their enrollment at PCS. Full disclosure: I am now a trustee of PCS.

Ironically, when I recently visited PCS to spend some time at the school before a board meeting, my visit coincided with a teacher-parent meeting to review Peter's progress, which I was invited to join. His mother, Diana, had reached out to me several years before about her son's lack of progress in reading: "I know he's smart, but reading is something else again." She was worried about placing Peter in a special school for all the reasons I've discussed. Reviewing his history and current assessment, I felt that attending PCS would be beneficial. Diana very tentatively agreed, but not without concern. Three years later, Peter had grown so much in every sense: reading, self-confidence, maturity. A worried mother has become an overjoyed parent, bursting with pride at her son's accomplishments. One of the reasons I enjoy visiting PCS is the climate. Observing the classes, seeing the students interact with one another, and observing the teachers and students working together, I was impressed with the positive school climate for dyslexic children. At this meeting I could see how climate makes such a difference.

As often happens, a special teacher made all the difference. It was readily apparent that Peter's special reading teacher, Chris, had come to know Peter at a deep level, going to great lengths to find just the right books to interest him and motivate him to read. Not limited to one specific reading program, she had a toolbox of reading instruction techniques into which she could reach for the specific approach that would meet Peter's needs. She was what we hope all reading instructors can be: a true, reliable diagnostician. Chris saw the whole Peter, not just a struggling reader. Her goal was not only to bring reading to Peter but to bring Peter to reading—that is, to foster his desire to read and, eventually, his love of reading.

Peter was captivated by stories about the Knights of the Round Table, so Chris focused on this topic for special projects. Satisfying this interest helped motivate Peter in his other subjects as well, including math. Remarkably, when a friend asked him about his favorite subject, Peter's reply was "Reading—I love to read." "Can you imagine this from my dyslexic son?" his mother said. "Oh what a long way we have come. Now I know he has a real future, and what a great one at that." Peter is now attending a public high school, where he continues to do very well academically and socially.

PCS cares deeply and constantly pushes forward to ensure that each dyslexic student there is receiving the specific instruction needed. In its head, Judith Fuller, the school is led by an individual who has the compassion, knowledge, and commitment to dyslexic students to make a positive difference. Her experience includes being a kindergarten teacher in Los Angeles public schools. She also worked assessing students and teaching their teachers how to use needed reading interventions, along the way obtaining a master's degree in reading at UCLA.

Fuller has been at PCS for twenty-six years in a full range of roles, so she is able to address any issue that may arise from multiple perspectives. She began as a fifth-grade teacher, then taught the youngest PCS students (second grade). She moved on to a role as reading teacher and then admissions director prior to assuming the role of school head. Coinciding with her transition to Park Century, she learned that her younger son, then in first grade, had been diagnosed as dyslexic. Here is her response to my question, "How did your experience as a mother of a dyslexic child influence what you do as an educator?"

> It absolutely informs every single day of my life here, because there is not a story I hear from the parents that at some point in my life I haven't lived. I think that people here who haven't had that experience have a real sense of the care, but they don't understand the anxiety and grief, because you're grieving at the time when you are trying to figure things out and support your child and you're sending them into the battlegrounds every day to do battle at school without all the skills they need. It's a tremendous amount of anxiety and grief is all I can see, which you need to process at different times, at different points in their journey. And I think I recognize that in our parents. I try to share my experience and try to give them reassuring messages and then listen. They need to be heard.

I asked, "What do you think people most misunderstand about dyslexia?" Fuller let out an exasperated sigh as she replied.

> That there's a quick fix—just teach them to read. It truly involves so much more. Even if they become an accomplished reader, they won't read as quickly as some of their peers and will need extra time. They may have incredible ideas and be able to dictate a marvelous story, but sometimes the details and the word recall will impact them later on when they're very accomplished. So you can teach them to read, but there're still other persisting issues and you really need to know compensatory strategies. It's a lot more complicated than just teaching a child to read.

I am so upset about schools not identifying early and telling parents they are getting "the gift of time" for these little, young children who are suffering daily. These educators seem to forget that in the early years school is about learning to read and that's exactly what the children find so hard, and yet these students are not provided with appropriate instruction. And the area they are strong in, which is content, comes in the later grades, after their self-esteem has suffered such a hit that they no longer feel capable of learning. So for me, early identification, early intervention, is the key to success and the key to turning around an entire generation of children at-risk. My goal is to get the kids in earlier and have a turnaround for those kids in the lower grades, because if you catch them before they fall, they will succeed. But if you wait, as so many schools do, until they're in fourth and fifth grade and they have so many emotional difficulties to overcome outside of the learning issues, it's a big problem. So you can understand why my reading goals and my social-emotional goals go hand in hand. You can't have one without the other, because nobody is available to learn if they feel less than [someone else] or stupid. Unfortunately, most of the kids come to us in third and fourth grades, fifth grade, sixth grade, and so we have to work on supporting their emotions and reducing their anxiety before we can even start intervention, because they're not available to learn yet.

We dedicate the first month of school to anxiety reduction before we even start ramping up instruction. There's a whole period of time that goes into anxiety reduction and trust-building; then we begin instruction. And when we do instruction, we acknowledge that every child is different, and they have different availabilities to learn and progress, and they kind of grow like little helixes. They'll have a burst in one area, and then you have to go and spiral back and support another area. It's just not rocket science.

What makes the biggest difference to kids at Park Century and helps them have confidence? There are a couple of things. One, when they start to talk with one another about how they're smart but learn differently. Also, not feeling alone, not feeling different, learning the self-advocacy piece, and then having work at their level that they can master and feel accomplished.

Park Century is structured to maximize learning opportunities. For example, the classes are small: The whole group class has twelve to thirteen students with a teacher and an associate teacher. The instructional groups for reading and math vary but are often one-to-three or even one-to-one.

In speaking to Fuller, I recalled visiting an eighth-grade classroom at PCS years earlier and being very impressed with the students making videos of the brain. I wondered if this course was still taking place.

> We started having them in eighth grade study the brain, study how the brain works, study how *their brain* works, and then make a video about how they learn, what their strengths are, what their weaknesses are, what accommodations they need, and then we have them dissect a sheep brain. And then they write their own neuropsych report on how they learn, and it really helps the students to understand themselves, what they need to do to manage and be successful.
>
> No matter how long I see kids, when I have that opportunity to be with them in the beginning, when they come in and they're scared and shut down and anxious, to then see them be able to stand up in front of a group of people and say, "I have dyslexia and I have ADHD and I need extra time and to sit in the front and I'll be fine." Yes, and what is so wonderful, to do so without shame. That's the big thing.

PCS works hard to ensure its students a well-rounded experience. For example, starting in fifth grade, students have the option of joining sports teams, including soccer, basketball, volleyball, and flag football, that play in a non-LD league. The school has a rigorous monthly community service called Panther Packs, which is considered to be character education involving kindness, respect, empathy, and resiliency. One of these traits is selected and talked about, and then a service is selected to help others. For example, the school recently held a bake sale to support victims of the extensive recent fires experienced by so many in California.

Judith Fuller concludes:

> I have to say that no one intervention is always perfect. There's still for us a feeling that it's trial-and-error, and we're always trying to

get better. We have staffings, where every two to three weeks the entire instructional team meets and we talk about how our students are responding. Are they still motivated? Are they shutting down? Are they anxious? Are they stalled out? What do we need to do to support them? I would say the gift of PCS is that we're always willing to evaluate and always look at if we are providing everything a student needs to reach their uncommonly bright potential.

School officials emphasize that a student at their school will have just as much of a social life as at an independent school, if not more, because he is not anxious, he feels comfortable with who he is, and he's around people who are very much like himself. The experience is much more normalized.

FORMING A BRIDGE

At specialized schools, uniquely, teachers speak to one another about their shared students. There is *a bridge, a unified curriculum, joining a student's reading teacher with his other teachers.* Rather than each teacher and class being isolated in a silo, there is ongoing monitoring and an exchange of information among each child's instructors, resulting in updates according to each student's specific needs. *Students are grouped in their classes (such as social studies, biology) according to their reading levels.* Instruction is therefore customized. If, for example, a student in middle school continues to require reading instruction, this is just added on and does not take the place of a content class.

It is vital to keep moving forward—the perfect is the enemy of the good. I have been dismayed to witness far too many times the tendency to seek perfection before going forward. As many knowledgeable educators emphasize, it's important to be aggressive in moving forward, far better than having a student working at a low level because of fear of progressing. The teacher can always back off if the approach is too aggressive. For example, in learning about the rules for pronunciation, if a student has not quite mastered a final *e*, that's okay. Teachers can start overlapping the next level of instruction and move on. I once observed a one-on-one tutorial situation in Connecticut during which an incredibly bright young boy appeared to be tortured by continuous repetition he couldn't quite manage. *Oh my,* I thought, *he would be so much better off moving forward to the next step while occasionally reviewing this material.*

SUMMING UP: NEXT STEPS

In truth there is no one perfect school environment that will suit every child who is dyslexic and his or her family. The key in selecting a school is determining which school's profile best matches your and your child's priorities at a particular point in time. Your priorities will change as a child goes on in school. In grade one your priority may be for your child to learn to read; later on, when the child is in grade nine, the provision of accommodations such as extra time on exams and participation in a team sport may be most critical.

Once you have determined your priorities, it is important that you know not only about each type of school but about the specific schools you are considering. To help you collect and organize the relevant information, I have provided a list of steps to take and a checklist of questions to ask about each school under consideration. (Clearly, not every parent has to ask every question; you must choose the questions that are important to you and your child.) By reading through these questions, you may learn about the range of possible programs and services that a school could potentially offer to your child.

Ask as Many People as You Can About the School. Try to speak to parents whose children already attend or are graduates of the school. Parents whose children are no longer in the school but who have attended recently are usually the ones who are the most frank and informative.

Visit the School. You must do so before deciding on any school, public or private—parents first, then the child. Ask yourself,

- → What is the climate like?
- → Do the children seem happy?
- → Do the interactions between student and teacher reflect mutual respect and positivity?
- → Is there a sense of orderliness?
- → Are the teachers and administrators open and friendly, and do they welcome questions?

Find Out How the School Views Itself. Most schools have a mission statement revealing that.

➔ Does the school have a policy regarding children with disabilities?

➔ What proportion of children at the school have dyslexia?

➔ What has the school's experience been with such children?

➔ Does the school use the word *dyslexia*?

Observe Several Classes in Session. Focus on the classes your child might attend. Imagine how he would react, socially and emotionally as well as academically, to being in this group. During such a visit, *force yourself to keep in mind your child as he is, as opposed to your imagined, idealized child.* Spend as much time in the school as you can. It often takes awhile to get a feel for a school, and your impression may change over the course of a day as you see more.

Learn About the Students Who Attend the School.

➔ Is the student body diverse (socially, ethnically, educationally) or homogeneous?

➔ What is the most common reason students come to the school?

➔ What are the students' general abilities?

➔ What is the prevalence of serious behavioral problems among the students?

➔ Do the children live nearby?

➔ How long does the average child remain at the school?

➔ Where do the children go after they leave the school?

➔ How many of the students go on to college, and which schools?

Learn About the School's Academic Curriculum and Its Reading Program. Once you have a general sense of the school and its students, ask about the instruction provided in reading and in other academic areas. Try to obtain specific answers about the bricks (the program's content) and mortar (the explicitness, the intensity, and the quality of the instruction) that make up the reading program. You can refer to pages 275–

284 to refresh yourself on the essential elements of an effective reading program.

→ What specific reading programs (or methods) are used, and are they evidence-based? Are they noted by the What Works Clearinghouse?

→ Since effective programs teach children basic decoding skills (that is, phonemic awareness and phonics), sight words, and strategies for decoding words, including longer, multisyllabic words, ask if these are elements of the program.
  • Are these elements taught systematically and directly?
  • Are decodable booklets used for beginning readers?
  • How soon are students given connected text to read?
  • Do these children regularly read aloud under supervision?

→ How is fluency encouraged? Is it measured, and if so, how? Are specific comprehension strategies taught?

→ What is done to promote the growth of a child's vocabulary?

→ What is the expected progress and the actual progress in reading of students at different grades?

→ Is there a specific writing program?

→ How large are the classes?

→ How are the students grouped?

→ How is instruction individualized to fit a specific child's needs?

→ What approach is taken with older students who are bright but cannot easily read texts in different subject areas?

→ How is reading instruction integrated with other academic work?

→ How do reading instructors and subject-matter instructors communicate? Is there a bridge—a protocol—connecting the two groups?

→ Are there provisions for hands-on or experiential learning (that is, learning through doing)?

→ Are there opportunities for advanced work in history, math, or science?

→ How is technology used? (While it's important to have the newest tablets, apps, and software programs, it is even more important that these are used as an adjunct and do not take the place of teachers in providing first-line reading instruction.)

→ Are students able to use keyboards, Livescribe, Immersive Reader, or other technology for note-taking and for writing essays in class as well as at home?

→ How is a student's progress in reading tracked?

→ Is there a formal course or ongoing support for organizational and study skills?

→ How is a child helped to keep track of his daily homework and long-term assignments?

→ Are self-advocacy skills taught and practiced?

→ Is there a faculty member with expertise (or interest) who coordinates services for students with dyslexia?

→ Are students provided with accommodations, and if so, which ones?

→ What is the process for a student to receive accommodations?

→ Are dyslexic students allowed extra time on their examinations? How much time, and how is this determined?

→ Can computers be used for essay examinations, especially for dyslexic students who have difficulties with handwriting?

→ What is the policy concerning foreign languages? Are waivers granted?

Learn About the Faculty.

→ Are any of the teachers reading specialists?

→ How available are the faculty members for students who require extra help?

→ Does the faculty in this regard have an open-door policy, specific office hours, or availability after school?

→ Who is responsible for monitoring and tracking each child's progress?

→ How does the faculty communicate and update one another on each child's progress?

→ How are parents kept informed of their child's progress?

→ What is the turnover rate for teachers? How long have most of the teachers been at the school?

Learn What Extracurricular Activities Are Offered.

→ Are there team sports, and if so, which ones?

→ Do the teams play in interscholastic leagues?

→ What are the opportunities for exploration of other interests and skills?

→ Is there an art program?

→ Is there a photography program?

→ Is there a drama program?

→ Is there equipment and a studio for video recording?

→ Is there a school radio station?

→ Is there a music program?

→ Is there a school orchestra or band?

As important as it is to ask the questions you want to ask and to take in all the information you can, you must also value your gut feeling. Your overall impression of the school is important and may provide a clue about intangibles that are difficult to characterize but may be extremely meaningful to you and to your child. Schools can look very good on paper but somehow not have the parts come together in a satisfactory way. Just because you can't easily put your gut feeling into words, do not dismiss it. Trust your instincts.

SPECIALIZED SCHOOLS THROUGH THE EYES OF A RENOWNED DYSLEXIC SCIENTIST

You have been introduced to specialized schools through the eyes of school heads. Seeing a school through the eyes of an insightful dyslexic adult, one of the most revered scientists of our time, provides an additional, personal, yet highly applicable perspective. George Church, known for pioneering research in genetics and a dyslexic, reflects on what school was like for him and what he wishes for dyslexic children today.

It is important to get their child into a place where their differences are embraced rather than being a source of embarrassment. Sometimes the source of embarrassment is not people literally bullying them; it is more people trying to be almost too nice, and the child can see that they are treating him differently. Kids are so sensitive to being different.

I remember when I was growing up I must have spent huge portions of my thinking time trying to be the same as everybody else. I was constantly struggling to blend in. It wasn't at all natural. And so if you can get your kid into some environment where being the same isn't the be-all and end-all, that's a win. Remember, at certain critical ages, if you are not wearing the same shirt and talking the same way as everyone else, you sense the difference.

Either they make fun of you or they might try to be too nice to you—that makes it worse. And you almost can't win. One of the most interesting experiences I had was a day I spent at a specialized school for dyslexic students. I spent a couple hours with the fifth-graders to the twelfth-graders and their parents. I was just amazed that when I would walk into the room, there were all these questions and engagement and real enthusiasm.

And then I went to my old high school, and it took me half the time that I had with them to get them to open their mouths. And I'm thinking it's because the high school that I went to was an elite school called Andover, where the Bushes went. I think the kids felt like they had more to lose than to win by opening their mouths. If they opened their mouths they might say something stupid. But at the specialized school it seemed like the attitude was part of the culture—you have nothing to lose. They would ask questions—it was quite dramatic.

There really is something special about a specialized school. Everyone there is special, everyone there felt it was them, the school against the rest of the world. While in most schools, it's you against everybody else at school.

I asked Church, "If you had a wish about dyslexia, what would that wish be?" His response: "If you have a safe haven for them where it felt like the child with dyslexia was joining the X-Men from Marvel Comics rather than being special, that would be better." Church explained further:

If you just had a school that embraces dyslexia as a feature rather than a bug, it would be great. I know it's tough on parents when you have a long commute. But I think it's even tougher on the child when he is in a special class within a school. You know, that's worse.

I think having separate schools is better. Because when you're in a class within a school, you're still with the other kids for the lunches and the phys ed and various other things. And you feel different—yeah, I guess you don't want to be in that special class. You want to do everything you can to get into the regular classes. And the kids and their families usually do everything they can not to be classified.

I certainly did. I was willing to say I'm a slow reader, but I didn't want to be classified. I didn't want to be different. From about age, I'd say, eight to seventeen, you're just petrified that somebody is singling you out. In fact, you watch what they are doing with the ones they do single out and you say, *That's not for me.*

It wasn't pleasant, and I said I don't want to be that kid. And whenever they see one of these kids being picked on, they then generalize to themselves. And they say, *Well, I'm not quite as bad as that kid, but I can't let them see that's who I am. Just a second from now, that could be me.* I lived it. I mean when I'm looking at something from the outside. But I just felt so helpless I couldn't help them. I couldn't help myself; all I could do is try to blend in.

I mentioned to the distinguished scientist that if he minded being quoted, I would not do so directly. His response: "I'm no longer embarrassed." I couldn't help but observe, "You think very differently, and you do an incredible amount of things and do them well." Church's response:

I'm not afraid of thinking differently at this point. When I was— even at the worst time I was afraid of working differently. But I wasn't afraid of thinking differently. Except I just had to be very cautious not to say what I was thinking. But today I can feel more comfortable saying what I am thinking.

I commented to Church that I think one of the problems with dyslexia is, in a sense, that because you do not appear different, people don't notice anything about you. But then when you don't read well—you read slowly, which is most often the case—people assume, *Oh, slow reader, slow thinker.* Church's response:

> Especially reading out loud. That's a real disaster. It is all the baggage that comes along with it. Kids at those ages are very, very sensitive. They have almost nothing else in their life other than determining who or what is different and who the cool kids are. That's what they do. That's their lesson, that's their job. And these are the kids that know that they are not the cool kids and are spending a huge part of their time trying to justify their existence or become cool or hide or use various coping mechanisms, but it is not fun.

This sentiment is from a man who, in terms of fame, scientific contributions, bridging and integrating multiple disciplines, is honored and universally admired and yet still carries with him the wounds he suffered as a child and deeply strives to alert parents how to avoid such painful experiences for their dyslexic child.

25

DR. LAURA CASSIDY AND THE LOUISIANA KEY ACADEMY

Your daughter is diagnosed as dyslexic. You are intelligent, caring, and knowledgeable and have reasonable resources. Your daughter fails first grade. What do you do? You investigate to find a good fit for your child so that she will learn to read, enjoy learning, and have fun. That was the situation facing Bill and Laura Cassidy, two physicians in Baton Rouge, Louisiana, regarding their daughter, Kate. At that time Dr. Bill Cassidy was also Congressman Cassidy, serving in the U.S. House of Representatives (today he is Senator Cassidy). The Cassidys are good friends of mine. We first met when Bennett and I were invited to speak at the Congressional Wives Club in 2006. There we shared the exciting twenty-first-century advances in the science of dyslexia, and a light went on for Dr. Laura Cassidy.

> The science made so much sense, yet it wasn't being acted on in schools. I knew there were dyslexic kids like Kate struggling in public schools, not identified, not being given what they needed. I just had it in my mind that my experience with Kate should be able to offer other dyslexic kids some help. And really that's what led to the Louisiana Key Academy.
> I started at the Department of Education, speaking to the peo-

ple in charge of literacy. I learned that there was not a real division for dyslexia; it was more literacy, with dyslexia underneath that. I didn't really see anything specific being done for dyslexia, and I was opposed to the notion that you would delay identifying children until third grade. I decided to start a charter school specialized for dyslexia.

Let me further introduce you to Laura Cassidy, M.D., the founder of LKA. Dr. Cassidy is a general surgeon specializing in breast cancer. Although she had never started a school before, Dr. Cassidy's call to action was the knowledge that dyslexia existed but that none of the nearby schools were recognizing or addressing it. Starting the school was an involved process. In addition to obtaining state permission for the charter, there was a need to find trained teachers, a location, a building, and funding. Dr. Cassidy attributes much of the success of establishing the school to increasing awareness of dyslexia and getting more people involved. Her belief is that the more people hear about it, the more parents of dyslexic children or people who understand the issues will come to the fore. For example, many classroom teachers know that these children's needs are not being addressed. They want to help; they just don't know how to do so.

The school opened in August 2013. It started small, with only kindergarten and first and second grades. Reflecting the lack of early screening and identification of dyslexic children, the kindergarten enrolled only seven children. Dr. Cassidy found the second grade easiest to fill because there were lots of kids, like her daughter, who had failed reading in the first grade or who had struggled to read in the first grade. Overall, the school's total enrollment was less than one hundred boys and girls.

The children attending LKA come from diverse backgrounds. There are students from disadvantaged families and from middle-class families, both African American and white children. How much money a family has doesn't matter, because it's a free public charter school. Attending LKA is a relief for these dyslexic children. Whether it's reading aloud in front of the classroom or trying to retrieve a spoken word, dyslexic children have anxiety. When they were at other schools, most of the children were made fun of, or they knew they were different from the other kids. Attending LKA gives these children breathing room, where they find other children like themselves. There's no sense of shame, no sense of being different from the other children. Children at LKA are less anxious, reflecting feelings such as "I'm here, and I'm getting what I need to read, and there are a lot of smart kids like me here."

Says Dr. Cassidy:

> A great many of our students didn't want to go to their prior school, where they were bullied or made fun of, or even didn't want to go to school at all anymore. Here at LKA it's still school, so I'm sure there are some days they don't want to go, but for the most part, I think our kids know the teachers love them and they're here to help them succeed. They know that, and that changes a child's attitude about going to school. The most important thing for teachers to understand is that these children are bright, despite the obstacles they face with language, whether it's reading, spelling, or at times writing. We focus on Dr. Shaywitz's Sea of Strengths model. While you're helping them on this journey, you target their reading difficulty and also work to build up their sea of strengths in thinking, reasoning, and problem-solving. If you don't get a child until third or fourth grade, their self-esteem has suffered as well, so you're dealing not only with the academic component but the emotional component as well. Almost all the kids at LKA have previously either failed a grade, failed a course, or repeated a grade, and they're still reading on a kindergarten or first-grade level. That's significant emotional and academic damage to the child and his or her parents.

Striking and very impressive is the degree of integration of how concepts and new words are taught, so that words the child hears in science and social studies are familiar, having already been introduced in his language arts program. Critical to this integration, the teachers are in communication with each other; there's a leadership team. Every child has his or her own plan, and that plan can be changed weekly.

Reading is the fundamental subject, so at LKA there are small group sizes. Classes consist of either seventeen or twelve students, with these groups broken up further as well. As LKA has grown, classes of twelve are further divided into two groups of six. Currently the teacher ratio is nine to one, or six to one for the kids who need the most intervention. As the students get older they don't need such intense intervention, but they all get ninety minutes of structured language arts *every day*—not three times a week, but every day. Those who need it get additional time as well. Other subjects, such as science and social studies, are integrated with what's going on in the reading class. It's important that the curriculum is delivered at the child's reading level. The child may be in the third grade, but if he is reading at a first-grade level, then that's the level at which he has to be taught.

Dr. Cassidy emphasizes that her approach has evolved over the years from the time the school was just a sparkle in her eyes.

> When we first started, the focus was almost entirely on decoding single words, primarily I think because I (and the teachers) like to see the phonemic deficit addressed, and that's done through decoding. And it is very important. At the same time, as I've learned from Dr. Shaywitz's teleconferences with LKA, we've got to start introducing the appropriate connected text as well and to do so very early on. I certainly didn't know that at the beginning, but I do now. So when we started there was a fear of connected text. Today our students are building strong vocabularies and expanding their worldly knowledge through the introduction of reading little stories very early on.
>
> In prioritizing the importance of children being fluent readers, we first focus on teaching the children the basics of decoding. There is a lot of competition for the discrete amount of time you have in a classroom, and we have to make sure that we allocate a lot of time to working to get the children toward fluency.

Dr. Cassidy is concerned about how very many dyslexic children across the nation go unnoticed, undiagnosed, and unremediated. Dyslexics are spread throughout every school and every classroom. As is far too often the case, the teacher has twenty-seven to thirty-two and sometimes even more students in a class, and there is pressure to keep the curriculum moving. The end result is that no one identifies the dyslexic child. Dyslexia comes to light only when those children are all in one classroom in one school, like LKA. Then it's dramatic, and dyslexia becomes the central focus.

Sadly, in the typical classroom, dyslexic boys and girls remain undiagnosed and on the periphery of their teacher's concern. As a result, much damage is being done to wonderful, bright children who are struggling to read and going unidentified. Parents are told something meaningless, such as "It's developmental," and to wait and wait and wait some more. So you can see how welcome a school like LKA is. It started in 2013 with a small group of second-graders and progressively added a grade each year, so that in the 2018–19 school year, LKA enrolled 350 boys and girls in kindergarten through grade seven, and there was a waiting list. Summer 2019 saw the addition of an eighth grade and the initiation of a year-round school program at LKA.

An increasingly acknowledged problem is that children are not being identified early enough, in kindergarten, before damage begins. Here's Dr. Cassidy's description of what she has observed at LKA:

> Our smallest enrollment is in first grade, and we're getting the largest influx in third grade. This means these children have for the most part been sitting in schools, not identified for years and not getting what they should have gotten.
>
> The schools do not have the professional manpower/womanpower or knowledge of what they need, and so you can see how it happens. The children come to us. We know they're smart. They start to improve, but it is so slow and laborious. Everyone wants it to be faster. We want them to be faster learners, but we have to deal with reality.
>
> Dyslexic children make up 20 percent of the student population. And yet, shockingly, schools do not typically screen for dyslexia. Why? I think they're afraid of what they would do once they find out the children are dyslexic.
>
> Personally, I think for right now, in addition to specialized independent schools, a new model—a specialized public charter like LKA—is the best solution. If you can have the flexibility and let the children get what they need, from a cost basis and an excellent delivery basis, that is the best solution.
>
> My greatest surprise in working to develop LKA has been the lack of awareness of dyslexia within the community, and along with that the pressing needs of boys and girls who are dyslexic. Reflecting dyslexia's paradoxical nature, the children seem typical in every *visible* manner—they are cute, sweet, and often highly intelligent—so they go unnoticed as you meet them in the playground, on the street, visiting a friend. However, their frustrating struggles trying to pronounce a word correctly, spell, retrieve a word they are trying to say, learn a foreign language—these fill up

most of the space at home, where increasingly distraught parents try to find an answer or at least a school, a class, or a teacher who knows what is going on within their child and who offers a plausible and effective solution. LKA offers such a solution, as evidenced by its strong embrace by parents and students.

We have close to four hundred children whose parents decided it was a good thing to send them here, and most of our children stay. There are countless testimonies of the parents telling us that their child attending LKA has made a tremendous change in their family and in their child's life because of the boost in their child's self-esteem and the knowledge that there are other children like them. Of course, they also actually start to read as well.

The teachers are totally invested; they have stayed with us and do a really good job. I see that they really enjoy it. They are invested in the child's life, not just one score or one week. They see the big picture. They see the potential in the child's life and not just some moments in time on a score or a grade or a class. I know they're taught to look at these things, but these teachers actually see the potential in these children. Part of it is the professional development they receive; part of it is what you learn just from teaching a lot of children with dyslexia from year to year.

This statement rang a bell for me. As physicians we learn some from courses and books, but we learn the most from taking care of patients over time.

One little child in particular came in who couldn't read. Today he is a remarkable, profoundly changed child. He was smart and incredibly creative but wasn't able to communicate well, which he can now do with both his teachers and his peers. And he is able to read. So the hope of his parents is now dramatically increased, as is the child's self-esteem. Basically, he is no longer isolated. He's able to socially interact and have friends, which he was not able to do before. What encourages me the most is that we've changed this child's life just about 180 degrees. What concerns me is there are many more like him who remain unidentified and don't learn to read, and whose self-esteem is under constant attack.

The children begin the day with structured language arts for ninety minutes, and the average is six children to one teacher. That's where the children are working on the decoding and some oral reading for fluency. An important change has been that we realized that unless we scheduled time specifically for children to read connected text, there often wasn't time enough in the period for this.

We have computers, we have the computer lab. Our students are learning the keyboard. If they have difficulty with handwriting, we encourage keyboarding. They practice in their classroom, but we don't pull them out for therapy aimed at improving their handwriting. We try to keep the children in their classroom, where the appropriate educational method is delivered.

What is so important to the success of LKA is that all the teachers, our accountant, people at our front desk, they all respect and understand that these children have dyslexia and that they're bright. They just have to be taught in a way that recognizes the child's dyslexia. Our job is to figure out what it is a child needs to learn to be successful in life.

Our graduates are going to be able to read close to grade level, if not at grade level. We don't want them to go through life having someone need to read a sign to them, someone to give them an audio device, or for them to feel like they are not valid people because they have dyslexia. The goal would be for a child who leaves LKA to be a fluent reader, to be proud of who he is as a person and as a dyslexic, and to be able to advocate for himself and to live a fulfilling life. I feel like that's what we give them. We're giving them the tools and helping the parents have the tools.

I think most children with dyslexia know they're different. At LKA they know they are not alone but a member of a community of many children. Our children are so diverse in their economic background, their gender, their race, and geography as well. It's like we're a microcosm of the world, which is great.

We now offer art in a newly added specialized room. The children are very creative and love art because it gives them a way to express themselves without having to use language. There are also opportunities for the children to express themselves verbally—for example, in activities in our new drama room. Our goal for next year is to have a sports team: soccer. Of course, they have PE.

Right now the children are all excited about providing nominations and names for a school mascot. Our school community also boasts of an active student council, where we've had some of the older kids who when they came to LKA were struggling academically and shy, who now have chosen to run for office. They fully engage in the election process, posting their picture and giving a little speech.

Basically, our kids are just normal, regular kids, and when they are in a safe environment with people who care about them and see their potential, they know that. Kids are incredible. Whether it's their parents or their teachers, they know they're judged. And if they're judged as having potential and value, they are going to be okay. They may struggle on a test or not make an A, but that's not really what's most valuable for them to get out of school. It's to have the self-awareness that you can go on to contribute to society. That's what our students get.

We hear all the time, "My child no longer hates going to school." We don't give homework that would cause a struggle or confrontation between the parent and the child at home. There may be repetition of what you learned in school, but it won't be laborious homework that involves reading new material or learning spelling or writing a three-paragraph essay that they're not ready to do. They will learn this at LKA but not be forced to do so at home.

Dr. Cassidy has strong feelings about the tendency in many schools to avoid using the word *dyslexia* and instead to use nonspecific, vague, ill-defined, relatively meaningless terms such as *learning differences.* She emphasizes that dyslexia is a discrete, scientifically well defined entity, and that use of the word *dyslexia* conveys much important information about the child.

Laura Cassidy's ultimate goal is for LKA to be a school from first through eighth grade in which children are fluent readers by the time they leave. She extends her vision to helping the children transition to high school through LKA's Dyslexia Resource Center, so if her LKA graduates are struggling in high school, the LKA Resource Center is there to help the schools better understand dyslexia and provide continuing accommodations and remediation. This will have an added benefit of providing professional development for the teachers in the high schools serving the LKA graduates as well as other dyslexic students in these schools who have not had the benefit of first attending LKA. Recognizing that so many teachers in other schools have little knowledge of dyslexia, the LKA Resource Center offers post-graduate experience and courses about dyslexia to more widely disseminate twenty-first-century knowledge of dyslexia.

Many parents have told us that their children now want to come to school. They will read at home in front of their parents or siblings, which is a huge thing for these children. They participate in plays or other public events, and that's changed their self-esteem. I think the greatest testimony to our school, though, is that unless the parents are moving, all the students are coming back. It would be even better if more children were coming in earlier, in first grade, and some of those children could then leave even sooner.

In transitioning from LKA, two things need to be in place. One, the child needs to be fluent or as close to grade-level fluency as possible. Two, when the student transfers, it has to be to a school that keeps the child moving forward and understands what dyslexia is, so that the parent is not having to continuously fight for his or her child to receive what that child needs. If the child doesn't get what she needs and then regresses, not only academically but also in self-esteem, that would be tragic.

Right now, in truth, there are not too many high schools that understand dyslexia. We are having more acceptance in the community, and while we are not there just yet, I think that as we grow and as the children get older, we will get there. I am hoping through combined education and outreach people will increasingly understand.

We are, happily, making progress in having schools understand the key need to screen for and identify at-risk boys and girls early—in kindergarten or first grade. For example, we are working with one Catholic school that is now screening all their kindergartners and first-graders with the Shaywitz DyslexiaScreen™. We will take the ones who fall out as at-risk for dyslexia and further test each of these children. Then, for those whom we identify as dyslexic, we will talk to their parents about options, which would include attending Louisiana Key Academy as well as after-school tutoring at our LKA Resource Center. In addition, several of their teachers are going to go through teacher training starting this summer. Our goal is to equip that school so that they can take care of children with dyslexia. That's the goal.

PART VI

SUCCESS AFTER HIGH SCHOOL AND BEYOND

SUCCEEDING IN POST-SECONDARY EDUCATION

You've really worked hard, harder than most people could ever imagine. You are dreaming of the future. Perhaps you want to be a writer. Maybe a business guru. What about a doctor or a scientist? Are these realistic goals for you as a dyslexic? My answer is a resounding *yes!* Perched atop lists of the best in virtually every profession are dyslexics, including

> *Writers:* John Irving, Richard Ford, John Grisham, and children's authors Dav Pilkey and Patricia Polacco
> *Athletes:* ice hockey player Brent Sopel, former professional football and current professional baseball player Tim Tebow, Olympic ice-skating champion Meryl Davis
> *Business leaders:* financier Charles Schwab, economist Diane Swonk, entrepreneur Sir Richard Branson
> *Doctors:* cardiac surgeon Toby Cosgrove, psychiatrist Stuart Yudofsky, emergency medicine physician Karen Santucci, orthopedist Tyler Lucas, radiologist Beryl Benacerraf, anesthesiologist Fred Romberg
> *Scientists:* Nobel laureate molecular biologist Carol Greider, renowned geneticist George Church, Nobel laureate chemist Jacques Dubochet

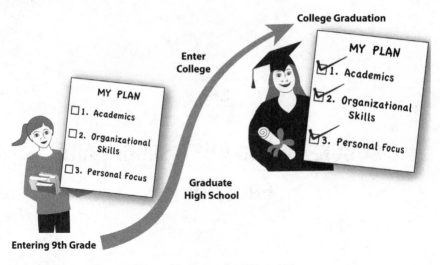

Figure 46. Getting from Here to There

Lawyers: David Boies, Gary Cohn

Public servants: Colorado senator Michael Bennet, former Connecticut governor Dannel Malloy, California governor Gavin Newsom

People in entertainment: superagent Ari Emanuel, producer Brian Grazer, actors Kiera Knightly and Orlando Bloom, director Steven Spielberg, writer-producer-director Brad Falchuk, actor, producer/director, and writer Henry Winkler

Each of these people experienced early difficulties involving reading. So the question is, how do you get from *here*—a dyslexic student entering high school—to *there*?

IT BEGINS WITH A PLAN

Just as when you want to build a house, you have to start with a plan and then go on to build a strong foundation, in making a successful journey from high school to college and a career, each dyslexic student has to make a plan that best suits his or her specific needs, skills, and hopes for the future. Planning for college begins years before the application process kicks in. It should start in ninth grade, and no later than tenth grade. Consider the planning process from the perspective of three broad areas:

→ *Academics,* including developing resiliency and the level of confidence and knowledge of dyslexia to comfortably and effectively self-advocate

→ *Organizational skills,* including actions and activities to undertake to sharpen a child's skills and attitude

→ *Personal focus,* especially on the inner self and possible conditions that might impact, and indeed impinge on, optimal learning.

ACADEMICS

Recommended academic-related considerations and goals:

→ *Increasing independence.* Once your dyslexic child is in high school, gradually encourage him to assume more responsibility for *increasing independence in decision making,* the sort he will require in college. His increased autonomy will reflect a growing base of solid knowledge of his dyslexia.

This is an ongoing process and not an arbitrary, sharp line of demarcation announcing immediate or sudden change. Rather, the goal is gradually, over time, to impart the knowledge of dyslexia along with the confidence in your child in order for her to be empowered to understand dyslexia and its impact on her, her strengths as well as the areas of difficulty. Together, along with thinking about her dyslexia, she should think about her interests, her areas of enjoyment, her ideal future career, and the implications of each.

→ *Developing self-confidence leading to self-advocacy.* As a dyslexic student gets to know who he is, feels comfortable in his own skin, knows that while he may not be a fast reader or a super speller, he is bright and has much to offer the world, he grows in his self-confidence, his independence, and in his ability to self-advocate.

The goal is for these elements to come together while the student is still in high school, if not before. I have met many dyslexic students who have shared with me powerful, and often painful, stories of their experiences of being misunderstood and criticized by well-meaning but ignorant teachers. All of these students light up when they recount the moment when they felt not only comfortable but impelled to stand up and advocate for themselves.

Having endured much during their school years, dyslexics develop *resilience* that many of their classmates who sail through school often lack. It can emerge in many forms, as Izer's mom describes:

Izer has an incredible belief in himself, which he has earned. At Gow, the specialized boarding school for dyslexic students that Izer attended, he would be up until three or four in the morning his whole senior year, working very hard. He does that, puts the time into it. When others in his frat are partying, he buckles down and does what he needs to do. Reflecting what he learned before coming to college, Izer realizes he has to work harder than the other kids to get through. . . . His experience working incredibly hard and succeeding established a work ethic that served him extremely well throughout college at the University of Pennsylvania and will continue to serve him as he now pursues graduate study and throughout life.

The positive role of resilience in a dyslexic person's life was crystallized for me by Adam, a young neurosurgery resident. Bennett and I were just closing the last slide of a PowerPoint lecture on dyslexia one sunny afternoon in Woods Hole, Massachusetts, when one of the attendees shyly came up to the podium. He wanted us to know that he was dyslexic and that without question he was confident in succeeding. As Adam explained, a neurosurgical residency, which requires eight very intense years, has perhaps the highest dropout rate in all of medicine. Adam shook his head and uttered with calm assurance, "Not me. I know I will be there to receive my certificate as a graduate of the neurosurgical residency program. Why? As a dyslexic, I have met and had to conquer and climb over so many obstacles put in my way, I have developed what I can only describe as resolute resilience, and honestly, nothing stands in my way now. If I take on something, I will succeed. It's a good feeling."

→ *Sharpening academic skills.* Review your child's reading fluency, vocabulary, ability to write essays and reports, math skills—anything he will need when you are not there to coach him and review assignments with him. Ask yourself, what skills does he still need to improve so that he is able to function more independently in college? Depending on the college, there will probably be some support, but not nearly as much as most dyslexic students receive when they are living at home during their high school years. Speak to his teachers, analyze his report cards, review any testing carried out by the school or privately. Ask yourself, would he benefit from more support, for example, in math or writing? Ask your son's teachers if they think there are things your son or you can do to help solidify his academic skills.

→ *Learning about assistive technology.* Would he benefit from any text-to-speech, speech-to-text note-taking? Are there software and apps that might help? Technology requires time to harness; a learning curve is associated with using different apps, and glitches will occur. So it is best to begin practice or tryout periods with any technology the summer before the senior year of high school, so together you and your child can determine if it is really helpful and if the time required outweighs potential benefits.

→ *Receiving appropriate accommodations.* Given that your child is dyslexic, there is an extremely high probability that he is a slow reader despite having good, and often very high, intelligence. Is he receiving all the necessary *accommodations* that will level the playing field—extra time on tests, exams given in separate, quiet rooms, use of a laptop in class and on tests? See pages 438–465 for a discussion of specific accommodations and their roles in supporting a dyslexic student.

In order for your child to request the most appropriate and effective accommodations in colleges and universities, she must be aware of her diagnosis of dyslexia, understand how it impacts her, and know that dyslexia is persistent. It does not go away over time. Your child will go away to college; her dyslexia will go with her. A common complaint from many college resource offices is that students come to see them later than they should. The student's explanation (or hope): "I thought I no longer had dyslexia." They arrive on campus magically thinking they no longer require support, or saying that they want a fresh start and will not use accommodations. Most people don't realize how much students who are dyslexic wish they did not need to use accommodations. The persistence of dyslexia in virtually everyone who has the condition, including bright college students, means that it is essential that they know what support and accommodations they will require in college, knowledge best acquired prior to college entry.

→ *Focusing on high school course selection.* Your child needs to develop an awareness of what the prerequisites are for areas of academic specialization. Is she prepared? Is she going to, and is she ready to follow a high school path focused on AP courses such as English language; English literature; U.S. history; biology, chemistry, or physics; or calculus AB or calculus BC? It is never too early in high school for your child to speak to her advisor and to make contact with a college advisor (if your school has such an individual) to discuss plans and prerequisites and her readiness (and desire) to pursue these pathways. If your child chooses to take AP courses followed by the AP exam, she must apply for accommodations in a timely manner. The college advisor should be helpful in this regard.

CAUTION: Be aware that school personnel often underestimate a dyslexic student's ability to pursue advanced work in certain areas, misjudging the student's high intellect.

Higher-Level Mathematics. If your child expresses an interest in science, technology, engineering, or math (STEM) but seems to have had trouble with early arithmetic or learning, consider highly regarded economist Diane

Swonk's experience in math. Early on in school, Swonk, who is dyslexic, couldn't memorize the multiplication tables or add a column of figures, but later on, calculus and algebra were, as she describes it, "a piece of cake." For Swonk and many other dyslexics, difficulties in performing early calculations or memorizing the times table are poor indicators of later ability to carry out complex mathematical analyses. Parents should not allow schools to prevent their dyslexic child from pursuing higher-level academic classes in a range of disciplines because of misunderstandings of aspects of dyslexia. For example, some educators have mistaken the term *slow,* as in slow reading (lack of fluency), to indicate overall slowness in mentation, which is typically not the case. One dyslexic student I know of was not allowed to take advanced algebra because she required extra time. The child changed schools and went on to excel in advanced math classes.

Creative Writing. At times a teacher may confuse a dyslexic child's poor spelling with an inability to write sophisticated essays or highly imaginative stories. John Irving was a C– student in English who struggled to produce a final draft that reflected his imagination and writing, rather than his poor spelling. Too often teachers, and parents, as well, see what is easy to see, pouncing on spelling errors that seem to leap off the page, and allow that to deter any focus on or consideration of the deeper, less visible, complex creative elements making up the child's essay. Don't allow this to happen to your child. Children should be judged separately for spelling and grammar and for creativity and imagination. Such an approach encourages the teacher to appreciate the rich creative thought that often lies just beneath the typos and spelling errors.

Medicine. As a middle-school student, Dr. Tyler Lucas met with his eighth-grade advisor, who asked what Tyler was interested in as a career. The response was quick and clear: "I want to be a doctor." The teacher shook his head, looked Lucas straight in the eye, and declared, "No, Tyler, medicine is not for people like you, people who are dyslexic." Dr. Lucas knew better than to accept the biased view of his teacher. Although I'm not sure what his advisor is doing today, I do know that Tyler is now a well-respected and highly successful orthopedic surgeon practicing in New York City.

In thinking about your child's future, keep in mind the whole child. Try to see beyond the glaring misspellings, bad penmanship, and awkward typos and look at her ability to reason, connect the dots, ask out-of-the-box questions, and create highly imaginative stories, characters, and games. If you or her teachers limit your vision to the superficial, you will restrict her opportunities and relegate your child to a lesser life than she deserves. In reviewing many dyslexic Yale students' personal histories, I am struck again and again by how a given successful student struggled and was often underestimated but, most typically with the strong support of her parents or a teacher who became her unwavering champions, went on to accomplish what years earlier would have been deemed the impossible. Don't underestimate your child's ability. Don't relegate her to second-class status when she, with your ongoing support, could excel at high academic levels.

Foreign-Language Courses. Dyslexia genuinely and powerfully interferes with a student's ability to master foreign languages. As I've noted, having to learn a foreign language involves a dyslexic's weakest area: getting to the sounds of spoken language. Much depends on the nature of the foreign language offered by the school. For example, Spanish, having a much more predictable linking of sounds to letters, may be more palatable to dyslexic students than French, in which, as in English, a letter can have multiple pronunciations. Some students find learning Latin prefixes and suffixes (which are also used in English) helpful in enriching their English vocabulary. Important, too, is *how the language is taught* and *what the expectations are.* For example, is English spoken in the class, with the focus mainly on learning vocabulary and watching films in the second language, or does the course accelerate rapidly to the point that the class is conducted almost entirely in the new language?

To try to accommodate a dyslexic student's predictable difficulties in foreign-language learning, you might ask if the course can be taken pass/fail. If this is not the case, I would strongly recommend requesting a partial waiver of the foreign-language requirement in favor of taking a course in the culture or history of a country that does not speak English. While this accommodation is becoming more common in colleges, high schools, for the most part, are still lagging behind. Hopefully this will change in the near future.

> *CAUTION:* Be aware that students enrolled in an International
> Baccalaureate program in high school must take a foreign lan-

guage (excluding Latin) and typically are not permitted language waivers.

In my experience, the initial offerings of a foreign language in high school are often approachable and not very demanding, even for dyslexic students, but again, this very much depends on the school. Many students have found it best to take the language course during the summer, when it is the only course they have to focus on, but only after first making sure that the school will accept this for credit.

ASL. An alternative that some students have pursued is to take American Sign Language (ASL), if offered by their school or elsewhere during the summer, when the time required is not an issue. CJ, a ninth-grader with dyslexia, was required to take a foreign language. Thinking that not taking a foreign language in high school would limit his college options, he searched for an alternative. He found an ASL summer course offered at a local community college and decided to enroll. Each class was held for three hours, three times a week, for six weeks. Only ASL was used in class.

CJ found learning ASL fairly easy. In each class he was given forty-five to fifty vocabulary words to learn. His teacher would sign in front of the class and he would write down the word or phrase that was signed. The next time the class met, these words were reviewed and then a new set of forty-five to fifty words was introduced. This pattern of instruction was fairly consistent, and CJ learned over eight hundred words during the summer. He found learning sign language easier than learning English vocabulary because ASL is so visual: hand gestures depict words and their meaning.

While CJ completed the course with an A, he is not able to sign fluidly. He is capable of expressing his ideas if he signs slowly, but he struggles to keep pace with anyone who knows sign language well. He believes that his ability to sign can be equal to that of his classmates and he plans to take ASL again next summer, and the one after that, in the hope of receiving a foreign-language credit.

CJ's experience with ASL stands as a positive recommendation for those students who are considering a foreign-language option. Some students who enroll in ASL courses actually find they enjoy it and soon find themselves thinking about minoring in ASL, along with a future career that would take advantage of their knowledge.

To sum up, when considering academic planning, none of us has a crystal ball to guess our child's or student's future. In thinking about a child's potential, what we can and should do, however, is not give credence to misleading and superficial clues and ignore more powerful but often less visible clues. Here, I am referring to a child's ability to think, to reason, to generate great ideas. As children, economist Diane Swonk and financier Charles Schwab were not linear but more abstract thinkers—a rare and enviable ability, which may be lost on many, including parents and teachers. Don't underestimate your child or his or her future possibilities. Aim high, have confidence in your son or daughter, convey that belief, and make sure that any academic paths are not prematurely closed off in high school.

Beware of well-meaning schools and faculty who may prematurely lower expectations and route children to a less challenging and ultimately less rewarding pathway. Instead, share with your child the stories of the many successful adults who also struggled early on in school and who, through belief in themselves, hard work, and persistence, were able to gain confidence in themselves and go on to realize their dream. View the Yale Center for Dyslexia & Creativity's public service announcement (PSA) and smile as you learn how a very special sixth-grade teacher, Mrs. Bennewitz, was able to recognize her student Patrick's dyslexia as well as his high ability and gave him that confidence. As a result, Patrick graduated from Georgia Tech, formed his own successful company, and now holds more than twenty-five patents. (Link to the PSA at dyslexia.yale.edu/dyslexia/dyslexia-psa/.)

ORGANIZATION

The transition from high school to college can be a shock to the system if a student is not prepared for the new responsibilities, not only in academics but those that are part of living on his own without parental support and reminders. Because the student is the active participant in this, I'm going to address her directly.

You might consider this your *Personal College Prep 101*. All students experience a kind of disequilibrium when they first enter college, and dyslexic students often more so, because for the first time they are truly on their own, without the physical presence of parents and obvious school support. Happily, much of this can be prevented and certainly lessened by preparation that goes on during high school. You might even think of high school as a prep school in this regard.

How you manage your time and work as a college student, how you make sure to organize and prepare for both the academic work and the activities of daily living, is vitally important. If it seems I am veering from the subject of dyslexia, it's only because time management is a part of overcoming it. To begin at the beginning, let's take a look at getting up in the morning. Now is the time to practice getting up on your own on school mornings, if you don't do so already. Alarms are easy to access on your smartphone, and many students find this works well for them. The important thing is to figure out what works for you.

→ *Set an alarm.* Once you are awake, is everything ready to go? Are the books and papers you need for today loaded into your backpack?

→ *Make and keep a schedule.* On a calendar, list your class meetings, when assignments are due (including longer-term projects), after-school events, and so forth. Get in the habit of actually using it. Look at it at the beginning of the day, enter new assignments as they are given out, and mark those that are completed. Develop a model that you are comfortable with and will use.

→ *Use helpful apps.* Start to become familiar with a growing number of apps designed to help college students make and actively use to-do lists. A very popular app favored by students and faculty is Wunderlist. Use it!

→ *Make time for yourself.* Once in college, students find it extremely helpful if they can *adhere to a relatively regular*

Figure 47. Taking Responsibility

schedule, one where a fairly consistent time is built in to work out, do yoga, run, dance, or engage in any hobby or amusement you choose. Having a constant look-forward-to time to get away from the anxieties and worries and be free to let loose and do something for yourself should be part of a daily routine. Getting into the habit of taking time to go out of your room and exercise will be welcome, indeed essential, once you are in college and you find yourself sitting in a room or library with hours of work to do. It will allow you to relax, feel better, and work more efficiently. Start doing this in high school.

→ *Try to master large, long-term projects.* The reports and projects you do in high school will become even more complex and expansive in college. High school offers a terrific opportunity to learn and become more adept at the organizational skills needed to approach and master such projects.

A major principle here is to think about the goal of the project and the components needed to complete it. Think of taking one bite at a time. I refer to the process as *thinking* and *doing.* A useful approach is first to conceptualize the project and divide it into meaningful components, listing each section or topic. For the *doing* part, make a time line and mark on your calendar proposed dates for completing each section. For each step you want to list *what* needs to be done, *when* it will be done, and *how* it will be completed. *As the project moves along, any external support should begin to be phased out.* Note-taking programs (described on pages 432–433) can make a major difference. OneNote or Notability allows you to easily group notes relating to specific topics. As each stage of the project is completed, treat yourself to some reward, like a favorite snack or time to play a computer game or briefly shoot hoops or do yoga.

→ *Use flash cards.* Here's a suggestion that Allison, a brilliant dyslexic young woman, shared with me: She uses *flash cards to help plan and organize* her work. You can use Post-its or index cards for the same purpose. One approach is to divide the topic into major areas and have a flash card for each and then arrange below each major card a series of cards, each containing a different subcomponent of that specific area. Once you're in college there are going to be lots of reports to do, and if you have already mastered the approach, you'll have a template to work with no matter what the topic is. Many dyslexic students also find making the flash cards and then reviewing them over and over again very helpful in making material stick. In preparing for a test, students visualize the cards and then during the exam find these images easy to retrieve.

Once you are in college, you will be very happy you did the necessary prepping.

PERSONAL

Switching back to parental concerns, high school is an ideal time to clarify and resolve any remaining issues concerning your child's acceptance and understanding of her dyslexia, and that it is a lifelong condition.

→ *Understanding the benefits of an early diagnosis.* Be reassured that your child brings with him to college the major advantage of early identification. He knows who he is and has adapted to his dyslexia. This typically means he has had support not only to learn but also to try out and optimize helpful learning strategies. While he is still in high school, he has the opportunity to work on activities that will build his confidence and chances of success in college. Earlier-identified students enter college having learned that they are not dumb. Most important, they have the benefit of having growing confidence in their abilities, a confidence that makes a world of difference once they enter college. They embrace the mantra *This is the way the world is, and this is the way I have to operate*—and they do!

→ *Fully understanding dyslexia.* Talking about dyslexia should be an ongoing conversation over time. Make sure your child understands what dyslexia is, that it is an unexpected difficulty, that dyslexia is a paradox with both weaknesses and strengths, and that it is a lifelong condition. Reviewing the yellow card (see pages 98–99) detailing the contrasting symptoms of dyslexia when your child is still in high school is often a great conversation starter. You also want to make sure she has her copy of this card.

Many have found it helpful to role-play, with the child and parent taking turns asking each other about how dyslexia has affected them. This helps to solidify the student's understanding of herself, providing very helpful practice in answering a question she will be asked on many occasions when she is in college—by the disability office staff or a professor—and even afterward, perhaps when applying for a job. At these times, she will be glad to have a high level of understanding of her dyslexia.

→ *Clarifying any other diagnoses or concerns.* It would be wise to make an appointment for your child to have a complete physical examination, including vision and hearing tests, during her junior year in high school. In this way, if a problem is detected, it can be addressed and treated prior to her going to college. The fewer the distractions, the better.

→ *Clarifying psychological issues.* There are several conditions that co-occur frequently with dyslexia, including attention-deficit/hyperactivity disorder and anxiety disorder. This is the time to be sure of the presence or absence of these conditions, and if present, to ensure optimal intervention, including

  • *reviewing medication your child is taking* for ADHD or anxiety to make sure it is necessary and that she is on an optimal dosage. This is a time when medication adjustments can be made, their effects observed, and any needed changes implemented before college. Of course, any changes should be discussed with the prescribing physician.

  • *considering nonpharmacological therapies.* If your son is anxious, you may want to consider whether other treatments may be helpful, such as cognitive behavioral therapy (CBT). CBT is aimed at changing a person's current way of nonproductive thinking and behavior, a change that often is transformative, producing positive improvement in his outlook and mood. If your child appears stressed, another approach is "mindfulness"— basically a state of awareness, of paying attention to what is happening at the moment without being judgmental. (See page 424 for a more in-depth discussion of anxiety and mindfulness.)

All these steps can make a huge difference for the dyslexic student entering college.

As you get ready to leave for college, you may want to keep in mind the thoughts of a wonderful, brilliant young woman, Hallel, as she prepared to leave for her freshman semester:

Next week I will be going to college and the person that is going could only be at the place she is with your help and diagnosis. [Hallel was recently diagnosed at our center.] Thank you so much again. You and your husband helped me change my story. Thank you doesn't really begin to describe how thankful I am that you came into my life. With your diagnosis, I was able to stop telling and believing a story that I felt was true at its very core. I was able to believe that another story could be told. You have changed my life and allowed me to see a different future—a future I am so excited to discover. Thank you for helping me to see myself. Now at the edge of the next adventure it's really nice to be reminded of the steps, challenges, and triumphs it has taken to get here.

SELECTING A COLLEGE

As your child progresses in high school and enters his last two years, thoughts turn more seriously to college, specifically to his choice of school(s) and the upcoming high-stakes standardized tests such as the PSAT, the SAT, the ACT, and AP achievement tests.

A BIT OF PERSPECTIVE ON COLLEGE ADMISSIONS

College choice involves many types of issues and preferences. My focus here is on specific issues that should be considered by a student who is dyslexic. Overall, you should aim high while at the same time giving yourself a broad range of choices. If you have an academic and personal record of achievement (even with a few inevitable disappointments here and there), don't be afraid to apply to rigorous schools, including those in the Ivy League.

My own institution, Yale University, understands dyslexia and welcomes deserving students who are dyslexic into its undergraduate, graduate, and professional schools. When I first came to Yale, I was invited to meet with the admissions committee and did so regularly. At these meetings I would review the admissions packets of students applying to Yale who may have in some way indicated or whose records suggested that they had a learning disability, most often dyslexia. My role was educational: to bring to the attention of committee members the rapidly growing scientific knowledge of dyslexia and to ensure that aspects of dyslexia such as slow reading or needing additional time on tests were not misinterpreted as an indicator of cognitive limitations. The admissions dean and committee members came to appreciate the unexpected nature of dyslexia as well as the challenges it presents. The admissions process was successful because the committee was determined to admit the best students and the university was committed to providing the support that would ensure that dyslexic students' performance on tests reflected their ability rather than their disability. Yale has the same high rigorous admissions standards for dyslexic students that it has for every other applicant. Having visited other universities, I have come to appreciate that the key to Yale's success with dyslexic students is its genuine understanding of the paradox that is dyslexia. The university does not misinterpret and get waylaid by such characteristics of dyslexia as slow reading. Rather, Yale takes into account the whole person, including the often incredible strengths of deep intelligence, out-of-the-box thinking, persistence, and grit that the dyslexic applicant brings to the school.

Yale, too, has been very fortunate to have extraordinary, visionary leaders such as Linda Lorimer, who in her roles as university secretary, university vice president, and most recently senior counselor to the president vigorously supported and continues to strongly support the university's commitment to excellence. Although she has now retired, her advice and perspective continue to be sought out.

Happily, admissions officers in colleges and universities around the country are demonstrating increasing understanding of dyslexia, including the potential of these students to succeed in higher education, and indeed in life. Our Yale Center for Dyslexia & Creativity has been honored to help foster this increased understanding by sponsoring two conferences, one at Yale, the other at Stanford, both titled "Diamonds in the Rough: Why Every Institution of Higher Learning Should Want to Attract and Admit Dyslexic Students." The audience for these events consisted of the deans of admission from colleges across the country who benefited from hearing from such eminent dyslexics as Cleveland Clinic CEO and renowned cardiac surgeon Dr. Delos (Toby) Cosgrove; then California lieutenant governor, now governor, Gavin Newsom; and financier Charles Schwab, each of whom described his early struggles with dyslexia and his current successes. The attendees also heard about the results of the Yale Outcome Study, which demonstrated that when followed into the working world after graduating from Yale, dyslexic alums performed extremely well in virtually all areas.

Hearing one after another dyslexic student and adult compellingly describe how they had succeeded in college was quite meaningful to the attending deans. Greg Buckles of Middlebury and McGreggor Crowley of MIT, along with many other attendees, indicated that they gained a new and deeply valued understanding of dyslexia, including insight into how best to evaluate the potential of dyslexic students to succeed at their respective schools.

STEPS IN SELECTING THE RIGHT COLLEGE

Dyslexics, like any other students, will want to consider location of the school (urban versus rural), school size, and courses and majors offered. There are also special factors that enter into the college selection process for students who are dyslexic.

RESOURCE OFFICE FOR STUDENTS WITH DISABILITIES*

It may surprise some to hear that a majority of dyslexic students, and I, too, would *rank the disability or resource office right at or very near the top of the list of special considerations in selecting a college.* This office and its staff can make a meaningful difference in the success or failure of a dyslexic student at a particular college. When functioning optimally, such offices serve as a welcome and highly effective bastion of support for a dyslexic student.

Given its special role, what should a potential student look for in assessing the resource office at a particular college? As a first step, check out services for students with disabilities at your top-choice schools. If you have difficulty finding the resource office, go to the school's website and type in the word *disability*, or simply Google the school with the search word *disability*. To quote Judy York, former director of the Yale University Resource Office on Disabilities, "If the school doesn't advertise a disability office, beware. Be scared." Indeed, some schools still do not have a formal office. If the school doesn't identify an office, that suggests that the college doesn't value it or give meaningful support to it.

* Resource offices are undergoing a transition in some universities. For example, as of September 2019, the Yale Resource Office on Disabilities is the Office of Student Accessibility Services. The goal is to create a more inclusive university.

Izer's experience of support and personal interest at Penn contributed to his positive experience at the school. As he put it while he was an undergraduate,

> They're great. They treat me as if they are my mom. Pam, my disability staff person, always greets me with a hug. "How are you—anything else you might need?" She sort of pushes me, just because she really wants me to succeed and do well. That's the sort of rapport I've gained with Pam.
>
> At the beginning of my freshman year, sometimes if I forgot I have to sign up to get accommodations for my exams, she would actually take the time to email me and say, "Izer, I haven't seen that you are signed up."

This attitude from the resource office serves students well. As one resource staffer explained to me, "As long as you come in, you show that you want to be successful, you're going to work hard, you gain." As Izer noted,

> It's good to visit the person who is in charge of your accommodations just to check in and say hello. It sends the message, *I'm here, I'm actually here. I really want to do this, to succeed, I want to get as much help as I can.* Then it looks good on your behalf, and they're more willing to help you.

→ A very clear message emerges: *Engage in personal interactions with resource personnel as often as you can.*

You should look for the following in judging a resource office's ability to meet the needs of a dyslexic student:

→ *Resource office location.* Accessibility is critical. Is the office conveniently located so that you can easily drop in during the day as you go to and from your classes or dormitory, or is it at the periphery or edge of the campus so that it is not easily visited? If you are dyslexic, you do not want to attend a college where a resource office costs you time instead of saving you time. Penn and Yale are exemplary in this way. *The location of the disability office, central or peripheral, sends a message about how the institution regards disability services.*

→ *Office staff and policies.* Learn how big a staff the office has and what services it provides.

→ *Make a visit.* The resource office should be one of your first stops when you visit a campus. Set up an appointment beforehand to meet with a staff member. You do not want to simply drop in and find the staff fully occupied. You must learn about what accommodations are available and how they are obtained, and if and how foreign-language waivers are available.

→ *Determine whether this office will work for you.* Are the people at the resource office welcoming? Is the atmosphere warm and friendly? Would you feel comfortable going there? Does the office look personal and inviting, or does it more resemble a visit to the DMV? Ask students you see at the center how they would sum up their interactions with the staff. Can you have one-on-one interaction with the same person over time if you want? Often the size of the office is a direct function of the number of students attending the school. For example, at Brown University the resource office houses a staff of five; the undergraduate population of Brown is about 6,400. Seventeen men and women staff the disability office at Columbia University in New York City, which has 14,000 full-time students.

SCHOOL CULTURE

Just as school culture is important for dyslexic students attending K–12, it is equally important to these students when they move on to college. When meeting with the disability office, be sure to ask about the culture of the school, especially the culture of the faculty. Would you want to disclose your dyslexia in this environment? You want to know how accessible the professors are, since in the right culture they can play a major role in your life. Professors often devote what some disability office staff have described as "an inordinate amount of extra time, voluntary time, working with students"—meeting after class, going over assignments, guiding the student's approach to the material, supporting the student in multiple meaningful ways.

As Emma, a highly successful dyslexic student at and now graduate of Washington University, says about Cornerstone, the disability services program at her school, "You need to have a good support system for accommodations. If I didn't have Cornerstone, it would have been horrible. They're so friendly and really fast about responding and taking care of things I need. Cornerstone has made a huge difference for me."

STANDARDIZED TESTING

In addition to learning about the resource office and school culture, you will want to consider whether the school offers *optional standardized testing*. Does the school *allow the student to decide whether or not to submit his or her scores on the high-stakes standardized tests such as the SAT or ACT or to take the test at all?* A surprisingly large and increasing number of fine colleges and universities offer optional testing requirements. Growing scientific evidence indicates that students who choose to submit these scores and those who choose not to, do *not* differ significantly in their GPAs and graduation rates.

In the study "Defining Promise: Twenty-Five Years of Optional Testing at Bates College (1984–2009)," presented at our "Diamonds in the Rough" conference, longtime prior Bates admissions head Bill Hiss and coauthor Kate Doria presented data indicating that the number of students applying to Bates who chose not to submit standardized test scores has risen from around 30 percent in 1990 to over 40 percent, and in some years close to 50 percent, of applicants. The difference in GPAs between submitters and nonsubmitters in that report is a minuscule 0.03 of a GPA point (3.16 compared to 3.13). The difference in graduation rates is 1 percent. To quote the authors, "In a word, in a college with very demanding academics, nonsubmitters have earned the same grades and graduated at virtually the same rates as submitters." In a far larger follow-up study involving 123,000 students attending thirty-three different institutions, Hiss and coauthor Valerie Franks replicated these findings: There were virtually no differences in GPAs or graduation rates between those submitting and not submitting standardized test scores. These findings are obviously relevant for dyslexic students, who often do not perform well on high-stakes standardized tests because of difficulties obtaining appropriate accommodations. As of 2019, over one thousand four-year colleges and universities did not use the SAT or ACT to admit bachelor-degree students. You can go to www.fairtest .org/university/optional to view the most up-to-date list of test-optional schools.

If you are not planning to apply to a school offering the choice of optional testing, you will have registered for either the SAT or ACT, or both, and requested the accommodations necessary for you to demonstrate your ability. The often unexpected delays in reviewing and granting such accommodations should make it a priority for you to get your application for accommodations in as early as possible. The College Board revised the SAT in 2016; a major change was to make the math section text-heavy. This has the unfortunate potential consequence of having students' math scores reflect their slow reading rather than their math ability.

Some students have found it very helpful to review as many questions from prior exams and review books as possible, while others rely on private tutors or specialized test prep companies such as Kaplan and the Princeton Review for directed practice. Review and practice are the key and should be incorporated into plans beginning the summer before the student's junior year, or early in that year. (Please see pages 471–485 for a discussion of changes in requesting accommodations on major tests, including PSAT, SAT, LSAT, and USMLE.) For those who worry about applying to colleges that continue to require SAT or ACT scores, keep in mind that many schools look beyond these scores. As former Connecticut governor Dannel Malloy, who is dyslexic, will tell you, "I was lucky some schools were willing to take a chance on me. My SAT scores were abysmal."

VISITING A SCHOOL

Having done your research and selected possible schools, the next step is to visit each of the schools. I have already discussed the essential visit to the disability office and speaking with a staff member. But of course you will want to see more than one aspect of a college where you hope to spend the next four years. Virtually all schools offer a formal tour, typically led by students attending that college. This gives you a chance to literally see the school and begin to gain a sense of its history and traditions, and to breathe in the atmosphere. Many students have told me they learned the most from an informal visit they carried out on their own. As one student said to me, you are able to get the backstory, a truer vision of the school. During such an informal visit you should speak to a number of students and ask questions about their experiences at the school. It is especially valuable to sit in on classes. This often requires little more than approaching the professor before class to say that you are considering applying to the school and want to see a class in progress. That way you can obtain a sense of whether the professor seems to know or have some kind of relationship with the students. For a dyslexic student, it is especially important to have one-on-one time with professors and have the professors know your name—all of which is much more likely in a class with a smaller number of students.

TYPE OF SEMESTER SCHEDULE

While most schools follow a traditional two-semester school year, some are geared to a trimester (such as Dartmouth) or even a quarter system (such as Stanford). It is important for you to know that things move much more quickly in the trimester and quarter schedules—especially in the quarter system—since each semester is shorter. Reading assignments must be completed quickly, tests come often, and much is covered in a relatively short time. If you need materials from the disability office—for example, text that is digitized to allow accessibility for text-to-speech programs—there is a lot of pressure to have this completed in a hurry. Quarter systems allow very little leeway for getting behind in homework; by the third week of classes, students are often already taking midterms. Having such a fast-paced schedule increases the difficulty and pressure for dyslexic students, though I know of several who have attended and graduated from each of these institutions and been very satisfied. You should not necessarily rule out attending such an institution, but do consider this cautionary note together with other features of the school in determining if the school is the right one.

WHAT KIND OF COLLEGE WON'T WORK FOR A DYSLEXIC STUDENT?

In general, dyslexic students do best in a smaller school, where they can make connections with their professors, meet with them, and give them the chance to get to know them personally. Students who are dyslexic would do well to avoid a very large, impersonal college, a college where each student may be just a number and just another test file and where there is such a heavy bureaucracy that it takes an inordinately long time to get a response to any request. A student who is dyslexic thrives within a smaller community of people who want to help and try to understand the problems he or she may have. Dyslexics, who themselves are often quite empathetic, do better in a more personal, caring environment than in a rote, mechanical one. That said, there are students who have found great support in large state universities, and others who have felt lost and not very well supported in smaller colleges. As in everything else in life, it is the individuals you meet and form a relationship with who can make all the difference in the world.

THE APPLICATION ESSAY

Should you mention your dyslexia or not? While there is no hard-and-fast rule, a growing number of students are deciding to talk about their dyslexia on their application. Dyslexia is part of who they are, and as one student wrote, "My diagnosis of dyslexia has enabled me to understand myself and to know how I best learn and to take advantage of this self-knowledge and go on to succeed in my studies." One student acknowledged that "asking for help is one of the hardest things to do, but once I understood my dyslexia I learned how to adapt and that, if needed, asking for help is a positive thing. Learning not to fear asking for help allows me to grow even more as a student who seeks knowledge."

In formulating your essay, think about providing examples of how having and understanding your dyslexia has been a positive influence in making you more determined and unafraid of temporary reversals, and has taught you that with determination and unfailing effort you can succeed in just about any challenge brought before you. As one dyslexic young woman shared with me, "Dyslexia has taught me to be efficient, hardworking, and not to take anything for granted. I am a realist when it comes to knowing what I have to do and for making sure I do it."

Here is the last paragraph of the college application essay of Blair Bucshon, now a sophomore at Cornell University, in which she synthesizes and powerfully expresses what being diagnosed with dyslexia has meant to her. It has the positive thrust all such essays should have.

> Last year, I was diagnosed with dyslexia. Finally putting a name to the problem has helped me to better understand how my brain works and the advantages I have because of dyslexia. I know the barriers that stand in my way and can conquer them rather than fighting an invisible enemy. I now recognize that while growing up with dyslexia was difficult, it was not a totally negative experience. I discovered different ways of solving problems that my peers were not able to see. I hope to give others with similar struggles the confidence to succeed like I have. I am blessed and consider myself unbelievably lucky to have the life I do, but I also believe that everyone has a personal journey and hardships to conquer. The

struggles of my first eighteen years have taught me how to persevere and the importance of believing in myself when times are tough. Now that I understand my weaknesses, I will never again hide in the back of a classroom, or lecture hall, or board room. Instead, I vow to myself that I will always be the one to stand up first, so that I can conquer my fears and make my best self better every day.

Legally, schools are not permitted to ask if you are disabled or the nature of your disability, so it is not mandatory that you disclose. However, as one admissions officer at our "Diamonds in the Rough" conference said, "It helps when you identify that you are dyslexic, because it often explains the applicant's story. That's a real benefit to the student."

28

IN COLLEGE: GETTING STARTED

Be prepared. Beginning college is a big change for all students and even more so for students who are dyslexic, such as Elise.

Elise explains:

> You don't realize how comfortable you are in your old routines until they're taken away and you're in a completely new environment. For example, you're not going to be home by yourself doing homework, but you're going to be in the library doing work with people who don't have dyslexia. You become aware of things you as a dyslexic have to do that nondyslexics don't have to do. At home you didn't see that others didn't have to do what you must do—get a handout and have to go get it scanned; other people don't have to do that. You didn't see them just whipping it out and reading it very easily.
>
> It comes more at the beginning of the semester, when you have so much more work to do. You have to look at the books and find out how you're going to get them transformed from text to speech, and just all the extra little things that people who have dyslexia have to do to get the alternative text. So when you're going to the

resource office and all your friends are having dinner, you're like, *Well, this sucks.*

Yet at the end of the semester Elise got two A's and two A–'s.

TRADEOFFS: ACADEMIC AND SOCIAL

There are only so many hours in the day, and as you breathe in the excitement of being in college, there is so much you've been thinking about doing. You're determined that even if schoolwork takes you much longer than it takes your classmates, you still want to have new experiences, make new friends, be a part of a group, socialize, attend events, and perhaps try out for a team. The trick, of course, is balancing all these activities. (This is one reason it is so important to get yourself as prepared as possible during high school.) As Julia, a college freshman, told me,

> This semester, my second at college, has definitely been hard, but socially it's been better. It's a weird tradeoff. I joined a sorority. I know that in college I wanted to be around people more and have a bigger group of friends than I did in high school. It's so exciting! I'm doing all these things that I would never have expected I would do, which is what I also had as my goal to do in college.

Vito is a sophomore and captain of the crew team at Iona College, where he devotes four hours each day to crewing. His advice to incoming dyslexic students very much aligns with Julia's:

> I would say do a sport, because if you don't do a sport, then all it is is academics, and then you're missing out on so much that college is all about. So do a sport, do a club, make friends, do something. All the time. You won't be sorry.

Here's what you can do to make your college experience work for you. Be sure to:

→ *Reach out to the resource office early.* Once you've finalized your college choice, make it a priority to contact the school disability office. A good time to do this is *early summer,* before school begins. Once school is in session and students begin streaming in and course selection begins in earnest, the college scene becomes quite hectic, with much confusion and very busy staffers. Introduce yourself to the staff and meet with them. This is a good time to help the office get to know you, your educational experiences, your dyslexia, and which accommodations have and have not been effective. Ask what documentation is required and what method—mail, email, in person—is preferred. Provide the office with the report that established your diagnosis and include recommendations for accommodations. Ask any questions you may have. Learn if you need to get further testing; this can be completed prior to the start of school. The office might recommend accommodations you have never considered that turn out to be quite helpful.

Find out what the resource center can and cannot provide for you and, equally important, what is expected of you. The kinds of information you are seeking might include:

- → how to arrange extended time for tests and what specific accommodations for tests are provided in addition to extra time
- → whether the school offers foreign-language waivers and the process for obtaining one
- → how to apply for a note-taker for your classes
- → how best to arrange for text-to-speech, if you find it helpful
- → whether you can bring snacks or drinks to tests that, with extended time, can go on for hours
- → whether you can use a laptop in class and on tests. What about the use of a calculator?
- → whether the school offers tutoring or coaching or help with organizational skills.

Here's a checklist that you can use when meeting with the resource office staff. Add other items that you have an interest in.

ACCOMMODATION	HOW ARRANGED	DATE REQUESTED/ APPROVED	NOTES
Note-taker			
Extra time			
Separate quiet room			
Use of a laptop—class			
Use of a laptop—exams			
Text-to-speech			
Foreign-language waiver/ substitution			
Use of a calculator			
Writing assistance			
Tutoring			
Coaching			
Organizational skills			
Other			
Other			

Policies regarding accommodations in college are pretty straightforward. If the school receives federal funds, that institution is required to provide reasonable accommodations to a student with a disability. The student's evaluation is reviewed to determine if it is complete, timely, and meets the school's accommodations guidelines. There is no firm published rule concerning how recent an evaluation must be. In a case involving accommodations and Boston University, a judge ruled that an assessment taken as long as five years before should be acceptable. According to the Association on Higher Education and Disability (AHEAD), documentation of the need for accommodations can take a variety of forms, including the student's description of her own experiences with test-taking and/or documentation from reports, school records, assessments, or an Individualized Education

Program (IEP) or 504 plan (see pages 468–470 for details). The student should give serious thought to relaying which specific prior accommodations were effective and which were not.

Should a student have to retake a test or have an evaluation for the first time, be aware that this can be quite expensive, often running to thousands of dollars. Some schools provide an evaluation at no cost. Other schools offer financial support. Many schools offer neither and the student must cover the cost himself.

ACCOMMODATIONS

Although there is variability among colleges and within specific majors, overall the most common accommodations provided are

→ *A note-taker.* The process is straightforward. Another student in the course is recruited as a paid peer note-taker. Most often it is the professor (not knowing the identity of the recipient) who makes the announcement. When completed, the notes are sent to the resource office, which in turn distributes them to the recipient. Confidentiality is maintained at all schools so that no one other than the recipient and resource office staff is aware of who is receiving notes. Typically the note-taker uses a laptop, ensuring that the notes are legible and easy to read.

In the near future many classes may well be recorded. This optimizes a professor's time, eliminating the need to repeat a lecture while allowing students to play it back as often as needed. Lectures may be recorded visually as well. Recorded lectures would obviously be of special benefit to students who are dyslexic.

In some institutions the future is already here. A service recently established at George Washington University (GW) in Washington, DC, is known as Lecture Capture. Professors can request to be enrolled and scheduled in a Lecture Capture classroom, where their lecture is automatically recorded and distributed to their students. Some of these classrooms also possess video recording capabilities. GW uses the technology wisely, with both students and professors benefiting. For example, students can bookmark these lectures so that they can easily return to sections for further review or clarification. Professors can monitor how many students viewed a specific lecture, how often they viewed the video, and how much of it they watched. Professors feared attendance would be adversely affected and are relieved that this has not been the case. In fact, they feel that students are getting a much deeper understanding of the materials more quickly.

→ *Extra time with a separate, quiet room for exams.* Once your evaluation has been reviewed and approved by the resource office, obtaining extra time for exams is straightforward. Additional time ranges from 25 percent to double time; typically time and a half is recommended and provided.

→ *Use of a keyboard (basic laptops with Word and no Internet access).* Use of laptops particularly, and at times only, for exams is becoming increasingly accepted. Both professors and those who grade the tests appreciate being able to read what is written easily rather than having to deal with bad or hurried handwriting. An additional advantage is that the test responses placed on the computer are saved, ensuring that those exams will not be lost. There are advantages for the student, too: She can cut and paste and reorganize what she has written and do so much faster than by hand.

Use of laptops in class is an entirely different issue. Professors are increasingly disallowing them during lectures, especially if they take place in a large lecture hall, because they have found students abusing the privilege by sending and receiving emails and looking at Facebook. You can understand how professors find that to be rude and discouraging. However, exceptions may be made for dyslexic students who need their laptops for taking notes. As in other areas, there is great variability among schools, and even within schools among professors.

→ *Text-to-speech recorded books.* Federal law and regulations, including the Chafee Amendment to the U.S. Copyright Law, Section 504, and the IDEA, support and allow the production and use of recorded printed material to be converted into audio for students with print disabilities, including dyslexic students. Furthermore, Section 504 and the Americans with Disabilities Act (ADA) both prohibit discrimination against those with a disability. Under these laws, states and local districts must ensure equal educational opportunity to students with a recognized disability by providing *accessible instructional materials*, often referred to as AIM. The resource office of most colleges has a mechanism for converting text to speech.

Entering freshmen find that college requires a surprisingly large amount of reading. There are a number of approaches to dealing with this onslaught, including

- making sure not to register for too many heavy reading courses in one semester
- developing the knack of skimming and understanding that not every word in an assignment has to be read
- using text-to-speech software. Digitized texts are available from the publisher, but given that students are often rushed from the time of shopping to the time of finally registering for a course, it simply takes too long to obtain the digitized text from the publisher before classes begin. Yale's functional adaptation to this predicament has been to send the books off to the school's printer, where they slash the binding and high-speed scan the pages into a digitized format. Students seem to be very satisfied with this arrangement.

At the University of Pennsylvania, if the book is available in audio format, the disability office will help a student approved for text-to-speech get access to it through the publisher. If not, the office will convert the textbook or PDF into audio format. The office will also convert a pamphlet or similar material into an audio format. Cornerstone at Washington University and the Office of Accessible Education at Stanford University have streamlined the process a bit further. At Stanford the software is distributed so it is all on the mainframe computers and readily accessible to the students. These institutions are exemplars of what students can and should expect from universities across the country. Not all colleges are on board with services such as these, but many more than you might think are. Services for students with documented dyslexia are moving forward at an incredible pace, and the vast majority of institutions are working to understand and respond to the needs of these students.

Be aware that students have very differing opinions of the usefulness of text-to-speech software. Some students suggest that its availability allows them to begin to deal with the tremendous and often unexpectedly intense reading load they are faced with as a college student. However, as one resource office professional said, "Many students who are early-diagnosed don't gravitate to it because it's one more thing to add to the mix of their life. They don't see the advantage of it." Some students describe "zoning out" while listening. Others may use the software only for some courses, arguing that it takes too long.

One disability office professional told me,

I know that we have a larger number of students who are approved for this than actually use it. Some may use it when freshmen and sophomores and then, as classes change and the nature of what they do changes, find they have less use for this technology. Or, in classes that don't do as much reading or are project-based or discussion-based, then maybe that is the reason behind why we don't see as many students using text-to-speech technology.

My advice, based on speaking to professionals and students, is to give this technology a try, and do so prior to college entry.

FOREIGN-LANGUAGE WAIVERS AND SUBSTITUTIONS

Many dyslexic students have had unnecessarily stressful experiences in trying to have their high school understand their need to be granted a partial waiver of the foreign-language requirement. Overall you will find a much more positive experience in college, where a great many (but not all) schools grant such waivers or substitutions. It is now part of the culture at Yale to be open to such requests. The first step is for the student to submit a request to the Resource Office. This request is reviewed and then sent to us at the Yale Center for Dyslexia & Creativity for further review and, if needed, assessment. We review the student's history and any prior evaluations within the context of deep knowledge of the impact of dyslexia on the ability to access the sound system of a second language. I think our knowledge of the scientific underpinnings of dyslexia and its clinical impact allows us to feel quite comfortable in granting such partial waivers when indicated. A letter with our recommendation is sent to the dean at the head of the Committee on Honors and Academic Standing.* Once approved at this level, the student is instructed to contact the dean of his college to formally request the waiver. Together the student and his college dean then write a letter of request to the Committee on Honors and Academic Standing, which formally approves the request. The registrar is notified, because the graduation requirements in the student's transcript have to change. At this point, the student meets with her residential college dean and selects the substitute course that would be approved. The student cannot arbitrarily select her substitute course; it must be approved by the residential college dean.

At some institutions, the change in requirement is referred to as a "course substitution" instead of a waiver, but a similar process is used to obtain the substitution. First the request is made to the resource center, which helps the student apply to the Committee on Academic Standing to make the final determination. Most of the students can manage the first level of a foreign-language course, which often primarily asks the student to memorize vocabulary words. However, dyslexic students run into trouble when they have to speak and understand the language. Do your research. Some schools, like Stanford, very rarely waive the requirement. In some engineering programs, including the one at Cornell University, foreign language is not required.

It is important to know that obtaining such a partial waiver is a lengthy, detailed process. At the same time, the climate can be welcoming and understanding, with the end result that the student receives his waiver and can enroll in a course(s) that will be accessible to him, and even enjoyable.

Additional accommodations include use of a calculator in the classroom and on some exams, essays instead of multiple-choice tests, and using pencil and paper rather than answering questions on a computer screen.

INFORMING A PROFESSOR ABOUT A STUDENT'S ACCOMMODATIONS
Once a student has been approved for specific accommodations, the next step is informing the professor. To begin the process, the student goes to the

* The Yale Center for Dyslexia & Creativity's recommendation is regarded as objective, substantiated proof that the student is most likely to experience difficulty in this class.

resource office and signs a release. This release gives the office permission to notify faculty of the approved accommodations. The nature of the student's disability is kept confidential; the faculty is *not* given this information, even if they request it. A professor is informed via a form letter that the student can drop in the professor's mailbox or hand to him or her after class; the student can also take it to the professor during office hours as a letter of introduction. Most often students hand the letter to their professor after class. However, I would suggest using the handover as an opportunity to explain that you are dyslexic and why you are requesting this specific accommodation. Students are best served by interacting with the faculty member in an ongoing way, developing a rapport. I strongly advise you to personally inform your professor at the very beginning of the semester that you are dyslexic and will be using accommodations. He will get to know you that way.

Professors enjoy teaching, and especially getting to know their students. They always have office hours, and even if on some occasions you can't make their office hours, you can contact them and ask if they have any time to meet. The response will usually be a welcoming, "Sure, let's find a time together."

CHOOSING A MAJOR AND SELECTING COURSES

One of the key decisions you will have is choosing a major (always keeping in mind that a major is not fixed in stone). When I ask students, "What do you, as a student who is dyslexic, wish you had known before you entered college?" I consistently hear, "I wish I had been more confident in taking risks." Dyslexic students often feel they need to prove themselves, or are too frightened to take a risk. For instance, they choose business rather than art, which they truly want to pursue. Based on their prior experiences in school, far too many dyslexics feel they are less worthy and must somehow choose the most respected major. One such student, however, had enough courage to switch his major from neuroscience to something he was good at and that really interested him: film. As a result, he is not only much happier, he is excelling academically, and far more active socially. For the first time he is relishing his college experience.

Students typically register for four, sometimes five, courses each semester. My very strong recommendation for students who are dyslexic and entering college is to register for three courses, not more. There is so much change and so much adjustment called for from freshmen that time must be available to figure things out, to adapt, to determine how your new college life can best work for you. At most schools a student can register for three classes and still be considered full-time. You will benefit from starting out slowly, allowing yourself to feel comfortable, and gain confidence by getting a good grade. Once you gain that confidence, your life will be a lot easier.

In selecting courses, the goal is to try for a mix of demands—not too many heavy reading classes, and at the same time not a course load full of very demanding writing or term paper courses. Remind yourself, too, of what finals will be like. Ask yourself, *Will I have to face one final after another, especially with multiple-choice questions, or can I ease the stress by taking some classes with finals and others that may require a final project or paper?* Many dyslexic students have found they achieve much better results with writing papers than with answering multiple-choice questions. As one student said, "I am horrible at multiple-choice, because it requires so much reading." Dyslexic students can be especially hard on themselves to prove they can compete. That's not the wisest way to proceed. In selecting courses it is always helpful to do a bit of detective work and find out about potential professors' reactions to students' need for test accommodations. At Yale, the Nobel laureate economics professor Robert Shiller at times may actually come to the Resource Office when dyslexic students are taking final exams to be physically present and to answer any questions they may have. I am told his presence makes a tremendous difference, helping to bridge the divide separating students who require extended time and who need to take their test in the Resource Office from the remainder of the class, who take the final all together elsewhere.

For courses that a student feels a bit unsure of, it's useful to inquire if the course may be taken pass/fail. Once again, different colleges have different policies. At some institutions a student can take no more than two pass/fail courses. In some instances taking a course pass/fail is considered not an accommodation but an arrangement worked out between a student and her advisor. This arrangement has to be approved by the college administration.

OTHER SERVICES

In addition to the accommodations we have just discussed, schools offer a range of additional services. The University of Pennsylvania has the Weingarten Learning Resources Center, an umbrella service center providing a range of academic support services and programs to students through its two divisions. One, the Office of Student Disabilities Services, is geared to students who self-identify with a disability such as dyslexia. The other, the Office of Learning Resources, offers to all members of the university community professional instruction and support in strategies for reading, writing, studying, and time management, among other services. Students who are dyslexic often benefit from the services provided by each of these divisions. Professionals here, as elsewhere, have found that students with dyslexia arrive on campus thinking they are cured and don't need support when they very much do need such support. New students who present themselves to Student Disabilities Services are referred early on to the Office of Learning Resources to support and strengthen strategies they already have developed. The learning process itself is a focus; students are instructed in how to get organized, maintain their concentration, and best demonstrate what they know.

I was especially impressed by the reading support offered at Penn, even more so after I spoke to Dr. Myrna Cohen, recently retired executive director of the Weingarten Learning Resources Center and the instructor in charge of the program focused on teaching students how to be more efficient readers. Penn has a very heavy reading load, and an important goal is to make sure students are being strategic about the courses they're choosing every semester. At the same time, Dr. Cohen helps students develop strategies for efficient and effective reading. Basically she urges students to be thoughtful about their reading, continually asking themselves, *Why is the professor assigning this reading in this specific course?* and addressing this central question by considering the whole course—for example, the syllabus, the professor's philosophy, and class notes—as part of an integrated message or purpose.

Another helpful model is found at Iona College in Westchester County, New York. There the Samuel Rudin Academic Resource Center, akin to the Weingarten Center, provides a range of services to both the full student body and, more specifically, to students who are dyslexic. The College Assistance Program (CAP) is a unique, highly specialized tutoring and support service that I became familiar with when a student I know, Vito Rocco, entered Iona. Vito, who is very bright and dyslexic, is incredibly enthusiastic about the role CAP played in making his first year at Iona successful, both personally and academically. CAP, to which students must apply for admission and for which they are charged a fee separate from tuition, offers two one-hour tutorial sessions each week; if needed, more can be arranged. The tutors at the conveniently located facility are professional learning specialists whom the student meets on an ongoing basis so that a relationship develops and the tutor gets to know and understand the student's needs. A range of skills are addressed, including reading, math, writing and editing, note-taking, and test preparation and test-taking. Student and tutor form a highly interactive dyad whose focus varies depending on the student's current needs. For example, as Vito explains,

> one day I have a paper due for this class. I'll bring the paper in for my tutor to look at and advise me. He'll compliment me because I'm not too shabby at writing. But then if I make a grammar mistake, he'll be there, because everybody makes grammar mistakes, and he'll explain how to correct it. So it's not like holding my hand; it's more of pointing me where I have to go. I like that.

Vito gives another example from his history class.

> I would read the textbook with my professor [tutor]. During the actual class, we had to read little passages that were online. So I would work with the tutor to read those passages with me and take notes on that. Then when we would have in-class discussions on this material, I would crush everybody on those because I actually read it. All the kids just don't read it. And I was like, "Ha, you can read and you don't read."

DISCLOSING THAT YOU ARE DYSLEXIC

Each dyslexic student must make his or her own choice on disclosure. Vito's response was typical of many: "If they ask if I'm dyslexic, then I don't lie, but I don't bring it up." When I asked Sara if she had met any students at Yale who are dyslexic, she said, "I have one other friend whom I talk to about dyslexia, but for the most part, I think people are pretty quiet about their dyslexia."

In contrast, Izer, at Penn, was open about his dyslexia. "All my fraternity brothers and friends know I am dyslexic," he said. "I'm very open about it. I feel that by being very comfortable being dyslexic, it actually helps a lot. I feel if you don't say anything or keep it hidden, people may say something behind your back, or like, *Why did you misspell this, why was it like that?* While if you tell them, especially your friends, it's more acceptable. *Yes, he's struggling with something else that we don't have. It's understandable.*"

What are the reactions when students share that they are dyslexic? They mostly reflect ignorance. Some students even assume dyslexic students read backward.

What about when you are going out with someone—do you tell the person you're dyslexic? Here's a typical response: "Since I'm confident about it and I don't make it seem like it's holding me back in any way, I do tell the person. . . . I explain the whole story: 'No, I'm fine, I can read, I've gone through training on how to learn my own way,' and it's always been fine. This has never been something that has been the downfall of a relationship." Others feel more like, "If the situation is appropriate, I'll share my dyslexia. Really, it's not something that says I'm different or have a terrible problem. It's who I am."

FIRST-TIME DIAGNOSIS IN COLLEGE

Some students in college have never been diagnosed with dyslexia but may indeed be dyslexic. How might you decide if you, or your child, or your student, is dyslexic? As you have read, suspect dyslexia in an intelligent person who since childhood has had spoken-language difficulties, including word retrieval; who learned to read late, reads slowly and with effort, and is a terrible speller; who struggled through foreign-language classes in high school; and who found herself having to work longer hours than peers or not being able to go to after-school or weekend activities in order to keep up with academic demands. In college there are typically four distinct areas that prompt a visit to the resource office:

→ *Inability to finish tests on time.* "I'm incredibly frustrated because I know the material, and yet when it comes to tests, I can't seem to finish when everyone else hands in their paper. As a result, I am doing very poorly in several courses in which I know the material quite well."

→ *A miserable experience in foreign-language class.* Students describe trying very hard in their French class, for example, and just not being able to follow the class. Yet they have classmates who are able to think, speak, and answer questions in French with ease.

→ *Inability to keep up with the demands in some but not all classes.* Typically, unidentified dyslexic students find themselves falling behind generally only in courses that have an especially heavy reading load. Comprehending the material is not the problem; being able to complete all the reading is the difficulty.

→ *Test at-risk for dyslexia on the Shaywitz DyslexiaScreen™, Adolescent-Adult Form.* There is now a brief ten-item, evidence-based screener that you complete and immediately lets you know if you are or are not at-risk for dyslexia (see pages 406–407).

If you are or know a student who is experiencing at least one of these difficulties, or who has been screened as at-risk for dyslexia, a visit to the resource office at the college is in order. Do not wait. With a diagnosis comes self-understanding and help.

GRADUATE AND PROFESSIONAL SCHOOL

Once you are comfortable in college, you begin to think, *What's next?* The questions begin to build as your dreams and hopes for the future rise up and your longstanding doubts and fears about your own abilities surface once again. Is pursuing postgraduate study the right choice for you as a person who is dyslexic?

First, you should know that the accommodations available to you in college are also available in graduate and professional programs. If you are harboring anxieties about entering the fields of law, medicine, business, or writing, you should be reassured (see pages 489–523). I am often asked, "I don't do well on standardized tests like the GRE. How will this affect my admission?" Here's how this affected one dyslexic scientist, Carol Greider, Ph.D., winner of the 2009 Nobel Prize in Medicine and professor and director in molecular biology and genetics at Johns Hopkins University: "If UC Berkeley had done the same thing that many of the other schools did, which was to apply a cutoff [on the GRE], then I wouldn't have gone to graduate school and made the discovery of telomerase and won the Nobel Prize." Instead UC Berkeley was wise enough to look past her low GRE scores and see the brilliance within her. Dr. Greider did her Nobel Prize–winning research as a graduate student at Berkeley. When successful dyslexic college and professional school graduates share their experiences and strategies for success, their most important conclusion is that you mustn't underestimate the critical role of perseverance and tenacity.

Those two attributes are reflected in the findings from our recent study. In my experience working with admissions officers at various institutions of higher learning, the major concern expressed about admitting dyslexic students is how they will do. Can dyslexic students manage the workload? Will they be successful following graduation? From interacting with dyslexic students at my own university and from experiences speaking around the country, I knew anecdotally that dyslexic students could and in fact did succeed at rigorous universities. As a scientist, I knew that the only way to prove it was to design a study that would test the hypothesis that dyslexic students could succeed at one of the most rigorous of universities, Yale.

The study, *The Yale Outcome Study: Outcomes for Graduates With and Without Dyslexia,* was straightforward. Focusing on Yale students who had graduated five or more years before, we asked questions about both their Yale experience and their work and life experiences since graduating. Our study sample was equally divided between Yale students who were dyslexic and those who were typical readers. The alums completed a 150-question survey asking about their academic experience at Yale, level of self-esteem at Yale, employment after college, work ethic, and learning style.

The only statistically significant difference in the academic experience of the dyslexic and nondyslexic alums was more "time needed to complete schoolwork," which was reported by more dyslexic respondents. There were no significant differences in "difficulty and enjoyment of classes," "self-perceived confidence, motivation, and intelligence," or "academic stress." Responses to survey questions about participation in college life were reassuring; the dyslexic group indicated that they did not spend any less time on extracurricular activities. Compared to their nondyslexic fellow Yale graduates, the dyslexic alums showed a trend toward higher overall self-esteem when they were in college.

The same positive pattern extended into the postgraduate years in responses to questions about their employment and their work ethic and perseverance. Overall, no statistically significant differences were found between the two groups in terms of income, future outlook, relationship satisfaction, life satisfaction, or confidence in goal achievement.

Reading some of the statements of these Yale grads helps to shed light on how these young men and women were able to be so successful or at their perception of it. For example, there was a recurring focus on their strengths:

> Dyslexia taught me how to take advantage of what I am good at and learn to manage my weaknesses.

Many came to view dyslexia as responsible for their success:

> Because I read slowly, I tend to focus more on what I am reading,
> and I think that helps me absorb the material better than some.
> In my work, it is very useful to fully grasp and understand the
> material I am reading, so in many ways, my slow reading may help
> my job performance.

Perseverance was noted by almost all the dyslexic respondents. For
example:

> I don't give up easily.

> Dyslexia has helped me in the long run because I've always worked
> harder than my peers.

A different way of thinking was expressed as well:

> I find I look at the bigger picture and can solve problems.

Finding so few significant differences between dyslexic and nondyslexic
Yale alums led us to conclude, *"When intellectually gifted students with dyslexia are given academic opportunity at a rigorous institution, they can succeed
academically, professionally, and personally."*
The Yale Outcome Study is the first to examine outcome at an institution of high intellectual rigor and to do so in multiple spheres. These are
extremely powerful results, perhaps a "game-changer," as one of my associates
proclaimed. The data are unequivocal: Dyslexic students can succeed if given
the opportunity. It is critically important that this information be shared
with both high school admissions counselors and college admissions officers.
Dyslexic boys and girls and their parents need to know this important and,
to some, surprising new information. Yes, a dyslexic student can aim high
and succeed. The data clearly show that given intelligence, perseverance, and
self-awareness, there is no limit to what a dyslexic can dream of achieving,
either in postsecondary education or in the workplace.

29

HELPING ADULTS BECOME BETTER READERS

There are ample opportunities to improve the reading abilities of adults who are dyslexic, and a wide range of career possibilities are open to them. To give you an inkling of what is possible, I am going to introduce you to a man who is very dyslexic and is now in the midst of a career neither you nor I would think a dyslexic adult could be successful in, much less one in which he could advance to the very top.

Meet Brian Bannon, a man in his forties who had difficulty catching on to reading as early as kindergarten, which he repeated. His concerned mom had him tested in first grade and he was placed in special education, where he remained throughout his schooling. The family did not have great means; his father worked in water well drilling and his mom in a flower shop. Brian described his mother as being really engaged, his true champion. She strongly encouraged his involvement in theater, and, as Brian told me, helped build his self-confidence. And yet,

> no matter what she did, not being able to learn like other kids and as quickly was something I carried with me for such a long time—just the feeling of being stupid, even when you have all the advantages of an engaged parent and a lot of caring teachers.

I was probably the only kid in special ed who had a diagnosed learning disability. Most of the kids basically had behavior problems. And what's really sad about that is that I'm the only person in that group of kids who graduated from high school.

I was the first person in my family to go to college, and only because I was a swimmer; I got scholarship offers. That was so unexpected for me. I never saw myself as being good at school, so I was going to go directly into a trade. Never in a million years would I have imagined . . .

Even though I had a lot of help, I never read at grade level—but I had a desire to learn. I remember I would sit in our little neighborhood library and I would imagine all of the knowledge in those books. But I never really thought it would be possible to be able to absorb all that knowledge and information I really wanted.

In eighth grade there was a teacher who was really trying to help me. He said for all history tests we could do open-note tests. I would study for hours and hours, make notes, bring them into class—and still fail the tests. I even tried to game the system—I would transcribe all three chapters and literally come into the test with all three chapters stacked up, and still fail the test. I just couldn't read my notes fast enough to answer the questions in time. And yet all my work taking notes was a real turning point for me. I learned that I had a level of discipline and a level of work ethic that was far beyond those of my peers.

Once I got into college I started to realize that all those years bumping up against the wall—trying one thing, trying another—really instilled a mindset and habits and discipline that defined my ability to chart out a career and be successful. I graduated at the top of my class in college.

From 2012 to May 2019, Brian served as commissioner of the Chicago Public Library, with eighty member libraries and one of the best public library systems nationally and internationally. This was Brian's title until May 9, 2019, when the New York Public Library named him as the inaugural Merryl and James Tisch Director of the library system. The New York Public Library, the largest public library system in the United States other than the Library of Congress, has ninety-two locations and 53 million items. Brian Bannon deservedly reached this pinnacle in the public library world by employing his sea of strengths, recognizing the great strengths surrounding his weaknesses and putting them into play. In Brian's own words:

> Once you are able to get over that hump, what you learn is that on the other side the high-ordered thinking is there, the discipline is there, and also the ability to work harder than everybody else who's there. All of a sudden things open up.
>
> I barely imagined graduating from high school, much less college. It was only the factors of a supportive parent, caring teachers, and a university saying, *We'll pay for your college if you come and swim for us* that even got me through the door.
>
> It was a teacher, my wonderful college history professor, who spurred my interest in library science. We were studying social justice and social justice movements and I got really interested in these questions around how people acquire knowledge and information to create action in the world. It was information as a tool for social change, information and knowledge as a tool for advocacy and self-empowerment. Then she said, "Brian, have you ever thought about going on to graduate school and doing work in library science?"
>
> I laughed at her. Library science? She responded, "You know, people have a narrow idea of what libraries are like today, and you should look into it." I was really interested in this set of questions—from my own experience of feeling shut out from the world of knowledge and information because of the challenges I had reading. I started reading and learning more about what libraries are really there to do. It opened me up to an intellectual idea about information as power. Looking back to my own experience as a dyslexic person, if you can't acquire knowledge and information that is all around us, you're really at a huge disadvantage in life and in the world. So that's what got me interested intellectually in libraries.

Brian took over the Chicago Public Library at age thirty-seven, after gaining experience at a series of libraries in Seattle and San Francisco, where his thoughts about the role of libraries in modern society coalesced. He discussed these with then Chicago mayor Rahm Emanuel, who had reached out to him about a potential leadership position in Chicago. You can just imagine the excitement rising as these two civic-minded men discussed what the mayor was trying to achieve: improve learning outcomes in this city, support neighborhood growth, strengthen economic development, and encourage workforce development. The two shared a sharp focus on the library as the vehicle to support community-based learning and unlock the creativity of neighborhoods as part of a strategy to make Chicago competitive in the information economy.

I was thrilled listening to this extraordinary individual fully engaging his creative thinking and his experiences as a dyslexic to help unlock the potential of all students, particularly those who are dyslexic. Brian argues that, yes, we must help support dyslexic students because our hearts tell us it is the right thing to do. At the same time, he emphasizes that it is also the smart thing to do.

> If you think about a city like Chicago, if 20 percent of Chicago kids are dyslexic, and we know that the brain has the potential of operating at a really high level, unlocking innovation, invention, it's those ideas and innovations that we're keeping our society from if we don't figure out how to make those connections. If we're not helping unlock the creativity and ingenuity and the innovation that exists in the minds of people in our society, we're going to miss out big.

Brian Bannon represents an inspiring example of how often and unexpectedly dyslexic men and women think deeply and at a level above most of us. All around us there are other Brian Bannons, with great potential they are unaware of, people who wish to advance in their work and in their reading ability. There is no deadline or age limit for when a person can learn to read or become a better reader. Research attests to the plasticity of the human brain and its ability to reshape itself in mature adults, as it does in children. There are many dyslexic adults who are out of school and working, or in retirement, who simply want to learn to read or have more focused goals related to their work, or to pursue a high school equivalency diploma. Such adults face a serious challenge, but one that they can meet successfully.

Lack of literacy is a major national concern. Thirty million adults (14 percent) in this country score *below* basic reading level in prose literacy skills—skills needed to comprehend and use connected texts—even though 45 percent graduated from high school. These adults are extremely inefficient readers, reading about sixty correct words per minute, *a rate expected of first- to second-graders.* Interestingly, surveys report that many limited readers perceive themselves as reading well. This may explain in some part why only a small percentage of adults (less than 10 percent) who have limited reading ability are enrolled in adult literacy programs.

Any adult, no matter his age, who cannot read or who struggles to read should participate in a literacy program. It will turn his life around. Literacy skills are critical in providing options for training, jobs, and careers. In trade schools, for example, students are expected to read course manuals written at the tenth-grade level or higher. In fact, in the coming years the vast majority of new jobs will require a twelfth-grade reading level. The ability to read impacts the quality of everyday life. Printed materials we read every day require a surprisingly high level of reading; a Friendly's menu, instructions on an ATM, and the Internal Revenue Service EZ form are all written at a tenth-grade level. At least a ninth-grade reading ability is needed to fill out employment applications, so low literacy influences trying to find a job regardless of the person's capability to do the job. To allow someone to enter adulthood and remain a very limited reader is to sentence him to second-class citizenry.

You or your spouse, your parents, grown children, or friends don't have to be stuck: You can rise above these difficulties. On a national level, almost 75 percent of the children of parents who have low literacy also are poor readers. In our Connecticut Longitudinal Study, in which children from age five are followed well into adulthood, we found that as adults, almost half of the dyslexic men and women (compared to approximately 20 percent of nondyslexics) had no education past high school, and that those who were dyslexic had triple the unemployment rate of nondyslexics.

In a recent paper Bennett, I, and our colleagues Daniel Herrera-Araujo, John Holahan, and James Hammitt used a strategy known as willingness to pay (WTP) to examine how dyslexic and nondyslexic adults (about thirty-three years old at the time of the study) valued improvements in cognitive skills. Specifically, we asked the now adult Connecticut Longitudinal Study participants questions to elicit their willingness to pay for hypothetical treatments that would raise five cognitive skills—reading speed, ability to read aloud, memory, ability to pronounce names and places correctly, and ability to retrieve facts and information rapidly—to "very good." These skills were to be improved by participating in a hypothetical intervention that required engaging in two one-hour training sessions each week for two months. The cost was to be paid as a one-time fee and would not be covered by insurance. For each of the hypothetical treatments the participants were first asked if they would participate in the intervention if the cost was $1,000. Those who responded yes were then asked if they would participate if the cost were $2,000; those who responded no to the first offer were asked if they would participate if the cost were $500.

We found that the rates of WTP were similar in four out of the five cognitive skill areas for both dyslexic and nondyslexic readers. However, dyslexic individuals were willing to pay significantly more than nondyslexics to improve their skill in *reading aloud*. This result provides scientific validation for the frequent stories I hear from dyslexics of all ages about how difficult it was and continues to be to read aloud. Clearly dyslexic adults continue to feel burdened by the effects of their dyslexia even into their fourth decade.

Adults who are poor readers and for whom English is their first language are for the most part dyslexic. Some were never diagnosed as dyslexic and received inadequate or no specialized reading help. Many of these dropped out of school, while some went on to graduate, although with extremely limited reading ability. Still others were identified as having a reading problem but received limited help. They learned to decode, perhaps, but did not receive the supervised reading practice necessary for fluency, so their reading skills are tenuous, and they do not read for pleasure or information. Some of these students dropped out of school and are now aiming for a GED diploma. There is an urgent need to recognize that these reading difficulties are most likely due to dyslexia and that the necessary systems to screen, identify, and effectively intervene were not in place to provide the needed help. As a consequence, many adult men and women who are bright and had the potential to benefit from evidence-based instruction were overlooked by society, including the educational system, and are now suffering the consequences. Those children who are left behind often follow a pathway filled with the consequences of low-level literacy and school failure and eventually drop out.

It is now possible for an adult to accurately, efficiently, and confidentially find out whether he or she is at-risk for dyslexia using the Shaywitz DyslexiaScreen™, Adolescent-Adult Form. This brief, ten-question, evidence-based screener is completed by self-report to immediately determine if he or she is at-risk for dyslexia. The questions are read aloud (by a computer or by a tester) to the adult being screened, who answers "Never," "Sometimes," or

"Always" to each question. Once all the questions are answered, the person receives immediate feedback on whether he or she is "at-risk for dyslexia" or "not at-risk for dyslexia." Information on the screener can be obtained at www.PearsonClinical.com/Shaywitz.

POSITIVE STEPS TO BETTER READING

A dyslexic adult can take positive steps to improve his reading and his life. It is well within your grasp if you have a strong desire to learn to read or to be a better reader. Several programs offer at least one reading year's worth of improvement for each six months of instruction, depending on the intensity of instruction and the severity of the reading problem. While addressing the problem early is ideal, it is never too late to learn to read.

GETTING STARTED: PICKING THE BEST PATHWAY

There are two complementary pathways you can pursue:

→ Look into the availability of an *adult basic education program* in your vicinity. Such programs typically help to prepare those who do not have a high school diploma for a GED exam or function as a bridge to programs and courses given at a community college.

→ If your reading level is below what is necessary to benefit from the adult education program's curriculum, you should locate a more basic program, referred to as an *adult literacy program*. These generally aim to teach students who require more intense and sustained instruction in reading. Following graduation from such a program, many students go on to attend adult education programs and to obtain their GED.

There are two national, online, Zip Code–based directories of adult literacy programs: the National Literacy Directory, http://www.nld.org/, which also has a toll-free number, and America's Literacy Directory, https://www.literacydirectory.org/. The U.S. Department of Education's Office of Career, Technical, and Adult Education has also launched a very popular mobile Web portal for adult learners to connect with online learning sites, https://

lincs.ed.gov/. You can also visit your local public library and ask for a state directory (some libraries also serve as sites for literacy programs). Often students needing adult literacy instruction are referred by community or social agencies or church groups.

PLACEMENT LEVEL

Once you have selected a program, the next step is to determine your placement level within that program. As is the case for children with dyslexia, there has been a sea change in the approach to adults with reading problems. Rather than a global test of reading, more up-to-date literacy programs give adults a placement test to determine their *exact* reading level and to detect specific gaps in their reading development. While children who are poor readers have not learned the foundational skills, adults, in contrast, may have acquired a smattering of some skills while they totally lack others. So placement testing is especially important for older dyslexic readers, since each is invariably starting from a different point. Keep in mind that the goal is to match the instruction as closely as possible with the adult's needs. Frustration at a slower-than-wished-for pace or being engulfed by low self-esteem and worries about the future often cause a student to drop out of the program. Teachers (and adult students themselves) tend to overestimate an adult's reading ability and are often surprised to see the results of the testing. The Comprehensive Adult Student Assessment System (CASAS) is frequently used for assessing placement. Other assessments are available as well; for the names of these, you can consult the Federal Register, which lists approved tests: https://www.federalregister.gov/documents/2015/08/12/2015–19847/ tests-determined-to-be-suitable-for-use-in-the-national-reporting-system -for-adult-education. Ideally, the reading instruction is evidence-based, systematic, and delivered in small-group settings.

ADULT ED PROGRAMS

Adults in particular do well with group instruction (with a well-trained teacher), and adult ed programs typically have teacher:pupil ratios of 1:16. It's engaging, highly motivating, socializing, and effective. Adults benefit from learning from one another. Optimally this instruction occurs as often as four times a week, for one and a half to two hours each session. Some successful programs meet twice weekly but for longer periods, such as three hours. Consistent attendance is critical, since it is one of the few factors associated with better progress.

ADULT LITERACY PROGRAMS

At times prospective students apply to adult ed programs but on evaluation are found to score below the level required to benefit from the program and are referred instead to an adult literacy program. Here the teacher:pupil ratio is much lower, typically 1:4, while for those needing basic literacy it may be as low as 1:1. Students like the more nurturing type of environment, one that doesn't look or feel like a classroom, which often evokes unpleasant memories. Many attending these programs are unemployed, on public assistance, or at low income levels. Programs are considered completed when the student can read at an eighth-grade level. Those who complete the program tend to be older, and to compensate for their reading difficulty they seek to obtain information from magazines, the Internet, television, radio, and family and friends. Studies have found three kinds of obstacles preventing completion of adult literacy programs: situational (such as lack of child care), institutional (such as scheduled class times), and dispositional (such as self-esteem issues).

For adults, issues of intensity and duration of instruction are particularly sensitive. Adults work during the day or may have young children—many students in adult reading programs are between sixteen and thirty—and it is often difficult for them to come to class more than twice a week. Long intervals between classes and lack of practice reading at home represent real barriers to progress. Adult students can expect to increase their reading one grade level for every hundred or so hours of instruction. Clearly, the more hours per week, the faster the improvement. So maximizing the amount of instructional time in class and particularly of practice at home is incredibly important.

The time it will take to complete the program depends on the level at which the student began and how often she attended class. If a student starts at a basic reading level and comes to class twice weekly, it will generally require three to four years; if the student can attend only once a week, she will not be ready for completion for five to six years. Typically students come once a week for a two-hour session.

I am often asked by potential students what the characteristics of a successful adult literacy program are. In one of the most successful programs I know, students attend a two-hour session four evenings a week. Intensity is stressed, and students may not miss more than thirteen days annually (in addition to a four-week vacation). Progress is often remarkable, some adults achieving a twelfth-grade or higher level of reading proficiency. While highly successful, the intensity and frequency of the sessions are unusual, for two important reasons. First, adult literacy programs are vastly underfunded; for the most part, they are led by trained volunteer instructors, of which there are far too few in number. Second, the population of struggling readers does not often have the luxury of attending classes frequently, for the reasons I've suggested.

READING PROGRAMS USED FOR INTERVENTION

Reviewing the What Works Clearinghouse reports, I was unable to find reviews of reading programs designed specifically for struggling adult readers. However, effective programs focused on a broader age range have been reviewed by the WWC. From this group, for teaching adults I recommend the Wilson Reading System. In addition, adults wanting to practice their reading skills will find helpful *Lexia Reading SOS (Lexia Strategies for Older Students)*, a computerized reading program. Lexia provides practice in a wide range of reading skills, including making letter-sound linkages, applying strategies for multisyllabic word analysis, and learning the meanings of roots, prefixes, and suffixes. Lexia is intended as a supplement, not as a total reading program.

It is very common for students to have work-related concerns. Carmine, a talented forty-three-year-old appliance repairman, had to improve his reading skills when his company began using new computer technology. Today, repair work is often based less on mechanical aptitude than on the ability to read. Most new washers and dryers are manufactured with computer chips that provide detailed printouts of an appliance's inner workings. As a result, all diagnostic work is based on the ability of the repairman to read and interpret schematics and other written information provided by the computer.

Although a high school graduate, Carmine was barely able to read. With the support of his company, he attended classes diligently for three years and is now reading at a high school level. Becoming a reader has changed his life. He has received several promotions at work, and his everyday life has benefited in many ways as well. At a recent breakfast with his colleagues, an article in a local newspaper lying on the table caught his eye. Since it concerned a friend, he began reading it aloud. As he describes it, "Reading aloud in front of my colleagues was a very exciting thing for me to do." He is now able to read and select a birthday card for his wife, fill out medical record forms in his physician's office, and complete entry forms to participate in his beloved hobby of motocross racing. Previously he would go to the track, ask for an entry form, and with much embarrassment take it, fold it, place it in his pocket, and with head down quickly walk out the door, hoping that "none of the guys noticed." His wife would complete the form for him at home later. This has all changed; now he can "read any word." He continues to attend the night classes, where he often teaches new tutors how to pronounce the sounds of different letter groups. He envisions attending community college one day and taking courses in computer repair.

One adult I came to know enrolled in a program in large part because she wanted to be able to read bedtime stories to her new granddaughter. Another student longed to be able to read aloud in her Bible study class. Each succeeded in reaching her goal.

If you know someone who could benefit from further reading instruction, there are a number of things you can do to help. For example, there are more materials than ever for adults with limited reading skills, including a weekly newspaper, *News for You* (www.newreaderspress.com/news-for-you), published by the New Readers Press, the publishing division of ProLiteracy International, and Thumbprint Mysteries, a collection of thirty easy-to-read books written in conjunction with the Mystery Writers of America, available from McGraw-Hill Contemporary Publishing (openlibrary.org/publishers /mcgraw-hill_contemporary). More and more libraries now have special collections of books that are of high interest to adults and can be read by those with lower-level reading skills. Ask your local librarian about them.

The General Education Development (GED) diploma is meant for students who did not graduate from high school but want to pursue further education or jobs that require a high school diploma or its equivalent. Obtaining a GED diploma requires passing a rigorous examination, revised in 2014 to align with Common Core State Standards (CCSS) for high school as part of the Department of Education career and college readiness expectations. The goal is a highly positive one: to help ensure that students have the skills and knowledge needed to enter a range of professions and occupations and attend colleges and universities. Three possible scores can be obtained:

> GED Passing Score: at or higher than the minimum needed to
> demonstrate high school equivalency–level skills and abilities
> GED College Ready: demonstrates the skills and abilities
> needed to enroll in credit-bearing college courses

GED College Ready + Credit: may be eligible for up to ten
hours of college credit

The 2019 GED consists of four content areas: reasoning through language arts, mathematical reasoning, science, and social studies. Reflecting the
imprint of the CCSS, the tests incorporate what are regarded as "clear sign-
posts indicating what is most important for college and career readiness." The
goal is to provide students with what is required for postsecondary academic
success and to obtain the "key college- and career-ready knowledge and skills
[that] are closely linked to being able to get the training necessary to earn a
living wage in high-growth industries." The hope is that by integrating the
college and career readiness standards into adult education programs, students will be prepared for and able to enter postsecondary training without
the need for remediation.

To repeat: The GED calls for tenth-grade reading skills and academic
knowledge. Students preparing for the GED may enroll in literacy classes
first to bring their reading levels up and then register for practice classes
focusing specifically on the components of the GED. A GED practice test
is available from the GED Testing Service. Accommodations are granted for
those who require extra time. If the individual has previously received accommodations from a standardized testing agency, there should be no need for
retesting. If an audio version of the test is requested, this usually requires
additional time and should be requested as an accommodation. The GED
has taken on additional significance given that many jobs, including those
generally regarded as low-skilled, require a high school diploma or GED,
with many industrial jobs requiring a preemployment assessment such as a
Career Readiness Certificate (CRC). For more information, visit the official
website of the GED Testing Service, www.ged.com.

Adult education teachers can often customize their instruction. A growing
number of companies (and unions) are sponsoring targeted reading instruction at the workplace itself. Participation is highly motivating; instruction
focused on the most common words found in work-related materials can
allow a worker to function more easily at his job or help him to move on
to the next level, as Carmine did. These programs are short but intense,
ranging from thirty to sixty hours over an eight- to ten-week period. Tom
Sticht, who developed highly successful literacy programs for the army in
the 1970s, believes that such work-related programs are especially effective
because of their relevance to the person's life goals. A student quickly expe-

riences tangible results, such as reading a cookbook if he wants to be a chef or a plumbing manual if he wants to repair pipes.

LIVING LIFE AS AN UNDIAGNOSED, VERY BRIGHT DYSLEXIC ADULT

So what is it like to be a bright dyslexic adult, one who was not diagnosed until he was well into middle age and who continues to struggle with reading to this day?

Perhaps the worst impact of having a low level of literacy, no matter how intelligent the person is, is that it robs him of his self-esteem. As I have noted often in these pages, many in our society mistakenly confuse slow reading with slow thinking, when nothing could be further from the truth. As a society, we have a responsibility to screen and to diagnose dyslexia early, since we now have the tools to do so. The urgent need for it was brought home to me recently. I am very proud of my university, Yale, for the depth of its understanding of dyslexia and its resultant actions. Our Yale Center for Dyslexia & Creativity serves as a referral center for students needing assessments. As part of this process, a student had been referred who was registered for an online course at Yale which required him to come to Yale once a month, and whose professor thought he might benefit from an in-depth assessment. I review each of these evaluations, and as I was going through this particular one, I was struck by the student's level of intelligence—highly superior, in the ninety-eighth percentile—in contrast to his very low scores in phonological processing and reading. Here was a classic example of the definition of dyslexia.

He was fifty-four years old. I phoned him, identified myself, and shared with him the results of his tests, which showed that he was dyslexic—and brighter than 98 percent of the population. "Are you sure you have the right person?" he asked. I assured him I did. I didn't hear anything from him for several seconds, so I asked, "Are you there?" "I am, but I'm crying," came the reply. I started shedding tears as well.

We met in person a few weeks later when he was on campus. Here's what he shared with me:

> Yes, I have very low self-esteem. So many people told me I wasn't
> very smart. Early in school I was placed in a class with intellectually

deficient children, and all we did was crafts and things like that. It made me feel horrible. I can still remember as a little boy going to Sunday school every Sunday and they would go around the room and you would have to read. It was a requirement, and I was humiliated each Sunday when I would be called upon to read. And it didn't go away. I remember even into college, if I knew that we were going to have to read a section, if they were going around the room selecting people to read and we were going up and down the row, I would go to the bathroom. I would not be present when it was my turn to read—that was in graduate school, for my master's.

He had compensated to a fantastic degree. He was going for a doctorate-level degree now. He had essentially memorized words in his specialty.

Yes, but at a cost. I can't do it—read these words automatically. Reading makes me totally exhausted, and I will sleep for a long period of time, which my friends say is abnormal. There are so many things. I would love to be able to watch foreign films, but I can't read the subtitles fast enough. The guidance counselor in high school told me I needed to select a technical field because I read so slowly and with such great effort.

I would really like to be in a situation where I could express my creativity and my intelligence without worrying about the image I have to portray. I feel like I can't be myself, that I have all these personas I have to show to other people. I have to hide my dyslexia so that they don't think I am not smart. I have a work persona—an exterior, if you will—that's very stressful to maintain, but with you, when I come in here, I don't have any walls, I would tell you anything about me. It feels like a safe environment, and that's one of the things when I first came here to test, I said, for the first time in my life, I know that I am going to be totally honest with the tests. This visit has opened the door, because even though I thought I was smart, this has validated that it just wasn't my wish or dream; it is real. What would make me happy is to be able to go to work every day and help people without having to worry about how I'm going to be perceived. It would mean that I could be the real me and that would be okay.

The beginning of my career I worked in a tertiary-care trauma center. So the patients were really, really sick. After a while I was always the one leading the cardiac codes and doing that. I didn't have to read it. I could react faster than I can tell you.

There are so many things, so many work-arounds that I have had to do to allow me to move forward on my intelligence so that I would not be stopped by my lack of a strong ability to read. For example, the GRE. I got around the GRE—if I had to take the GRE, I would not be able to go to grad school. Here's how I did it—the MSN [master of science in nursing] program accepted me as long as I had a certain GPA, 3.5 or something. The same thing here, at Yale. I didn't have to take the GRE for the degree I am aiming for, doctor of nursing practice. If it had been the Ph.D. program, I would have had to register for and take the GRE. The irony is that I would function better in the Ph.D. program because I love research and I would love to write.

Dyslexia is a really important part of my life, but I have real concerns about disclosing it. People have their preconceived perceptions. If I had a wish, it would be to have been diagnosed as dyslexic way earlier and then received the reading instruction I would need to become a good reader, so that my intelligence and reading ability would be aligned together, and that I could truly be myself, inside and outside. How much easier life would have been for me. On the other hand, I am relieved and even happy to know that it is not too late, and that there are still things I can do to help me be a better reader.

While early is always better, it truly is never too late. Acclaimed dyslexic crime novelist Amanda Kyle Williams did not read her first book until she was twenty-three years old. As a child, she struggled to read even the simplest sentences. Teachers described her as anxious; her family thought she was lazy, slow, or stupid. She dropped out of high school and became addicted to cocaine. She did not discover that she was dyslexic until she was twenty-two. "The diagnosis changed every way I felt about myself and every way I felt about my chances in the world," she said. "I lived in that undiagnosed, unsupported world for many years. When you grow up feeling dumb, being told you're not trying or not smart enough to do what is asked of you in school, it marks you. It just does." Sadly, Williams passed away in 2018.

MORE HOPE ON THE HORIZON FOR ADULT DYSLEXICS

A major effort is now under way to develop breakthrough technology—software—for mobile devices such as cell phones and tablets that would bring about significant improvement in the user's reading skills.

In 2014, Bennett and I were invited to join the advisory board of the Adult Literacy XPRIZE. As described by the XPRIZE Foundation, the XPRIZE is "the world's leader in designing and managing incentive competitions to solve humanity's grand challenges." Adults with low literacy, primarily dyslexic adults, are the targeted beneficiaries of this grand challenge, along with a sizable group of people whose native language is not English, the English language learners (ELLs). One hundred fifteen teams competed, and as I write this, the grand prize has just been announced and is shared by two teams: Learning Upgrade (San Diego, California) and People ForWords (Dallas, Texas). These two teams, along with two other teams—AmritaCREATE, Amrita University (Amritapuri, Kerala, India) and Cell-Ed (Oakland, California)—were also selected to compete for the second phase of the competition, to test the practical applications of the apps in communities with a high prevalence of dyslexic adults. AutoCognita (Hong Kong) was also one of the finalists. These apps generated the greatest gains between a pre- and post-test to both native English-speakers and English-language learners who initially were reading English at a third-grade level or below. Specific details are provided in the Appendix.

Even in our increasingly technical society, it's still all about reading: instructional manuals for the workplace, for computers, for cars, for putting together children's toys; directions for taking medications, for making up baby formulas, for safety, for voting; messages via mail, email, the Internet, pager, and fax. Learning to read opens the door to a brighter, better future, and given the effective programs to teach reading at all levels of reading ability and at all ages, it really is never too late to learn to read.

PART VII

MAKING IT WORK FOR THE DYSLEXIC CHILD AND ADULT

ANXIETY, ADHD, AND DYSLEXIA

Virtually everyone who is diagnosed as dyslexic, whether it is a Yale student, an Uber driver, a physician, a middle-school boy or girl, a mom, or a retired senior citizen, comes with a full load of anxiety. In fact, it is rare to see a dyslexic who doesn't also have anxiety or, for that matter, ADHD. It is important to be aware of these very common comorbid conditions, for they too must be recognized, diagnosed, explained, and treated if the dyslexic individual is to have an optimal outcome. Anxiety and ADHD so commonly occur along with dyslexia that it is critical that once someone is diagnosed as dyslexic, anxiety and ADHD must be asked about. Both anxiety and ADHD are well described and well delineated, and each can be treated effectively. In the following sections we review each of these conditions that commonly occur along with dyslexia.

ANXIETY

Sam was a second-year medical student attending a prestigious school. He was diagnosed with dyslexia in middle school and struggled particularly with standardized tests and oral examinations. Sam was able to persevere due to his own resilience but also because of the incredible support he had from his family, including his mother and father and two brothers. Sam remembers his long and sometimes arduous journey to get where he is. It made him so anxious not to know when he would be called on to read aloud or when he would have to go off to a special room to take his exams. He remembers the stares of the other children in middle school and high school while he took his examinations in a separate room with extra time. He remembers hearing his fellow classmates complain to the teachers, "That is not fair. Why does Sam get extra time and we don't?" Sam could not comprehend this, since he wished he didn't need to take the extra time and could be like everyone else. He also remembers being told by one of his admissions counselors that he would never get into a medical school. "There are other good professions," he was told. He remembers one of his medical school teachers patronizingly imploring the rest of the class, "Would someone please tell Sam what is going on?" Sam used all these comments for motivation.

Sam was diagnosed with anxiety. He was a shy child growing up but had friends and a wonderful childhood. At college, particularly in stressful situations, he experienced panic attacks. During these attacks, lasting fifteen minutes at most, he had a choking sensation, heart palpitations, and a feeling of impending doom; he felt he was having a heart attack and was going to die. The anticipation of these attacks made him even more anxious. Performance situations such as written or oral examinations would often trigger the panic attacks, as would meeting individuals in social situations. Group study projects or oral examinations or public speaking made him so anxious that at times he would try to avoid the situations altogether. When he couldn't avoid them, his anxiety would become so great that his hands would begin to tremble, he would sweat, and he would lose his focus and concentration.

When this anxiety became debilitating, Sam sought professional help from a psychiatrist. Through his discussions with his psychiatrist and through reading, Sam learned that anxiety is a normal reaction to stress and can actually be beneficial. In some situations, however, anxiety can become excessive, and while the person suffering may realize the anxiety is excessive, he may also have difficulty controlling it and it may negatively affect his day-to-day living. Sam was informed by his psychiatrist that the spectrum of anxiety disorders includes social anxiety disorder, panic disorder, generalized anxiety disorder, post-traumatic stress disorder, and obsessive-compulsive disorder. He was also told that given his symptoms, he most likely had both social anxiety disorder and panic disorder (PD).

Social anxiety disorder is the third most prevalent psychiatric disorder in the United States, following major depression and alcoholism. It is defined in the *Diagnostic and Statistical Manual of Mental Disorders,* fifth edition:

→ Marked fear, anxiety, or avoidance of social interactions and situations that involve being scrutinized or being the focus of attention, such as being observed while speaking, eating, or performing in front of others.

→ Fear negative judgment from others and, in particular, fear of being embarrassed, humiliated, or rejected, or offending others.

It often first appears in adolescence, at around fifteen or sixteen years of age. A great number of people are so affected by it that they cannot work or start or maintain professional relationships. There is also a high comorbidity of social anxiety disorder with major depression.

Panic disorder (PD) is also common, with a prevalence perhaps as high as 2 percent of the population. Like many anxiety disorders, a combination of biologic and environmental factors is involved in PD. The onset is often in late adolescence and early adulthood, and PD occurs more often in women than men. The core symptoms involve short, discrete periods of intense fear accompanied by somatic symptoms such as a racing heart, sweating, and shortness of breath. It is defined in the *DSM* fifth edition:

→ Recurrent, unexpected panic attacks which involve an abrupt surge of intense fear or discomfort that reach a peak within minutes and include four or more symptoms (including palpitations, sweating, trembling, feeling short of breath, feeling of choking, chest discomfort, nausea, lightheadedness, fear of losing control, numbness, and chills).

Sam's problems with word retrieval, so common in dyslexic readers, were especially exacerbated by his anxiety. In the top portion of Figure 48, a typical reader uses his intact higher-order thinking abilities, as you have previously seen (pages 58–60), to develop a concept and identify the abstract semantic representation of the word that represents this concept. For typical readers the next step in this two-step process—uttering the word—is performed smoothly. For the dyslexic reader (bottom portion of Figure 48), the first step, using his cognitive ability, is performed well, but the second

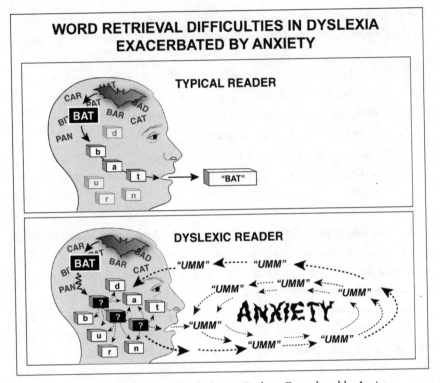

Figure 48. Word Retrieval Difficulties in Dyslexia Exacerbated by Anxiety

step is disrupted and made even worse by anxiety (for further details, see pages 58–60).

Sam had always been very resistant to taking psychiatric medications, believing it was a "sign of weakness" and that he should be able to "be strong and power his way through it." However, when his psychiatrist posed the question, "Sam, if I told you that you had high blood pressure or diabetes, would you be reluctant to start those medications?," Sam knew it was more the "stigma" of being on psychiatric medication that bothered him than the medication itself, and he agreed to give it a chance.

The medications effective in treating anxiety (as well as ADHD) generally work by acting on neurotransmitters, chemical messengers that either stimulate or inhibit neuronal impulses in particular areas of the brain. The core first-line medications employed include the selective serotonin reuptake inhibitors (SSRIs), such as fluoxetine (Prozac), paroxetine (Paxil), sertraline (Zoloft), citalopram (Celexa), escitalopram (Lexapro), and fluvoxamine (Luvox). Other commonly used medications include the serotonin and nor-

epinephrine reuptake inhibitors (SNRIs), including venlafaxine (Effexor), the norepinephrine-dopamine reuptake inhibitor (SDRI) buproprion (Wellbutrin), and the alpha-2 adrenergic receptor antagonist mirtazapine (Remeron). These medications are all equally effective in treating anxiety; there have been no head-to-head long-term studies demonstrating superiority of one medication over another. In selecting an anxiety medication, it is important to consider the history of a family member being effectively treated on the medication, other medications the patient is currently taking (to avoid negative drug-drug interactions), the possible side effects, and how long the medication stays in a person's system.

Sam was placed on the SSRI sertraline (Zoloft). All SSRIs can take up to four weeks to become effective and have similar side effects, the most common including gastrointestinal disturbances, such as nausea and diarrhea, and sexual side effects, including anorgasmia (failure to have an orgasm) and decreased libido. There is also potential weight gain and some sedation with this class of medications.

Another class of medications that are employed in the treatment of anxiety disorders include the benzodiazepines. The common ones include diazepam (Valium), alprazolam (Xanax), clonazepam (Klonopin), and lorazepam (Ativan). In addition to concerns about tolerance and dependence, side effects include drowsiness and potential cognitive impairment at higher doses. Benzodiazepines are immediately effective, which is particularly helpful when individuals are having a panic attack.

The psychiatrist discussed all this with Sam and explained another class of medications, beta-blockers, which are helpful in controlling the physical symptoms of anxiety, such as sweating, trembling, increased heart rate, and blushing when giving a talk or taking a test. These symptoms are believed to be the result of epinephrine release, which is inhibited by the beta-blockers. These medications do not affect the cognitive aspects of anxiety and affect only the physical symptoms. Propranolol (Inderal) is a popular beta-blocker. Its onset is immediate, and potential side effects include decreased heart rate, fatigue, weight gain, and depression.

Sam was also informed that medications were only part of the treatment for anxiety disorders. An equally important treatment involved undergoing individual behavioral therapy, particularly cognitive behavioral therapy (CBT). As you read earlier, CBT is aimed at changing a person's current way of nonproductive thinking and behavior, a change that is often transformative, producing improvement in outlook and mood. It is typically provided

by a psychologist with special training in the methodology and is backed by evidence of its effectiveness in children and adolescents, particularly for anxiety and depressive disorders.

Cognitive behavioral therapy generally involves eight to twelve structured sessions divided into exploring the patient's cognitive thoughts and behaviors. The patient first is made aware of his distorted thinking and then taught behaviors to help alleviate them. Sam's distorted thoughts included believing that he was always being scrutinized and was the center of attention when he entered a social situation. He believed that the other individuals in the social situation were judging him negatively. In CBT, Sam was asked by his therapist for evidence to support these automatic thoughts. Sam learned that these cognitive distortions were called "catastrophizing." His therapy sessions also included "graded desensitization," during which he would gradually go to different social settings with increasing anxiety potential. He initially went to a coffee shop by himself and then rated his anxiety. Several days later he went to a small party where he knew everyone, and again rated his anxiety. After several weeks he attended a large gathering where he knew no one and again rated his anxiety. The goal of this behavioral therapy was to expose Sam in a stepwise progression to gradually more unfamiliar and larger settings to decrease his anxiety symptoms.

Sam underwent CBT and was started on sertraline, with the addition of clonazapam to be used as needed if he experienced a panic attack. He was also given Inderal, to be used before performance situations such as giving a presentation or taking a test. He initially found the sertraline to be sedating but after several weeks experienced no side effects from the medications. He did not like alcohol, so it was easy for him not to have to worry about not drinking while on the medications. He found the combination of the medications and CBT particularly effective for his test-taking anxiety during oral presentations and standardized examinations.

Since Sam was dyslexic, his test-taking anxiety was exacerbated. At such times, he especially found practicing mindfulness, based on an ancient Buddhist concept, very helpful in diminishing his anxiety. Mindfulness can be promoted by meditation, but it is not synonymous with meditation. Mindfulness is achieved through a variety of activities, including yoga, tai chi, and meditation, which have in common the goal of trying to bring mental processes under greater personal control through training attention and awareness. The hoped-for (and often observed) result is a sense of calm, clarity, and focus. There is clinical evidence that mindfulness decreases anxiety and increases a sense of positivity.

The principles of mindset can help ameliorate anxiety as well. As discussed in Carol Dweck's insightful book, *Mindset: The New Psychology of Success*, mindset refers to your attitude about yourself, your abilities, and your future. The growth mindset championed by Dweck is based on the belief that change can be brought about by the individual, which is welcome news to dyslexics. If you harbor the belief that change is possible and that effort translates into positive change, then if you work on improving a quality or characteristic, your effort will eventually pay off. Think of the successful dyslexic writers, athletes, doctors, scientists, attorneys, public servants, and those in entertainment noted earlier, who struggled when younger and who through continuing effort are now at the very top of their respective fields. Dweck reminds us that when NASA was asking for applications for astronauts, "they rejected people with pure histories of success and instead selected people who had significant failures and bounced back from them." Dyslexics, all too familiar with failure, have developed a growth mindset and succeeded.

While anxiety and dyslexia were a part of Sam, he made sure they never defined him. He was able to graduate from one of the top medical schools in the country and is now a board-certified physician doing what he loves to do.

ATTENTION-DEFICIT/HYPERACTIVITY DISORDER (ADHD)

Attention is intertwined with dyslexia for two important but very different reasons. First, as you have read, dyslexic readers are not automatic; to them reading remains effortful. I have referred to them as "manual" readers. This means that they have to dip into and soon use up their reservoir bucket (see pages 136–137) of attention in order to read, so that they become attention-deprived. As a result they appear to have an attentional deficit, so that after reading a short time they become easily distracted. Second, there is a very large overlap or comorbidity between ADHD and dyslexia.

Sophia was in the seventh grade and had been diagnosed with dyslexia. Her reading had improved dramatically with the help of tutors and the provision of extra time on tests, and her grades had improved as well. At the same time, her teachers had noticed that she was becoming inattentive in class and was easily distracted. They described her as "not being focused" and "not being able to follow instructions." These behaviors were also noticed by her parents at home. When they asked Sophia about her day, her mind would wander and she would take a long time to answer their straightforward questions. Both her teachers and her parents noticed that she was often sidetracked when asked to complete an assignment and would take a long time to do it, if she completed it at all. Sophia was disorganized at home, often losing or misplacing her schoolwork. After many months without any improvement, her parents, with a recommendation from the school, sought a consultation with a child psychiatrist.

The psychiatrist spoke at length with Sophia and her parents. With the parents' consent, she also spoke with officials at the school and gave Sophia two rating scales to help with the diagnosis of attention-deficit/hyperactivity disorder: the ADHD rating scale and the Conners rating scale. After reviewing all the information, she told Sophia and her parents it was most likely that Sophia had a form of ADHD, predominantly Inattentive Type. Sophia's parents were told that the co-occurrence (or comorbidity) of dyslexia and ADHD is quite common. In our own studies, when we considered a sample based on children referred to a clinic for learning disability, 40 percent of the children with dyslexia were also found to have symptoms of ADHD, and over half the children with ADHD were found to have dyslexia.

The psychiatrist explained to Sophia and her parents that ADHD is characterized by three cardinal symptoms: inattention, hyperactivity, and impulsivity. Clinicians have come to recognize that the hyperactivity and impulsivity seem to track together, and inattention often appears separately. Clinicians recognize these as three types of ADHD: inattentive, hyperactive/impulsive, and combined type (children with all three symptoms). Sophia most likely had ADHD predominantly inattentive type. This type of ADHD is usually not apparent until the child is eight or nine years old and usually is lifelong. The symptoms, according to the newest *Diagnostic and Statistical Manual* (*DSM* 5), include:

→ Often fails to give close attention to details or makes careless mistakes in schoolwork, at work, or during other activities (e.g., overlooks or misses details, work is inaccurate)

→ Often has difficulty sustaining attention in tasks or play activities (e.g., has difficulty remaining focused during lectures, conversations, or lengthy reading)

→ Often does not seem to listen when spoken to directly (e.g., mind seems elsewhere, even in the absence of any obvious distraction)

→ Often does not follow through on instructions and fails to finish schoolwork, chores, or duties in the workplace (e.g., starts tasks but quickly loses focus and is easily sidetracked)

→ Often has difficulty organizing tasks and activities (e.g., difficulty managing sequential tasks; difficulty keeping materials and belongings in order; messy, disorganized work; has poor time management; fails to meet deadlines)

→ Often avoids, dislikes, or is reluctant to engage in tasks that require sustained mental effort (e.g., schoolwork or homework; for older adolescents and adults, preparing reports, completing forms, reviewing lengthy papers)

→ Often loses things necessary for tasks or activities (e.g., school materials, pencils, books, tools, wallets, keys, paperwork, eyeglasses, mobile telephones)

→ Often easily distracted by extraneous stimuli (for older adolescents and adults, may include unrelated thoughts)

→ Often forgetful in daily activities (e.g., doing chores, running errands; for older adolescents and adults, returning calls, paying bills, keeping appointments)

The child psychiatrist also emphasized to Sophia and her parents that ADHD is a clinical diagnosis and cautioned that at present, while rating scales are helpful, there is no evidence-based test that will reliably diagnose ADHD. They were urged to be skeptical of any measures claiming to identify ADHD, despite claims for EEG measures or experimental psychological measures such as continuous performance tests.

The psychiatrist told Sophia and her parents that good interventions have been developed and there is a long and successful history of therapy for ADHD, both using medications and behavioral interventions. The main pharmacotherapy treatment options include stimulants, atomoxetine, anti-depressants, and alpha-2 agonists. The behavioral treatment strategy for children and young adults with ADHD includes working with the child, the parents, and the school. The child and the parents strive to maintain a daily schedule, reward positive activities, set small but realistic goals, and use checklists to help the child stay focused. The school can help by providing accommodations, including extra time to complete assignments and taking tests in a separate room to avoid distractions. The school can also arrange to have the child sit close to the teacher and provide note-takers to help her organize assignments and work. Other behavioral treatment strategies for ADHD include physical activity as well as mindfulness.

The stimulants begin to work immediately. They include methylphenidate (Ritalin) and amphetamines (such as Adderall) and come in both short- and long-acting preparations. The short-acting preparations can last up to six hours, while the longer-acting preparations, such as methylphenidate ER (Concerta), amphetamine/dextramphetamine XR (Adderall XR), and lisdexamphetamine (Vyvanse), can last from ten to twelve hours. While generally safe, the stimulants can cause anorexia, growth retardation, jitteriness, exacerbation of existing tics, and increase in heart rate and blood pressure. Atomoxetine (Strattera), a *nonstimulant,* is another class of medications used to treat ADHD. Its mechanism of action includes increasing the amount of available norepinephrine, and unlike the stimulants it can take up to three weeks to begin to act. It can last up to twelve hours and can cause dry mouth, constipation, nausea, headaches, and palpitations. Antidepressants are a third class of medication options used to treat ADHD. Buproprion (Wellbutrin) is the most common among them. Its effects can last up to fifteen hours, though it can cause seizures, hypertension, and increased anxiety. The fourth class of psychotropic medications include the alpha-2 agonists, such as clonidine (Catapres), which work by indirectly increasing norepinephrine, and their benefit can last up to twelve hours. The side effects can include dry mouth, low blood pressure, and sedation. Clonidine is classified as a category C medication, meaning that there can be potential detrimental effects on fetal development.

After careful consideration and review with her parents and her psychiatrist, Sophia selected the long-acting stimulant Concerta and was prescribed 18 mg/day. She found the medication very effective in helping her focus; her teachers and parents noticed a positive change in her behavior at home and at school. Sophia tolerated the medication well but noticed a decrease in her appetite and feeling "jittery" when she first started the medication. Because she took the medication in the morning, it did not affect her sleep, and she took "drug holidays" on weekends, when there was no school, which she found helpful as well. Sophia worked with her parents and the school psychologist on behavioral techniques, including setting daily schedules and making lists to help complete assignments. She was granted accommodations by the school, including extra time on exams, which she was able to take in a separate room.

Recent reports suggest that stimulants may have beneficial effects on reading in children with both ADHD and dyslexia. Two small clinical trials have also demonstrated improved reading scores in children with both ADHD and dyslexia during treatment with the nonstimulant medication atomoxetine (Strattera). More recently, we examined the effect of atomoxetine on the reading abilities of a group of adolescents with dyslexia alone and another group with both ADHD and dyslexia. The study indicated:

→ Atomoxetine treatment improved the reading scores of children with dyslexia, either dyslexia alone or dyslexia and ADHD.
→ Improvements in the children with dyslexia were in critical components of reading, including decoding and reading vocabulary.
→ For children with both dyslexia and ADHD, improvements in reading scores were *distinct* from improvement in ADHD inattention symptoms.

These data represent the first report of improvements in reading measures following pharmacotherapy treatment in children with dyslexia (and also ADHD) evaluated in a gold-standard randomized, double-blind trial. We are currently examining the effects of atomoxetine on neural systems for reading and attention in children with dyslexia only and those with both dyslexia and ADHD.

THE ROLE OF TECHNOLOGY

Critical for the dyslexic child using technology is his or her facility in inputting and accessing digitized text.* *Teach your child to touch type as early as possible. First grade works for many children.* This means that the child must be adept at keyboarding. Arizona State University professors Steven Graham and Karen Harris, in the context of teaching writing, emphasize the importance of the student learning to use word processing: "Many schools still use 19th-century writing tools such as pencil and paper, even though scientific studies demonstrate that students in grades 1 to 12 show greater improvement in their writing over time when they use word processing to write at school versus writing by hand. . . . We obviously need to move writing instruction more squarely into the 21st century, making it possible for our students to take advantage of word processing and other electronic methods for composing." A number of good programs are available to teach keyboarding to children, including *Typing Instructor for Kids,* available from Amazon for both PC and Mac systems. It uses games to motivate children as they learn to type. Once the dyslexic child has mastered keyboarding, he will

---

* Technology continues to change at a rapid pace, and any specific programs mentioned may be superseded by better ones or updated versions.

have at his disposal a number of programs for note-taking, text-to-speech, and speech-to-text.

In the same vein, since text-to-speech is perhaps the most ubiquitous of the technologies offered to all students, especially dyslexic students, it is important to recognize the logical and sequential steps necessary for its use. These are:

→ Obtain digitized text or have text of interest digitized
→ Decide which text-to-speech app you want to use
→ Apply text-to-speech app to read digitized text of interest

Knowing these steps should help students to understand the various technological fixes or apps described in the subsequent sections.

While there are many advantages for the dyslexic child in using technology, there are potential downsides as well. This is especially important to remember when the child is using text-to-speech software. In studies of eighth- and ninth-graders, poor readers presented with text bimodally (onscreen, highlighted, while being voiced) improved in comprehension compared to students who were presented with the text visually. But a negative effect of text-to-speech software on reading was demonstrated in fourth- and fifth-graders, presumably as the consequence of passively listening to the auditory presentation rather than actively reading the text. These data support what I always tell my patients and their families: Text-to-speech software is often very helpful to older children and young adults in high school, college, and graduate school but must be used cautiously and sparingly in younger children, who must be encouraged to read as much as possible. As I have already said, the way children improve in their reading fluency is to read connected text as much as possible. Text-to-speech technology can limit the practice in reading that dyslexic children need so critically.

You and your child must grasp that, no matter what the technology, there is a learning curve associated with its use.

A speech-to-text tool such as Dragon, for example, requires a fair amount of practice and patience before it can be of help. Note-taking software and text-to-speech software—the most commonly used technologies—are usually much easier to learn. It is crucial that the dyslexic child not be discouraged. That's why I think it is so important for him to begin to take ownership of the use of the technology while he is in high school, or even before. Parents often assume a great deal of responsibility for their dyslexic child's high school work. They will not be there in college, and it is a good idea for the dyslexic child to become familiar with and use the technology on his own while still in high school.

TYPES OF TECHNOLOGY AVAILABLE

NOTE-TAKING

For dyslexic students in middle school, high school, and especially college, note-taking is of great concern. As I've said, students in the class are often paid by the college to take notes, which are then disseminated to dyslexic students, whose identity is kept confidential. Some popular examples of assisted technology for note-taking follow.

Livescribe Smartpen. This device uses a special pen housing an infrared camera which, when coupled with the special dot-matrix paper supplied by Livescribe (which acts like a GPS for the pen), allows the smartpen to capture everything the student writes or draws on the dot-matrix paper. These notes can be uploaded directly to a computer, tablet, or smartphone, where additional software can translate the student's written notes to digitized notes that then become searchable. A built-in microphone on the Livescribe Echo or, using Livescribe 3, the microphone on a tablet or smartphone, allows the student to record a lecture while taking notes. The recorded lecture is synchronized via Bluetooth with the student's written notes so that the student can refer to the audio if he has missed something in the lecture. Livescribe is compatible with Mac, PC, iOS, and Android systems.

OneNote. This is another popular note-taking program that many dyslexic students use. OneNote is free and packaged with Microsoft Office for PC, Mac, iPad, and Web versions, so it is readily available to most students. It was designed for students to take notes and has a number of very useful templates, including simple lecture notes, detailed lecture notes, lecture notes and study questions, math/science class notes, and history class notes. OneNote includes text-to-speech (*Immersive Reader*) and speech-to-text (*Dictate*) as well. OneNote backs up to the Web so notes should be accessible from any device.

Evernote (https://evernote.com). This is a popular note-taking program, too, that is accessible on any device. It is maintained in the cloud so that notes are available at all times, and it is very easy to learn. There are no predefined templates for students to use in class, but the program is so versatile that students should find it easy to create their own.

Notability. This is another good choice (download at the App Store), though at the time of this writing it is compatible solely with Apple products. A nice feature of Notability is that students can record a lecture as they take notes on it and the audio is synced with what's written. If the student has missed something, she can tap on the text input and access the recorded portion of the lecture. Notability allows inputting with both typing and handwriting using the Apple pencil.

LIBRARIES OF DIGITIZED BOOKS AND PERIODICALS

Bookshare (https://www.bookshare.org/cms). This library has the largest collection of accessible titles—at the time of this writing, over 500,000—and is entirely free to students with dyslexia. Students register with Bookshare and then can download the books they need for school using any number of text-to-speech platforms, including a platform provided by Bookshare. Using these programs, dyslexic students are able not only to listen to books with high-quality text-to-speech voices but at the same time see highlighted words, sentences, or even whole paragraphs onscreen.

Additionally, dyslexic students who have an individualized education program (IEP, see page 469) are eligible to apply for digitized textbooks and other school material through Bookshare (and other libraries, including Learning Ally) from the National Instructional Materials Access Center (NIMAC), a federal agency established by the Individuals with Disabilities Education Act (IDEA) that serves as a national digital-format repository for all K–12 textbooks, workbooks, and other classroom reading materials published after 2007. NIMAC is limited to K–12 textbooks, so it does not include textbooks for college, graduate, or professional schools.

Learning Ally (https://learningally.org). Started over sixty years ago to help blind World War II veterans, this resource now offers the largest collection of *human-narrated* audio textbooks and literature. At one time human-narrated tapes were the only source of text-to-speech, and some students still prefer the human voice. Learning Ally has more recently expanded its repertoire to include computer-generated voices, in what it refers to as VOICEtext Books. Learning Ally has a library of over 80,000 titles, mostly human-narrated books, and charges parents, college students, and adults $135 a year.

Project Gutenberg (https://www.gutenberg.org). This library offers over 60,000 free ebooks that can be read online or downloaded to Kindle. All books are out of copyright or in the public domain. Over time more and more books become available, and so the library is constantly expanding.

Google Books. Another very important source of digitized books, this project hopes to digitize the holdings of many university libraries and began with the digitization of the University of Michigan library. Over time all books out of copyright or in the public domain will be digitized and available via a Google search.

TEXT-TO-SPEECH

As you have learned, once a dyslexic student has accessed the digitized text he needs, the next step is to use one of the many text-to-speech apps designed for reading the texts. Previously, I mentioned the text-to-speech capability *Immersive Reader,* built into OneNote. It is now accessible on all Microsoft programs, including Word, PowerPoint, OneNote, Outlook, Teams, Forms, and Flipgrid, and is available on all PC, Mac, iOS, and Android systems. Some of the libraries, including Bookshare, offer free readers that can be accessed on either Apple or Windows platforms, including Android devices. One of the most popular text-to-speech apps that many dyslexic students find helpful is *Voice Dream Reader* (www.voicedream.com), available to both Apple and Android users. Voice Dream Reader is very versatile, can automatically download from libraries, including Bookshare, and uses a large number of increasingly humanlike digitized voices. A very nice feature of Voice Dream Reader is synchronized highlighting of words, sentences, and paragraphs, depending on the reader's preferences. It will also look up words in a dictionary. As noted earlier, it is possible that in middle school and high school, seeing the words smoothly synchronized with speech improves comprehension and knowledge retention. Similar to other text-to-speech programs, Voice Dream Reader allows the reader to pause and rewind; to change speed, font, size, spacing, and color for optimal readability; to highlight; and to take notes. Voice Dream Reader can read just about anything that contains text and loads from anywhere, including note-taking programs and any website. A companion app, *Voice Dream Scanner,* will digitize to your smartphone any document and make it available for text-to-speech.

Another very popular text-to-speech is *Kurzweil firefly* (https:www .kurzweiledu.com/K3000-firefly/overview). Kurzweil was one of the first systems developed to convert text to speech, and has been made more accessible using the Web. Kurzweil will operate on all Mac and PC devices, including iOS and Android systems. Many voices are available here as well. The system is quite expensive for an individual to purchase, but many high schools and universities have the system in place and make it available to dyslexic students at no charge.

Other text-to-speech programs that many dyslexic students have found helpful are *Read & Write Gold* (https://www.texthelp.com/en-us/products /read-write) and *Read: Aloud* (https://learningtools.donjohnston.com). Many high schools and colleges have purchased one of these programs, which have many similarities to Kurzweil, and each can be used on Apple, Windows, and Android systems. Another popular text-to-speech is *Natural Reader* (https:// www.naturalreaders.com/software), used on PC, iOS, and Android systems. One version of the system is free, and more versatile versions cost about $10. Upgrades of the voices are more expensive. The program is very popular with college students.

For dyslexic students, it is often helpful to use speech-to-text programs to get oral material into a digitized text format. Many cell phones use programs like this to answer emails and seem to work reasonably well. *Dictate,* packaged free with the Microsoft Office suite of programs including Word, OneNote, PowerPoint, and Outlook, is now available for use on Apple, PC, and any device using iOS or Android systems. Another commonly used speech-to-text program is *Dragon,* from Nuance, which has a number of versions, some with very specialized vocabularies for professional use, such as medical and legal. Read & Write Gold and Voice Dream also have speech-to-text capability. Even newer speech-to-text programs may change the landscape dramatically. Programs such as *Otter.ai, Descript,* and *Rav* now offer speech-to-text transcription very rapidly and very economically with a high degree of accuracy. Though the learning curve for speech-to-text programs may be intimidating, know that it is surmountable, and the programs may be helpful for dyslexic students.

OTHER HELPFUL APPS FOR STUDENTS WITH DYSLEXIA

Quizlet, basically an electronic version of flash cards, is very helpful for learning math, foreign languages, geography, and many other subjects. It's very easy to make your own set of flash cards for subjects you or your child might need, but a major advantage of Quizlet is that other users of it, primarily teachers, have made their flash-card sets available for anyone to use. As one very successful dyslexic college student told me, "Flash cards help so much, especially if you write them yourself, because it creates an image in your head. When I'm in an exam, if I don't remember it, then sometimes I'm able to close my eyes and re-picture the card, and that helps a lot."

CliffsNotes are known to all students, especially dyslexic ones. Now owned by Houghton Mifflin Harcourt, these study guides are available for numerous books and topics. Helpful for dyslexic students, the text that is downloaded is already digitized and can be inserted into any of the text-to-speech programs.

Another useful group of programs are those relating to *schedules and to-do lists,* and these are readily available on both Outlook and Google platforms. *Wunderlist* (https://www.wunderlist.com) is another app that students find helpful to organize all their to-do lists. This free app works with Apple, Windows, and Android devices.

SensusAccess (https:/sensusaccess.com), which is free, allows students to automatically convert documents into a range of alternate media, including audiobooks (MP3 and DAISY), ebooks (EPUB, EPUB3, and Mobi), and digital Braille. The service can also be used to convert inaccessible documents such as image-only PDF files, JPG pictures, and Microsoft PowerPoint presentations into more accessible formats. The student loads the file he wants converted, sends it off, and receives the converted file via email.

Textbook scanning services are of great help to many dyslexic students in high school and college. At the college level, this service may be performed by the office of disability services. Otherwise, a number of private vendors (for example, *Blue Leaf* book scanning, www.blueleaf-book-scanning.com) will scan the textbook or chapters so that the digitized book or chapters are accessible on the text-to-speech programs noted above.

Finally, especially for younger children, an increasing number of apps that help students learn *common words* are available. I am indebted to Professor

Emerita Elaine Cheesman, who noted many of these on her *App Chat* blog. The apps listed below are all downloadable from the App Store for iOS devices. They include apps such as *Fry words* and the more difficult *Fry words Ninje*, which help children learn the thousand most commonly used words in written text. An additional list of 220 commonly used words, excluding nouns, is featured in *Sight Words by Photo Touch*. Dr. Cheesman also noted several apps that are designed to help teach preschoolers the fundamentals of reading. These include *Handwriting Without Tears: Wet-Dry-Try*, *Partners in Rhyme*, and *OG Card Deck*. Just as a reminder, parents may want to consider the computer interactive programs effective for very young at-risk readers: *Headsprout*, *Ooka Island*, and *Lexia Core5 Reading*. These programs are more fully discussed on pages 215–216.

32

ACCOMMODATIONS: BUILDING A BRIDGE TO SUCCESS

If what I write here is already familiar to you, please bear with me: I feel accommodations are so vitally important to the dyslexic reader that what I say about them bears repeating.

If you are a dyslexic reader, accommodations represent the bridge that connects you to your strengths and in the process allows you to reach your full potential. By themselves, accommodations do not produce success; they are the catalysts for success. They grow in importance as a dyslexic progresses in his schooling. As he advances, his strengths—in thinking, in reasoning, in vocabulary, and in analytic skills—mature. At the same time, his academic challenges increase as well. Consequently, with time it becomes even more crucial for the dyslexic reader to access his strengths in order to compensate for his phonologic weakness.

EXTRA TIME

The most critical accommodation for the dyslexic reader is the provision of extra time.

→ *Dyslexia robs a person of time; accommodations return it.*

Studies carried out over the last two decades confirm a dyslexic reader's absolute *physiological* need for extra time. For him, additional time is obligatory, not optional. It is the dyslexic's unique constitution that makes it possible for extra time to exert a positive effect.

→ For the dyslexic reader, learning capacity is intact; he simply needs time to access it.

Accumulating scientific evidence shows that typical readers and dyslexic readers follow very different pathways to adult reading. For typical readers the route is smooth and orderly: Their phonologic skills increase with age, they become more accurate and more automatic in their reading, and they generally identify words without any need to rely on context around the words. By fourth grade typical readers are no longer using context to figure out a word. The dyslexic, however, must take an alternate, indirect, and demanding route. This secondary pathway will get him to the same destination, but it will take a lot longer. He learns to read accurately, but to achieve the same level of reading as his nondyslexic classmate, he must read much more slowly and with greater effort. The automatic route to reading is unavailable to him. Consequently, if he is to identify many of the words on the page, he must pause and rely on the support of his higher-level thinking skills. He must survey the context and get to the word's meaning by this slower and more indirect pathway.

David Boies is, as I've said, dyslexic. He tries not to rely on notes in court, because he is not automatic in his reading. If he were to need to refer to his notes, he would not be able to capture the words on paper instantly; he would need time. This is the imprint of a lack of automaticity in reading. For others, that lack sometimes manifests itself in unexpected ways. For example, John Irving finds it frustrating that he cannot rapidly locate his flight on the departure monitor in an airport, though he tries very hard.

I look at that thing and I am the only one who is competitive about it. I think, *God damn it, I'm going to find that city and flight and gate number before [my wife] Janet.* . . . I'm just killing myself to find the flight to Dusseldorf or Paris or Helsinki and Janet will look up and say, "Oh, it's B-9," like that. Instantly. I can't do that.

What his wife, an automatic reader, can do at a glance, Irving, a dyslexic, must do manually, and therefore much more slowly.

EXTRA TIME AND THE NEED FOR QUIET

Brain imaging studies have provided us with a new level of understanding of the path dyslexic readers must take. To review briefly, when typical readers encounter a word, they automatically integrate all its features: the way the word sounds (phonology), the way the word looks (orthography), and the meaning of the word (semantics). For dyslexic readers there is inefficient functioning, particularly of those systems in the left occipito-temporal region, the word form area, and as a result dyslexics have to resort to slower, less efficient neural pathways to manually figure out the word and get to its meaning. This process takes much longer and consumes a great deal of energy. As I mentioned earlier, the results of brain imaging studies of children and adults show different brain activation patterns in dyslexic and typical readers, a pattern referred to as the neural signature of dyslexia (see Figure 23, page 78). Instead of the distinct activation observed in the back of the left side of the brain in typical readers, dyslexic readers demonstrate an inefficient functioning in that region, with activation seen in the left and right frontal regions as well as the right posterior region (see Figure 25, page 80). An auxiliary team of neural systems is trying to take over for the inefficient primary reading system, and it requires extra time. Dyslexic adults who are able to achieve average levels of accuracy have to do so by resorting to these backup, slower, secondary neural routes to reading.

From these and other studies of students, older adults, and children, we know that, as Maggie Bruck, currently a professor of psychiatry at Johns Hopkins Medical School, said, "The patterns of deficits that characterize dyslexic children are the very same patterns that characterize adults with these childhood histories. Despite the fact that many of the students have been educationally successful, and despite the fact that as adults they have increased their level of word recognition skills, these data suggest that the primary deficits have not dissipated." Brain imaging studies in adults with dyslexia confirm Bruck's clinical observations.

Studies of university students who are accurate but slow readers—in the United States, the United Kingdom, France, and Italy—indicate that the phonologic deficit and the inefficient neural function persist. So, as shown in Figure 26 (page 82), if the dyslexic reader is going to decipher the print before him, he will have to rely on higher-level sources of knowledge—vocabulary and reasoning—and slower, secondary neural pathways for reading. The equation for adult readers who are dyslexic goes something like this: *higher-level thinking skills + context + extra time = meaning.*

The bright dyslexic reader's reliance on context is both absolute and unique. Children who are poor readers but not dyslexic—for example, those who have across-the-board language difficulties—are generally not benefited by context. They do not possess the verbal skills, particularly the vocabulary and reasoning skills, necessary to help identify unknown words. It is only the dyslexic who possesses a sea of strengths she can apply to the surrounding context to help figure out the mystery word. Since the direct route to meaning is not available to a dyslexic reader, he must apply his intelligence, vocabulary, and reasoning to the context around the unknown word in order to get to its meaning. This means taking a secondary reading pathway that requires extra time.

Highly able dyslexic readers have the desire to go on to higher education, but they face the barrier of timed standardized tests, on which they more often than not perform poorly in relation to their knowledge and ability. As I said earlier, they are particularly penalized by timed multiple-choice exams, which typically provide sparse context and time constraints. These tests are not a fair measure of a dyslexic individual's ability but rather a measure of his disability (see Figure 49 below).

John Irving and the late Stephen J. Cannell, another writer, both scored poorly on the verbal SAT. The renowned academic physicians Dr. Delos Cosgrove and the late Dr. Graeme Hammond scored so low on their Medical College Admissions Test (MCAT) examinations that if not for unique circumstances in each of their cases, neither would have been admitted to medical school. David Boies and the financier Charles Schwab join the illus-

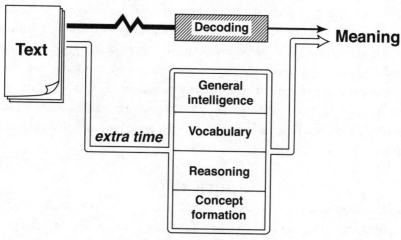

Figure 49. Using Context Takes Time

trious circle of highly gifted individuals whose performance on timed standardized tests vastly underestimated their abilities and almost kept each from pursuing and succeeding in his life's ambition. Dr. Carol Greider also had an underwhelming score on her GREs, which resulted in her acceptance by just two of the thirteen graduate programs to which she applied. She carried out her Nobel Prize–winning research during her graduate years at UC Berkeley. Apparently, as you read earlier, the admissions team there wisely saw qualities in Dr. Greider that were missed by the other program administrators, who were blinded by her low GRE scores.

SATISFYING THE GOALS OF A FOREIGN-LANGUAGE REQUIREMENT

You know by now that students who are dyslexic struggle with foreign languages. Since these students have yet to master the basic phonology of the language they have been immersed in since birth, they are hardly likely to become proficient in a new language when they are young adults. In secondary school, students can, and often do, slip by through rote memorization, diligence, and pleasant personalities, receiving passing grades without having the slightest ability to converse in that language. But in college students are expected to become proficient in a language, able to converse in it and read its literature.

At times students will come to realize the futility of their quest and request a partial waiver of the foreign-language proficiency requirement. Alternatively, an experienced instructor may be the first to appreciate the degree of a student's difficulties and recommend a waiver. As noted earlier, at Yale, after careful review, a partial waiver may be granted with the stipulation that the student is expected to take six course credits in the study of the culture, history, politics, or literature of any specific non-English-speaking culture or society. This model policy reaffirms the importance the university places on the foreign-language requirement and its desire to have students meet the spirit of the requirement rather than solely pass a course. This is a very useful and proper accommodation. Furthermore, it prevents needless suffering and waste of a student's time and energy while allowing him to focus on courses he has a chance of mastering.

DIGITIZED TEXTS

Just how helpful it is for dyslexic students to be able to listen to digitized or recorded text is exemplified by stories of dyslexic students who endured years of struggle before they began using text-to-speech technology. For example, as one relieved dyslexic college senior explained:

I could remember just about everything I heard, but read too slowly to keep up, and so much effort was put into trying to get the words out that I had little comprehension afterwards. Once I began using text-to-speech technology, everything changed. I started to feel much better about myself. I really was learning the material and I no longer made excuses to avoid doing my work because it made me feel bad. The audio changed my life. For the first time ever, I got straight A's.

Beyond schoolwork, dyslexia tears at the independence of the maturing adolescent. Having his texts digitized and available for text-to-speech technology helps give the student back his sense of independence. Accordingly, once a dyslexic child is in middle school or high school, plans should be made to have his textbooks digitized. Listening to the digitized books allows the student to participate in courses and to study at his level of understanding rather than be held back by his slow reading. Furthermore, digitized texts and text-to-speech technology introduce him to vocabulary words that he may not have encountered in his otherwise limited reading. Listening and following along with the text in front of him allows a student to actively dig into his reading, underlining, taking notes, and highlighting—important reinforcement activities that were never even considered when he was totally focused on deciphering the words on the page.

Building a bridge to success for the maturing dyslexic student requires three basic ingredients:

→ *Extra time and a quiet, separate room for examinations*
→ *Some form of waiver of the foreign-language requirement*
→ *Digitized texts coupled with text-to-speech technology for listening.*

In addition, there are other useful accommodations:

→ *Alternative testing formats* (short essays, oral reports, projects)
→ *Reliance on visualization* (figures, graphs, illustrations) rather than print alone
→ *Courses and grading that emphasize concepts versus isolated details* (content versus form).

To offer practical guidance for integrating these and other accommodations into a student's educational program, I am going to focus on Stephen and Gregory and on the plans I developed for each of them.

STEPHEN'S PLAN (HIGH SCHOOL AND COLLEGE)

Stephen Bennett was fifteen years old and about to enter his freshman year of high school. After he was evaluated at the Yale Center for Dyslexia & Creativity and diagnosed as dyslexic, I met with him and his parents and explained his diagnosis. For adolescents like Stephen who will soon be expected to function independently while attending college, it is particularly important to understand at a very basic level what dyslexia is and how it might affect his everyday life. Gaining this awareness and knowledge, Stephen would be better equipped to figure out how to deal with the impact of his dyslexia in a range of situations. He would gain autonomy and in the process become a powerful advocate for his own needs.

I gave Stephen the following summary outlining the predictable consequences of dyslexia and the most appropriate accommodations for each. Because students who are dyslexic learn best by first understanding the big picture, I began by providing Stephen with a conceptual framework summarizing the overall rationale and goals of the plan.

→ *Understand how you learn: your strengths and your weaknesses.* Visualize an encapsulated, circumscribed phonological weakness surrounded by a soaring sea of strengths (excellent reasoning skills, vocabulary, the ability to understand concepts).

→ *Remember the goal: Maximize your strengths, minimize your weakness.* Your plan is based on understanding that dyslexia is a problem accessing the basic sounds of spoken language that affects reading, writing, and speaking. Here are the five basic consequences of dyslexia for you and what you should do about each.

1. READING IS SLOW AND LABORIOUS

→ *Request extra time on examinations.* Extra time on examinations is a necessity. The amount of time cannot be determined from testing; it should be based on your own real-life experience. When you first request this accommodation, request time and a half. Based on how well this works for you, you can request more or less time on subsequent occasions. Since your decoding skills are not automatic, it takes you longer to read each word. In addition, many bright dyslexics like yourself use the context of a paragraph to help them identify a word that they cannot automatically decode. (See Figure 49, page 441.) This allows them to fill in what the unknown word probably means, even if they can't pronounce it. This indirect route to word meaning takes extra time. You absolutely need this extra time if you are to show what you really know. *Never feel guilty about requesting extra time. A dyslexic needs extra time just as a diabetic requires insulin.* A quiet, separate room and earbuds help to minimize distractions and allow you to concentrate as you read the test questions. Although you read words slowly, your comprehension is at a high level, because you can apply your skills in vocabulary and in thinking to figure out the meaning of a passage.

→ *Avoid taking too many courses with large volumes of reading.* You should limit yourself to no more than two *reading* courses per semester. If necessary, when in high school take summer courses. When you begin college, try to take a reduced course load the first semester (or even the first year), until you adjust to the academic and other demands of university life. You can take college summer courses or a fifth year in college as well.

→ *Pace yourself when you read.* Read for twenty minutes, take a break, exercise, return to your reading. Since your reading is fragile, read in a quiet room. As noted previously, rather than being an *automatic* reader, you are a so-called *manual* reader who must use up large amounts of attention during reading. Listening to background music such as jazz or classical—as long as it does not have lyrics—helps to block out or mask any distractions.

Getting the Most Information from What You Read. Many students do not know how to get the most information out of a passage they are assigned to read; they focus on all the details and remember few. But one approach to getting the most out of reading has worked for many students: encouraging *active* reading, which is the basis of remembering what you have read. This approach is called SQ3R. Each letter stands for a process that will help in your reading:

> *S—Survey.* Look over the chapter before you read it. Look at the title and subheadings of each section; read the introduction and the summary at the end. It helps to know what to expect from your reading. This puts you into a mindset for receiving the information.
>
> *Q—Question.* Change the headings into questions and then let these questions guide your reading. For example, in a biology textbook, if the section heading is "Functions of the Stomach," change that to "What are the functions of the stomach?"
>
> *R—Read.* Read the text to search for answers to the questions you have created. Highlighting or underlining the key sentences helps you to recall the information. Taking *brief* notes on the highlights of each section will also help you to remember the important points.
>
> *R—Recite.* Say your answers out loud.
>
> *R—Review.* After completing a section, go over the main points of your highlighted sections or your notes by reciting the information out loud and making sure you are correct.

This very simple process will help you organize your reading and gain more from it. There are other aids:

> → Dyslexic students frequently find *taking short notes on the computer* very helpful in their studying. In fact, many distill things even more, taking notes on their notes until they are left with a very small number of notes. Rewriting the notes is helpful in remembering and also creates a concise, abbreviated set of notes from which to study for a future examination. Choose from the several helpful note-taking apps (OneNote, Evernote, Notability, Livescribe).

→ *Active learning* along with *repetition and practice* is the key to remembering. Actively engaging with the material by highlighting, underlining, and even enlarging the print, if it is on a computer, are all helpful in making the material salient to you. Explaining or rephrasing specific ideas or concepts to yourself aloud and reading your notes aloud over and over again help you to keep information in your memory and to retrieve it when needed. Visualizing what you read may be especially helpful to you. Creating an exaggerated or ridiculous visual image, or any image that is meaningful to you in some way, will aid your memory. All these suggestions are consistent with current thinking about reading comprehension, which emphasizes that the goal of reading is not just to repeat by rote what the writer says but to go deeper into the text so that the reader makes inferential interpretations of what will happen and why. It is a good idea as you read to ask yourself questions having to do with your predictions of what will happen and what makes you think so.

→ *Obtain books in digitized form.* Obtain digitized versions of your textbooks from libraries such as Bookshare and Learning Ally. In college, have them scanned in by your resource office for disabilities. Ask your instructor for a book list *before* the start of the semester so that you can be sure of having the text digitized in time for class. Use text-to-speech software to listen to as you read the text. Hearing and seeing the words help you to pronounce and to remember difficult words. Listen actively. For example, you can highlight and underline important parts as you read along. Do the same for pleasure reading. This way you can get to enjoy the stories and literature you might not read in their printed form. Magazines are also available.

Modern text-to-speech programs allow you to select the rate at which the text is read and select words, sentences, and paragraphs to highlight. Options are available to change the font style, increase the white space between lines, and move the text around. A search function enables you to rapidly locate a particular word within the text. Specific sections of a text can be clipped out and saved as a separate text file. In addition, these programs provide bookmarks, dictionary functions, voice notes (you can dictate short notes attached to specific text sections and play them back when needed), and spell-checkers and will create an outline from any text. They also link up to and read Web-based emails and other Internet text.

Speech-to-text programs have options to read aloud and pronounce words that you have said or typed, so you can make sure you typed the word you intended. Other options include homonym support; for example, if you wrote, "I want to *heel* the sick," the program will read aloud the definition of *heel* and provide the spelling and definition of other like-sounding words. Once you see and hear the word *heal* read aloud with its definition, you can correct your spelling. There is a learning curve associated with text-to-speech and speech-to-text programs, and it may be helpful to locate someone you can turn to for expert advice.

Find Alternatives to Reading the Originals. If you have to read a book for a literature class, you might try to find an alternative version. For example, Shakespeare's *Hamlet* and Theodore Dreiser's *An American Tragedy** have been made into movies. Many students know that a very useful website for movies is www.imdb.com, which provides a complete catalogue of just about every film ever made so that you can check to see if an assigned book has been made into a movie. (Ideally you will watch the movie version as a preliminary to reading the novel, although there may be times when the text is so difficult or lengthy and your time pressure is so great that you are not able to read the novel in its entirety.)

Both Charles Schwab and David Boies told me that they learned a great deal from Classic Comics when they were younger. Now there are graphic novels and other graphic books that can serve the same function as comic books did for these two dyslexic readers.

In addition, a number of other options are available to the dyslexic reader.

---

* The most famous movie version is called *A Place in the Sun.*

For example, earlier I mentioned materials referred to as high interest, low readability, such as the Oxford School Shakespeare series available from Amazon. This series includes not only the complete Shakespeare plays but also clear explanations of difficult words and passages, a synopsis of the plot, summaries of individual scenes, and notes on the main characters. A similar series, Streamlined Shakespeare, is available from the publisher Gander (and via Amazon) and features works of Shakespeare rewritten so that someone reading at a fourth-grade level or higher will be comfortable. Many other high-interest, low-readability texts are available from a number of publishers, including High Noon Books, a division of Academic Therapy Publications (www.highnoonbooks.com). These series provide increasingly complex reading material in a very user-friendly format, incorporating both fiction and nonfiction. Using Lexia reading levels, your local library staff should be a great source of high-interest books at your readability level.

Watch the news on television, listen to it on the radio, and visit museums. In museums you can rent audio guides and listen to minilectures about the art. Even if you don't read a lot, there are alternative ways to get information and to learn about the world. Television, movies, YouTube, graphic novels, and even comic books are acceptable and encouraged if they are of help to you.

Preview Reading to Identify Words You Can't Pronounce. If you can't pronounce a word, the chances are you won't recognize it in a textbook or in a lecture the following day. Most text-to-speech software includes a dictionary. Many dyslexics get into the habit of using this dictionary to learn not only the meaning of unfamiliar words but their pronunciation as well. This is particularly important in courses such as biology and chemistry, where complex new terms often appear (such as *oxidative phosphorylation*).

Talk Through the Material with Your Teacher or Tutor on a One-to-One Basis. Since you have trouble getting information from printed material, you will benefit from meeting with your teacher (or a tutor). This usually works best after you have read the required pages. An active, one-to-one oral interchange will help you get at the concepts and remember them. Using new terms in such a dialogue allows you to get to know unfamiliar words and to recognize each more easily the next time you see it in print or hear the word in a lecture. You often will remember what you have *heard* much better than material you have *read*. In addition, try to join a study group when preparing for a test. Talking through the material and listening to the group often helps much more than sitting alone in your room trying to plow through volumes of material.

If You Know You Have Trouble with Multiple-Choice Tests, Avoid Them. You use context to get at words you cannot decode. Multiple-choice short-answer tests do not provide enough context to help you get to the meaning of difficult-to-decode words. Short essays are the best test format for you, allowing you to demonstrate the true level of your knowledge, and you should request them.

Avoid Speed-Reading Classes. They will not benefit you or any other dyslexic reader. Don't waste your time.

2. THERE IS A BASIC PROBLEM IN
ACCESSING THE SPOKEN LANGUAGE SYSTEM

→ *Obtain a partial waiver for foreign-language requirements.* Remember, dyslexia reflects a deficit in your most basic level of language processing: getting to the sound structure of words. Given your problems with learning your native language, you will invariably experience even more difficulty in acquiring proficiency in a foreign language.

→ *Visualize the material.* Visual imagery and visual study guides will be especially helpful for you. You will learn best when you can picture what is on a page or are able to convert the printed information into a visual format such as a chart or graphic. *Inspiration* software (www.inspiration .com) helps you to organize your ideas visually; with it you can create flowcharts, idea maps, tree charts, diagrams, and outlines. Inspiration has a very useful Web-based component, *WebspirationPro,* particularly addressing the needs of advanced high school and college students and a range of professionals. It is connected online and, using visual materials, permits a user to show her ideas, organize the related information, and create flowcharts and diagrams to indicate connections. Then, using Microsoft Word or Google Docs, the program will create a *written outline* reflecting the information gleaned from the visual schema. This is the answer to many a dyslexic's (and nondyslexic's) dream.

→ *Do not allow language problems to influence performance in other academic areas.* You may have problems on tests simply because of a language problem—you cannot easily or swiftly read the question or instructions. Having the problem read to you or listening to an audio exam can be very helpful in ensuring that a test measures your ability and not your reading skills. This is often the case when it comes to math word problems.

CAUTION: Beginning in 2017, the math SAT became language-heavy. This makes it difficult for dyslexic students who know their math very well but read slowly to have their score reflect their ability in math rather than their slow reading.

3. HANDWRITING IS LABORIOUS AND BARELY LEGIBLE

→ *Use your laptop or tablet and type everything you can.* Because you have trouble forming letters and words, you often write very slowly and illegibly. The best cure for this is word processing. Make sure you take your laptop or tablet everywhere you go (which you may already do). Use the computer for note-taking if the instructor speaks at a reasonable speed, and to write assignments. Always try to write your essays and tests on the computer. This will free you from the labor of writing by hand and help you focus on the content of your written work. Note-taking programs were discussed earlier. Remember that many colleges, if requested, arrange for note-takers.

→ *Record lectures.* Many dyslexic readers record their lectures using note-taking programs.

→ *Use speech-to-text to record your own essays and then text-to-speech to hear the essay.* As a student hears the words he has written read back to him, he has an opportunity to determine whether what he has written makes sense. These programs also have extensions allowing for word prediction and homonym correction. For example, if the student wants to say "from sea to shining sea" and instead writes "from see to shining see," the program will ask whether he means *sea,* referring to a body of water, or *see,* referring to the visual process.

4. GIVING ORAL RESPONSES ON THE SPOT IS SLOW AND LABORED

The same phonologic weakness that affects your reading can also affect your ability to come up with spoken words quickly. Oral language, too, depends on phonologic ability: You must go into your internal dictionary, access the appropriate phonemes, put them in the correct order, and then these will activate the muscles of articulation to form and then express the word. Ironically, the most difficult step, using higher thinking abilities to develop a concept and identify the abstract semantic representation of the word, is not a major problem. Rather it is the lower-level component, accessing and ordering the units of sound that represent the word and programing the articulatory muscles that utter the word, that is the troublesome component. The result is many *um*'s or misspoken words, along with much embarrassment on the speaker's part. (See Figure 13, page 59, and Figure 48, page 422.)

You may recognize these signs in yourself:

→ You have trouble expressing your thoughts in spoken words.
→ You mispronounce words or phrases.
→ You have trouble finding the correct word, often talking around a topic.
→ You have a problem reading aloud.
→ People say you tend to use imprecise words: *like, stuff, thing, you know.*
→ You tend to confuse words that sound alike.
→ You know what you want to say, but those words don't come out of your mouth.

If you have these kinds of issues, technically referred to as word-retrieval difficulties, that explains why you cannot come up with the exact answer when called upon in class, even when you think you know the answer. Once you understand the basis for your problems answering oral questions quickly, you can take steps to remedy the situation. Your teacher is an important partner in this; you must explain to her that your dyslexia impacts how you respond in class. Share the suggestions below with your teacher.

Give Prepared Short Oral Reports Rather Than Instant Oral Responses in Class. Given time, dyslexics can prepare and deliver excellent oral reports. Technology can help in an almost unending number of ways, especially in organizing and delivering such presentations. For example, you can learn to use *Power-Point* slides to prepare reports. This allows you to take advantage of your strong visual and thinking skills as well as to use the slides as an aid during the presentation. You will have to practice at home. Preparing a PowerPoint presentation (or simply making an outline on the computer) is also an excellent exercise in organizing large amounts of material and learning how to select the most salient points.

Phonological Slips Should Not Be Mistakenly Interpreted as a Lack of Knowledge. Knowing that you are dyslexic and understanding the impact of dyslexia on you as well as why you are experiencing difficulty can be enormously helpful. For example, if your oral response sounds off the mark, suggest to your teacher that it may be a phonological error and she should consider allowing you to explain. Dyslexics often have difficulty in finding the precise or specific word. You should be asked to elaborate or more fully describe what you are trying to say. Tell your teacher, "I misspoke. Let me explain what I mean."

Your Written Work Should Be Graded on Content Rather Than on Form, Especially Spelling. Many dyslexic readers conceptualize at a very high level, but as a result of a phonologic weakness they experience difficulty with handwriting and spelling. The same difficulties decoding words (going from letters to sound) show up when you try to carry out the reverse process, spelling, which relies on encoding sounds and transforming each sound into letters. Because you are unable to encode the exact model of a written word (that is, its spelling), you will also confuse homonyms. You will not notice that you wrote *heal* for *heel* or *sail* for *sale*. As mentioned earlier, many of the text-to-speech and speech-to-text programs offer homonym help. Spell-checkers are also a great help. Even with the clever software packages currently available, it still may be helpful to have someone proofread your written work. Because spelling errors reflect your dyslexia, essays should be graded for content and grades should not be lowered because of spelling errors.

5. FOR DYSLEXICS, LEARNING IS MOST OFTEN A TOP-DOWN APPROACH, GOING FROM MEANING TO FACTS

Select courses where the emphasis is on concepts, not on details. You will do much better in courses where the emphasis is on your strengths—understanding broad concepts and ideas, using reasoning and analytic skills—rather than on memorizing isolated bits of information. Rote memory requires strong phonologic abilities. Dyslexics best remember information that is attached to meaning. Before selecting a course, find out if you are expected to memorize specific facts or instead to demonstrate your knowledge through projects and reports. The latter will usually be much better for you. Before the course begins, don't be afraid to discuss with your instructor how you will be evaluated. Be your own advocate: Explain what you need and why.

Stephen and his parents met with the principal of his high school and presented the results of my evaluation and recommendations. They were pleasantly surprised by the principal's positive reaction. He seemed relieved; we had given him a plan of action.

The new information produced a significant turnaround for Stephen at school. He and his teachers were surprised and pleased at what a difference extra time and a quiet room made in his test performance. Some of his teachers now offered to provide Stephen with copies of their own lecture notes for him to preview before class, and his classroom contributions improved significantly. He substituted a course in French culture as a partial waiver of the foreign-language requirement. In high school he used his laptop for all his courses and was amazed at how helpful it was to type all his work. Seeing Stephen use his laptop in class resulted in several of his nondyslexic classmates also using laptops. Stephen was pleased, because this made him stand out less. Like many other dyslexics, Stephen wanted more than anything to appear like everyone else and not be noticed.

Stephen went on to college, incorporating my recommendations. He graduated with honors, went on to business school, and is now doing well on the mergers-and-acquisitions desk of a prominent investment bank while working with his company on helping to identify and support employees who are dyslexic.

GREGORY'S PLAN (MEDICAL SCHOOL)

You already know about Gregory, the bright young man whom I wrote about in *Scientific American.* He was experiencing frustration in medical school and knew it didn't have to be that way. I met with the dean of student affairs and told him that a few critical steps would make a significant difference. Operationally, this meant that Gregory's professors had to recognize and attempt to accommodate rather than penalize Gregory. In words that are now familiar to you, I told the dean that a dyslexic like Gregory cannot provide instant answers, but he can learn and learn very well if given the opportunity to do so. I said it was *important to separate the learning process from the evaluative process.* For example, during clinical clerkships, a first attempt or response is often regarded as the final evaluation. But Gregory had to be given an opportunity to learn without fear; if he responded hesitantly or incorrectly on the first try, it should not be taken as an indication of a failure to learn. *His evaluation must not confuse oral language facility with ability.* Accordingly, *Gregory must be evaluated on the basis of his knowledge and reasoning skills, not his glibness or speed of verbal response.*

I said that for each course Gregory needed to meet regularly with someone knowledgeable with whom he could preview and talk through the content. This would help overcome his problem with rapidly retrieving words. Faculty members, fellows, senior residents, and retired faculty members have served effectively in this capacity. I said it was critical that this individual *not* be a part of the evaluative process.

I also said it would make a real difference if there was an individual who would function as a faculty advocate and mentor to Gregory, someone who would represent his needs to the clinical course directors. Such a person should be a member of the academic faculty who carries the authority and respect of the institution, has an understanding of the nature and consequences of dyslexia, and is genuinely supportive. Having a respected faculty member run interference for the student makes a world of difference. The student may say the same things but with little effect.

Finally, I said that the evaluative process must be a reflection of Gregory's ability and not a measure of his disability. I told the dean about how unfair multiple-choice examinations were for Gregory and why. This had been demonstrated during Gregory's obstetrics/gynecology rotation. During the six weeks of this course, his performance was observed around the clock by a range of attending doctors and team members; the consensus was that he performed at an honors level. In stark contrast, he failed the multiple-choice written examination—a result he had anticipated and spoken to the course director about. The extreme disparity between his excellent clinical performance on the wards and the multiple-choice examination speaks for itself and clearly documents the inappropriateness of this type of examination to measure Gregory's knowledge.

A dyslexic medical student like Gregory who has had to adapt to and successfully cope with a chronic disability for virtually his entire school career clearly has a great deal to offer as a physician, including many of the most meaningful yet least teachable qualities: compassion, empathy, and sensitivity as well as intelligence and motivation. With the kind of support and understanding I indicated, Gregory would thrive.

Most of my recommendations were adopted, and Gregory graduated. He was then accepted into a very competitive residency, which he easily completed. He received the accommodation of extra time to take his specialty board certification and is currently director of a large medical unit in his specialty. He reads voraciously, not only professional journals but newspapers and many periodicals, including *The New Yorker* and *The Economist*. All the challenges posed by dyslexia and his overcoming them have made him an exceptionally strong leader. He is both a great problem-solver and fearless.

Our new level of understanding of dyslexia (and learning) and of the importance of accommodations is now being transformed—through the requests of individual students, through the insights of perceptive teachers, and through changes in public policy—into real and lasting changes in educational practices and policies. The scientific basis of, the urgent need for, and the life-transforming difference made by the provision of accommodations have been proven and are rapidly taking root in classrooms across the country. As a result of these much-needed adaptations, dyslexic students will now have the opportunity to develop their full potential. Not only will the individuals benefit, but society as a whole will be a winner, too.

ANSWERS TO THE MOST FREQUENTLY
ASKED QUESTIONS ABOUT ACCOMMODATIONS

HOW IS DYSLEXIA IDENTIFIED IN COLLEGE AND GRADUATE STUDENTS?

Dyslexia is a clinical diagnosis; it must be made by a clinician who knows the student and who is able to thoughtfully synthesize the pertinent historical information, clinical observations, and results of relevant testing. The diagnosis is based on a pattern of findings; it should never be made or ruled out on the basis of an isolated test result. Tests are just proxies. A person's lifelong history and the reality of how that individual reads, especially how he reads aloud, should not be neglected.

The hallmark of dyslexia is an *unexpected difficulty* in reading in relation to the person's other cognitive and academic abilities. In college (and beyond), the most valid approach relies on the person's accomplishments—academic performance (outstanding grades in math, philosophy, or chemistry), educational status (attending or a graduate of a competitive college, graduate, or professional school), or professional status (attorney, physician, writer, engineer, or successful businesswoman). Within this context, a college student, a law school graduate, or a physician who reads slowly and with great effort is manifesting an unexpected difficulty in reading.

Reading accuracy should not be used as a measure of reading proficiency in a bright, educated adult. By adolescence, a student who is dyslexic often will improve his ability to read words accurately so that measures of word accuracy are expected to be in or to approach the average range. Consequently, the finding of average word-reading skills in an educated adult dyslexic is not very helpful in determining if he is experiencing difficulty in reading. Scientific evidence shows that adults with childhood histories of dyslexia who appear to be able to identify words accurately are reading those same words differently from others; they are reading them more slowly and using different brain systems.

Lack of fluency is the only true index of dyslexia in a bright adult. How the person reads *aloud* is a critical measure of fluency: Does he stumble over words, hesitate, mispronounce, omit, or add words as he reads? Does he read with *prosody* (expression)? Prosody is a key criterion for fluency, often overlooked and underappreciated. Prosody provides insight into how well the reader *understands* what he is reading. Does he adjust his reading to reflect the meaning, reading faster or more slowly, raising his voice or whispering, depending on the message being conveyed by the words on the page?

ARE THERE OTHER PROBLEMS
THAT COULD BE CONFUSED WITH DYSLEXIA?

In a bright postsecondary school student or an accomplished adult, very few problems can reasonably be confused with dyslexia. If the pattern I have outlined in this book has been established in someone, she is dyslexic. Only dyslexia produces the clinical syndrome characterized by the paradoxical pairing of a phonologic weakness and higher-level cognitive strengths expressing themselves throughout the person's life. If a person has a history of phonologically based speaking, reading, and spelling difficulties going back to childhood and now reads very slowly despite signs of her cognitive strengths, that person is dyslexic.

Sometimes ADHD is confused with dyslexia; it shouldn't be. As you know, dyslexia reflects difficulty getting to the basic sounds of language; ADHD reflects problems with the modulation of attention and activity. The symptoms, the neurobiology, and for the most part the effective treatments differ. However, at times a person who is dyslexic may appear not to be paying attention to her reading because it is so demanding for her to decipher the words on the page. It's not that she has a primary attention problem but rather that reading, for her, requires an unusual outlay of attention. Having said this, it is important to remember that there is a very large co-occurrence of dyslexia and ADHD. (See pages 425–429 for a more detailed discussion of ADHD and dyslexia.)

WHAT IS THE BEST WAY TO ESTABLISH HOW MUCH
EXTRA TIME IS NEEDED FOR AN EXAM BY A PERSON WHO IS DYSLEXIC?

The only valid gauge is the person's own life experience. There is absolutely no test that can provide this information. Each accomplished dyslexic has developed her own route around her phonologic deficit. The specific strategies or alternate pathways that she has perfected over the years and found to work for her will determine how much extra time she requires.

HOW OFTEN SHOULD A PERSON BE TESTED?

Dyslexia is a persistent, lifelong condition; it is not outgrown. Once a student has been identified as dyslexic and has received accommodations, there is no logical reason to believe that there will come a time when he will not require this accommodation for timed tests in order to access his cognitive strengths. Scientific studies have clearly demonstrated that the thread of the phonologic deficit persists throughout a person's life; imaging studies demonstrate that even so-called compensated dyslexic students continue to call upon secondary neural routes to read, and these result in reading more slowly than their peers.

> → *There is absolutely no evidence to indicate that a person who is dyslexic as a child will as an adult magically metamorphose into a person who is a fluent reader and uses traditional primary brain systems for reading. Requests for repeated assessment are without merit. There is no evidence of benefit from asking students past high school who have recently been appropriately assessed and diagnosed as dyslexic to go through the expense and the psychological trauma—and the potential for misinterpretation—of a new series of evaluations.*

Many believe that the request for frequent testing of the dyslexic once he has completed high school represents an artificial barrier raised to discourage students from applying for accommodations. The requirement for additional time for a dyslexic reader is so fundamental that there is no tenable counter-argument. There is no plausible rationale for believing that a person who is dyslexic will no longer require extra time.

DO ACCOMMODATIONS POSE AN UNFAIR ADVANTAGE?

Researchers have compared the performance of dyslexic and typical-reading college students on timed and untimed tests. The results were consistent: Only students diagnosed as dyslexic actually showed a significant improvement in test scores with additional time. Those who worry about the "advantage" of extra time for a person who is dyslexic fail to appreciate that even with additional time, this individual will continue to feel rushed. *Extra time is not an advantage; it is an attempt to level the playing field.* Even with the additional time, a slow reader will continue to feel at least the same or more time pressure compared to the ordinary reader. No one wants to feel different, including dyslexic boys and girls, and if not absolutely necessary, they would much rather not require accommodations. For the most part, being treated differently brings about embarrassment and shame. Students request and accept accommodations because they have no choice if the exam is to be a measure of their ability rather than their slow reading.

DO PEOPLE FAKE DYSLEXIA
JUST TO GET THE PERKS ASSOCIATED WITH IT?

The belief that somehow parents, particularly middle-income parents living in the suburbs, are seeking a diagnosis of dyslexia to gain some imagined advantage for their child is insidious, unfounded, and malicious nonsense. Such harmful attitudes create a backlash that is harmful to children who are struggling to learn to read, and to their parents, who are struggling to help them.

The Blue Ribbon Panel, appointed to examine flagging on accommodation test results for standardized tests (see page 485), concluded that *rather than overidentification of suburban students as dyslexic, the problem was the relative underidentification of disadvantaged, especially minority, children.* The panel strongly urged that the major testing agencies make proactive efforts to inform these disadvantaged dyslexic students of their rights to testing accommodations.

HOW EARLY SHOULD ACCOMMODATIONS BEGIN?

Once a child is expected to write, to complete time-consuming class or homework assignments, and/or to take standardized tests, consideration should be given to providing accommodations.

→ *Teachers should be encouraged to grade a student who is dyslexic on the basis of her thinking and not her spelling.*
→ *If homework seems to go on for hours, the student should be allowed to reduce her homework assignment.* For example, she could complete every other question.

Reducing the time required for completing assignments is critical; it is demoralizing and self-defeating for a student in third or fourth grade to spend hour after hour struggling to complete her work at home. While text-to-speech technology is a helpful accommodation for older students, it must be used very cautiously and sparingly in younger children, who must be encouraged to read as much as possible.

When providing accommodations, a child's dignity and sense of privacy must be considered. Providing accommodations most sensibly and most effectively should become part of the planning discussions routinely held within each school, and not carried out as a spur-of-the-moment, and sometimes thoughtless and hugely embarrassing, gesture.

IS THERE A DOWNSIDE TO REQUESTING ACCOMMODATIONS?

Because of the all-too-frequent misunderstanding associated with the diagnosis of dyslexia, most adults and children prefer to keep their dyslexia private. In a recent article my associates and I discussed the pros and cons of asking for accommodations in professional school:

> The medical student with dyslexia is frequently faced with what seem to be two very poor choices: if the student self-identifies as having dyslexia and asks for what he physiologically requires—the accommodation of extra time—his instructors and other students might think he is not very smart or is trying to game the system. However, if the student does not self-identify, he will not receive accommodations to which he is entitled by law and will not be able to demonstrate his knowledge.

HOW DOES THE NEED FOR ACCOMMODATIONS ON EXAMS INFLUENCE THE KINDS OF WORK THE DYSLEXIC PERSON WILL BE ABLE TO DO ONCE HE IS OUT OF SCHOOL? WILL HE BE ABLE TO DO "THE JOB"?

Most dyslexics can and do succeed on the job. A dyslexic generally does not have a problem practicing medicine or law or engineering or writing novels or plays—if he can survive the hurdle of timed multiple-choice tests. These tests tap a dyslexic's weakness.

Having said this, dyslexics will be relatively more successful at some occupations than others. The jobs with which they will have difficulties tend to involve dependence on lower-level skills, with very little requirement for higher-level thinking abilities. A person who is dyslexic would not enjoy work entirely centered on clerical or filing duties.

In law, the ability to read quickly or to carry out rote mechanical skills is often confused with the ability to think and reason. Early on, few would have predicted that David Boies, who did not learn to read until third grade, would become the leader in a field that seems to be so dependent on reading. Yet Boies's life experience reinforces the dictum that it is not how fast you read but how well you *think* that counts. Boies's approach is to skim through the text until he recognizes what is important; then he slows down, hones in on this material, and carefully reads the critical facts. He has an uncanny ability to grasp exactly what is important. Once he has set his sights on the most meaningful target for further analysis, he can concentrate on this much-reduced volume of material and carefully analyze it.

Earlier I mentioned dyslexic physicians, including surgeons, who have thrived. What flies in the face of so much that we assume to be related to surgical expertise is that some of the most renowned surgeons failed anatomy in medical school. One of those surgeons told me that anatomy as usually taught in the first year of medical school is not related to surgical skill. "It actually has little to do with the practice of surgery," he told me. "Performance in a first-year anatomy course has more to do with the ability to memorize the names of assorted body parts by rote." Most surgeons are in agreement that "surgery has to do with thinking, with judgment, with knowing what to do, and not with knowing the names of particular structures." Hopefully, this information will caution medical educators to be more thoughtful in their career counseling. They should not assume that dyslexic medical students who have trouble memorizing anatomic labels cannot become outstanding surgeons.

Slow-reading lawyers, poor-spelling writers, and surgeons who struggled with anatomy in medical school—they all flout conventional wisdom. From these individuals we learn that reading slowly tells nothing about the ability to comprehend, that poor spelling has little to do with one's ability to write creatively, and that an inability to memorize or rapidly retrieve the names of anatomical structures does not portend one's skills in operating on those same body parts. Many school advisors would have counseled each one of these individuals against pursuing their chosen career, in which they have found such fulfillment while also contributing to society. I hope that educators and parents alike will encourage children who are dyslexic to pursue their dreams.

A CAUTIONARY TALE

I want to share with you what a wonderful young woman, Skye Lucas, whom you met earlier, when she was a yet-to-be-identified dyslexic struggling reader, learned about the importance of accommodations in higher grades. Skye sought to enroll in a high school that had some understanding of dyslexia, especially accommodations. The school she chose had indicated to Skye and her mom during a preselection interview that it was very dyslexia-friendly.

Virtually from the moment she arrived at the school, she was struck by how homework-heavy it was. She spoke to her teacher, explaining that she was dyslexic and asking, "What should I do if forty minutes go by and I haven't finished the reading due for tomorrow?" The response: "Just try to get it done in forty minutes. I don't see why you wouldn't." "That was really the first time I noticed that, basically, I was lied to about whether or not this school was actually friendly toward dyslexics about accommodations," said Skye. (Assurances cannot always be trusted. Parents of dyslexic children must always dig deeply.)

Skye was disappointed by the school's approach to spelling errors on history tests and the like. "Even if I had the correct answer, points were taken off if I spelled *separate* with an *e* instead of an *a*. It was very demoralizing. . . . I told my teacher, 'I'm not supposed to get points taken off for spelling. It's part of my accommodations.' As soon as I said that, she snapped back at me, 'Just because it's part of your accommodation sheet doesn't mean that I have to give it to you.'"

For Skye, extra time was critical, and knowing that her school had a policy that students had to request additional time at least two days before the actual exam, she always made sure to do just that. However, other dyslexics at the school at times were so filled with anxiety about their upcoming exam that they occasionally forgot and did not request the accommodation in time. Amazingly and sadly, they were refused additional time, although they had been approved for this accommodation and school personnel knew it was necessary. More often than not, these students could not finish the test and failed. You can only imagine how heartbroken and fearful they became.

The school also promised a learning specialist to help Skye, and she was understandably excited about this. "There was a learning specialist in the school—Merle," said Skye. "Her office was a closet hidden in the art room, separate from the school. It was difficult to get there, and it seemed that this was a covert thing that the school was trying to hide. Predictably, I pretty much had to explain to Merle what dyslexia was as well. This was kind of frightening and a presage of what my school year was going to be like."

This school wasn't going to work for Skye. The need to memorize dates and facts and to spell words exactly right left this very bright student feeling dumb. She and her family set out to find a school that understood both dyslexia and accommodations. This time they found it: Dwight School in New York City.

> They have a learning center called Quest. It is not in a closet; it's actually a full room of teachers who work with students, one that is located pretty central, next to classrooms. And best of all was the head of Quest, who really understood and strongly supported accommodations; he also knew how to speak about dys-

lexia and talk with a dyslexic. Both he and the teachers at Dwight weren't clueless about accommodations; they didn't try to fight me on accommodations. They were actually very friendly about my accommodations and didn't make me feel bad about needing them. They were understanding, which really made a difference. I received extra time on testing quizzes and I didn't have points taken off for spelling. You don't know how important something like accommodations are until you don't have them or [they] are given to you with visible resentment.

Skye is precise in what she would tell her old school she wished they knew:

I would tell them that they really need to be more emotionally supportive, especially in understanding what dyslexia is and understanding why a dyslexic would need certain accommodations. When I was at the other school, I really felt emotionally attacked by teachers 24/7. I think I was attacked because of my accommodations. I really feel my current school understands me, not only academically or in terms of accommodations, but also emotionally.

Figure 50. Penn, class of 2021

Listening to Skye and so many others I have spoken to, there can be no doubt of the critical importance of the school's understanding of dyslexia and the importance of accommodations, provided with emotional support. Yet even at this new, more understanding school, Skye was advised not to apply to the University of Pennsylvania because she would "never get in."

In the fall of 2017, Skye Lucas entered Penn as a freshman and was invited to become a member of the very special literary space known as the Kelly Writers House. As Skye's mother, Geralyn, shared with me, "Who ever thought a dyslexic who failed first grade would be recruited to the Writers House at the University of Pennsylvania?"

At the beginning of 2018, I was attending a birthday party for one of my grandchildren when my cell phone rang. Much to my happy surprise, it was Skye, who was calling to tell me two things. First, as she has done many times, she shared how she thinks of me very often and how my unwavering confidence in her has made such a positive difference in her life, propelling her to go forward fearlessly, and how my caring so much about her was the subject of her Bat Mitzvah speech. Second, Skye told me that despite her high school counselor's warning not to apply to Penn because she would never get in and, on the low probability that she were to be admitted, she would never be able to do the work at such a rigorous college, she was doing beautifully at Penn, and she joyfully added that she had received her first-semester grades: Her GPA was a robust 3.8.

Skye's story is one each reader of this book must keep in mind, whether you are dyslexic or a parent or a teacher of a dyslexic student.

33

THE LAW IS ON YOUR SIDE

Since its first description over a century ago, dyslexia has been recognized as an *unexpected difficulty in reading for an individual who has the ability to be a much better reader.* Unfortunately, in some circles the basic component of dyslexia—its *unexpected* nature—has been overlooked, including in its definition. Now its unexpected nature has finally been codified in the most up-to-date, scientifically validated definition of dyslexia in the recently passed First Step Act, which was signed into law as Public Law 115-391. Given its extraordinary importance and implications for those who are dyslexic, I am going to restate it here: "The term 'dyslexia' means an unexpected difficulty in reading for an individual who has the intelligence to be a much better reader, most commonly caused by a difficulty in the phonological process-ing (the appreciation of the individual sounds of spoken language), which affects the ability of an individual to speak, read, and spell." In addition, the law mandates a "Dyslexia Screening Program . . . a screening program for dyslexia that is (A) evidence-based with proven psychometrics for validity; (B) efficient and low-cost; and (C) readily available."

This legal definition was preceded by Senate Resolution 680, which was passed by a unanimous and bipartisan vote in the U.S. Senate in 2018. In addition to the definition included in the law, this seminal resolution empha-sizes that dyslexia is a paradox, so that an individual with dyslexia may have both weaknesses in decoding that result in difficulties in accurate or fluent word recognition and strengths in higher-level cognitive functions such as reasoning, critical thinking, concept formation, and problem-solving. The resolution also notes that "the achievement gap between typical readers and dyslexic readers occurs as early as first grade" and reemphasizes the critical importance of "early screening for, and early diagnosis of, dyslexia . . . for ensuring that individuals with dyslexia receive focused, evidence-based intervention that leads to fluent reading, promotion of self-awareness and self-empowerment, and the provision of necessary accommodations that ensure success in school and in life."

If passionate advocates for dyslexic individuals want them to benefit from scientific progress, we must join together to ensure that this wonderful progress is translated into action. We worry that people are fearful of change and of letting go of an out-of-date definition that has now been replaced by one reflecting new knowledge and a new level of understanding of dyslexia. This contemporary definition truly defines dyslexia, noting not only its unexpected nature but also the role of a phonological deficit; the incorporation of the Sea of Strengths model; its prevalence in relation to that of general learning disabilities and in the overall population; the critical role of early identification; the great importance of evidence-based interventions to support self-awareness and self-empowerment; and the role of accommodations to enable dyslexic students to look forward to happy and fulfilling lives. Because there is so much ignorance and confusion about exactly what dyslexia is, this up-to-date definition will allow parents, educators, and policymakers to better understand and educate boys and girls who are dyslexic. We have taken a major step forward in closing the action gap that so often surrounds dyslexia.

This up-to-date definition carries into action the plea made in my congressional testimony in September 2014 and to the Senate HELP Committee in May 2016, where I stated the necessary goal of bringing together twenty-first-century science and education and closing the action gap in translating scientific progress in dyslexia into policy and practice. This new law and resolution represent a win for every child (or adult) who is dyslexic, for his or her parents and community, and for our entire nation, but state laws still need addressing. Today over forty states have dyslexia laws, and these laws vary greatly. Some are not very accurate.

To help you plan an effective implementation strategy for your child, your student, or yourself, it's important first to review the three federal laws that affect dyslexic students.

→ The Education for All Handicapped Children Act (EAHCA), enacted in 1975 and then updated as the Individuals with Disabilities Education Act (IDEA) in 1990 and updated again in 2004, is designed to ensure that grade-school and secondary-school students with a disability are provided a free appropriate public education (FAPE) in the least restrictive environment (LRE) tailored to their individual needs. One of the law's pillars is that students with a disability are entitled to an individualized education program (IEP) that spells out the services to be provided. The law specifies fourteen categories to define students with a disability who should be guaranteed a free and appropriate public education. One of those fourteen is the category of "specific learning disability," which includes dyslexia. Children require an IEP in order to receive special education services.

→ The Americans with Disabilities Act (ADA), first passed in 1990 and then amended in 2008 (the ADA Amendments Act, or ADAAA), prohibits discrimination based on a person's disability. It is meant to level the playing field for people with disabilities, including those who are dyslexic. The ADAAA, discussed in detail below, is critical in ensuring that dyslexic students receive accommodations on high-stakes tests such as the SAT, ACT, and examinations for graduate and professional schools.

→ Section 504 of the Rehabilitation Act of 1973 (commonly referred to as 504) is frequently applied in cases where students may not qualify for an IEP yet may require accommodations. While a 504 designation may provide your child with accommodations, it does not provide the extensive protections and services for children found eligible for an IEP under IDEA, including interventions for dyslexia.

With this as a preface, let's review some of the most important and relevant components of these laws directly applicable to protecting and supporting dyslexic students. Because of its direct relevance to obtaining educational services for children with dyslexia, we begin with one of the most important components of the IDEA, the IEP.

THE IEP

The first step in the IEP (Individualized Education Program) process is determining whether the child meets the criteria for disability as defined under the IDEA. While procedures for determining eligibility differ in each state, the first step in the process is a referral. This typically is a written request for an evaluation to determine if the child qualifies as a child with a disability. The referral can be initiated by either her school or her parent and must be carried out within sixty days. Parents have the final decision on whether they agree to the referral or not. The evaluation is typically done by the school. If you believe the evaluation is not accurate, you have the right to request an independent evaluation. If you request this early enough in the process, the school may pay for this independent evaluation, though if there is a time constraint you may need to pay for this yourself. Make sure that the professional you select to perform the independent evaluation is someone who is comfortable using the word *dyslexia* and diagnosing dyslexia. Often this can be accomplished by asking the evaluator, "Do you diagnose dyslexia?" A negative or wavering response will tell you that this may be someone to avoid.

In order for your child to demonstrate that she is eligible to receive services under IDEA, she must meet two requirements:

→ She must have a disabling condition, dyslexia, which is included under specific learning disabilities.
→ The impact of the condition (her dyslexia) must create a need for special education services.

Once it has been determined that she meets these requirements, the school district must convene a team to develop the child's IEP.

→ There are well-defined rules about when the team must be convened (within thirty days) and who will participate at that meeting:
  • the parents of the child
  • a regular education teacher
  • a special education teacher
  • a school administrator who can design the program, knows about the general education curriculum, and knows the availability of resources

- an individual who can interpret the instructional implications of the evaluation
- sometimes the child, and, at the parents' or school's discretion, others who have knowledge or special expertise

→ The law requires that the IEP must contain specific information:
- the child's present levels of academic achievement and functional performance
- measurable annual goals
- an explanation of how the child's progress toward meeting the annual goals will be measured
- special education and related services and supplementary aids and services that will be provided to the child
- an explanation of the extent to which the child will not participate with nondisabled children
- appropriate accommodations necessary to measure academic achievement and functional performance on statewide and districtwide assessments
- the beginning date for providing services and modifications
- a statement from the IEP team if it determines that the child must take an alternate assessment in place of the statewide and districtwide assessments

IDEA ensures that parents play a central role in the IEP process, beginning with the referral, continuing with evaluation, and then proceeding to the creation and implementation of the IEP. My strong recommendation is for you as your child's advocate and champion to make sure you avail yourself of these legal safeguards and obtain the optimal services your child requires.

PRESERVING THE RIGHT TO ACCOMMODATIONS: THE AMERICANS WITH DISABILITIES ACT AMENDMENTS ACT OF 2008 (ADAAA)

The Americans with Disabilities Act Amendments Act of 2008 (ADAAA) is the most important federal law affecting your or your child's ability to obtain accommodations, especially on high-stakes tests. My philosophy about overcoming dyslexia is that *strength and victories come from the empowerment gained through knowledge.* Knowledge about the law and policy affecting the granting of accommodations is essential to that. There has been a cosmic shift in the law and the attitude of government agencies responsible for enforcing the law in favor of granting requests for accommodations to bright and academically high-performing dyslexic students. The more specifics you know, the more you will be able to forcefully advocate for your child or yourself.

It is important for you to be aware of what the law says, because it is often in conflict with what testing agencies claim. Over the years after the passage of the original Americans with Disabilities Act (ADA) in 1990, progressively restrictive judicial decisions posed a serious setback to all that had been achieved in gaining civil rights for the disabled, including those who were dyslexic. The dyslexic community owes a huge debt of gratitude to a group of especially caring and committed congressional representatives and senators who crafted the Americans with Disabilities Act Amendments Act of 2008. The ADAAA made clear Congress's intent in passing the law so that there would be no question when cases came before the courts as to what Congress had in mind.

THE DEPARTMENT OF JUSTICE PROTECTS THE RIGHTS OF DYSLEXIC APPLICANTS TO OBTAIN ACCOMMODATIONS

In 2016, the U.S. Department of Justice published an extremely important document that codifies the critical issues Congress sought to effect with the passage of the ADAAA: "The Final Regulatory Assessment, Final Rule—Amendment of ADA Title II and Title III Regulations to Implement ADA Amendments Act of 2008." Here are its most important elements:

→ In a change from previous regulations, dyslexia, and not learning disability, is considered the primary central entity. Throughout, the DOJ refers to "dyslexia and other learning disabilities."

→ The term *disability* is to be *interpreted broadly.* The primary object of attention in cases brought under the ADA should be whether covered entities have complied with their obligations not to discriminate based on disability.

→ Congress instructed that "the question of whether an
individual's impairment is a disability under the ADA
should not demand extensive analysis." Specifically,
comparison of an individual's performance of a major life
activity to that of most people in the general population
usually will not require scientific, medical, or statistical
evidence.

→ Types of evidence that are less onerous to collect, such
as statements or affidavits of affected individuals, school
records, or determinations of disability status under
other statutes, should in most cases be considered
adequate to establish that an impairment is substantially
limiting.

→ The Final Rule emphasizes that Congress clearly intended
the ADA Amendments Act to expand coverage: "It is our
expectation that because this bill makes the definition
of disability more generous, some people who were not
covered before will now be covered."

→ When considering requests for modifications,
accommodations, or auxiliary aids or services, the entity
should *give considerable weight to documentation of past
modifications, accommodations, or auxiliary aids or services
received in similar testing situations* or provided in response
to an Individualized Education Program provided under the
IDEA or a plan describing services provided under section
504 of the Rehabilitation Act of 1973 (often referred to as a
section 504 plan).

→ The ADA as amended expressly prohibits any consideration
of the ameliorative effects of mitigating measures
when determining whether an individual's impairment
substantially limits a major life activity.

→ Importantly, an example in the Final Rule illustrating
dyslexia shows a *discrepancy* between the person's age,
measured intelligence, and education and that person's
actual versus expected achievement.

→ One of the most important elements in the Final Rule is
the concept of *condition, manner, or duration.* These terms
are critical in interpretation of the rules of construction for
"substantially limited." Specifically, it is made clear that it

could be useful to consider, in comparison to most people in the general population, the *conditions* under which an individual performs a major life activity; the *manner* in which an individual performs a major life activity; or the *time it takes* (duration) an individual to perform a major life activity.

→ These three critical factors, *condition, manner,* and *duration,* provide significant support to bright dyslexics requesting accommodations. Giving deference to how a person reads is a major leap forward. The score a dyslexic obtains on a test does not reveal the extraordinary effort she must put forth to obtain that score—much more than most people who obtain the same score.

The Final Rule notes that the language regarding the "*difficulty, effort,* or *time* required to perform a major life activity" should prove extremely helpful to dyslexic individuals asserting a need for testing accommodations.

→ An individual with dyslexia may need to approach reading or writing in a distinct manner or under different conditions than most people in the general population, possibly employing aids including verbalizing, visualizing, decoding, or phonology, such that the effort required could support a determination that the individual is substantially limited in the major life activity of reading or writing.

In passing the ADA Amendments Act, Congress clarified that courts had misinterpreted the ADA definition of *disability* by, among other things, inappropriately emphasizing the capabilities of people with disabilities to achieve certain outcomes. For example, someone with dyslexia may achieve a high level of academic success but may nevertheless be substantially limited in one or more of the major life activities of reading, writing, speaking, or learning because of the additional time or effort he must spend to read, speak, write, or learn compared to most people in the general population.

The Final Rule notes that some courts have found that students who have reached a high level of academic achievement are not to be considered individuals with disabilities under the ADA, as such individuals may have difficulty demonstrating substantial limitation in the major life activities of learning or reading relative to "most people." The Final Rule does not support such a view. It states:

→ "When considering the *condition, manner, or duration*
   in which an individual with dyslexia performs a major
   life activity, *it is critical to reject the assumption that an
   individual who performs well academically or otherwise
   cannot be substantially limited in activities such as learning,
   reading, writing, thinking, or speaking.*" (italics added)

The Final Rule quotes Congress that the comparison of individuals with
dyslexia to "most people" requires a careful analysis of the method and manner in which an individual's impairment limits a major life activity. For the
majority of the population, the basic mechanics of reading and writing do
not pose extraordinary lifelong challenges; rather, recognizing and forming
letters and words are effortless, unconscious, automatic processes. The process of reading for an individual with dyslexia is word-by-word and otherwise
cumbersome, painful, deliberate, and slow—throughout life.

→ The Final Rule reflects congressional intent and makes
   clear that the *outcome* an individual with a disability is
   able to achieve *is not determinative of whether an individual
   is substantially limited in a major life activity.* Instead,
   an individual can demonstrate the extent to which an
   impairment affects the condition, manner, or duration
   in which he performs a major life activity, such that it
   constitutes a substantial limitation. The ultimate outcome
   of an individual's efforts should not undermine a claim
   of disability, even if that individual ultimately is able to
   achieve the same or similar result as someone without the
   impairment.

The Final Rule emphasizes that the testing results or grades of an individual seeking reasonable modifications or testing accommodations may not be
relevant to determinations of disability, in contrast to what testing agencies
often claim.

As we know, often standing between the dyslexic individual and the
demonstration of her intelligence and deep knowledge is her slow reading. Once a dyslexic reader is allowed extra time to complete a test, she is
able to demonstrate her high ability; without such an accommodation, the
test merely becomes a reflection of her disability. Scores on such unaccom-

modated tests often do great harm to dyslexic students, preventing access to higher education and desired careers. Perhaps of most concern, those students are left with the incorrect perception that they are not intelligent.

Judge Sonia Sotomayor understood the critical importance of extra time for a dyslexic reader. Hearing the case of Marilyn Bartlett, a dyslexic lawyer seeking accommodations on the bar examination, Judge (now Supreme Court Justice) Sotomayor found for the dyslexic lawyer, wisely opining: "Just as a person in a wheelchair can use an above-ground entrance to gain access to a building if a ramp is available, a dyslexic can draw meaning from complex text if permitted the time required to slowly decipher each word manually. To a dyslexic, time is his ramp."

New York Times reporter Jim Dwyer described the interchange between the judge and the plaintiff:

> The woman sitting in the witness box was presented with a printed page, and asked to read it aloud. She used two hands and her lips. One index finger tracked the words left to right across the page; the other moved down the lines, from top to bottom. She mouthed the words to herself before speaking them. She read the word "indicted" as "indicated."
>
> The judge, Sonia Sotomayor, glanced at the clock. It was 11:13. At the end, she had a question for the witness, Marilyn Bartlett:
>
> "What did you just read?"
>
> "I haven't got a clue," Bartlett replied.
>
> "Neither have I," the judge replied.

Although the passage was just 426 words, it had taken Bartlett—who had earned a doctorate in education and a law degree and had her verbal IQ measured as "superior"—eleven minutes to read it, the sentences so excruciatingly drawn out that no one could remember their meaning.

After listening to the testimony of the defendant's experts and listening to Bartlett read "haltingly and laboriously," the judge found for the plaintiff, determining that "*by its very nature, diagnosing a learning disability requires clinical judgment*" and is "*not quantifiable merely by test scores.*" Consideration of the condition, manner, or duration of how a person carries out the life activity of reading (or speaking or learning) is consistent with Judge Sotomayor's determination that the diagnosis of dyslexia is not, and should not be, restricted to a consideration of a score on a test. This landmark case highlighted the dangers of relying on automatic numerical cutoff scores for determining if a person is disabled or not. The Bartlett case is extremely important for bright students who are able to self-accommodate and score average or better on a test of reading but who at the same time are not able to read in the same condition, manner, or duration as most people in the population. This case is important, too, because it highlights that the diagnosis of dyslexia is a clinical one and is not, and should not be, made by test scores alone.

One would think that with the signing of the ADAAA and the explicit language expressed in the Final Rule, dyslexic students who are bright, work extremely hard, and earn grades that permit hopes of attending college or graduate, medical, business, or law school would have minimal difficulties obtaining accommodations on high-stakes exams. Unfortunately, this has not always been the case. Some deserving dyslexic individuals have continued to find it extremely difficult to obtain accommodations, even those who were diagnosed early on in school and have received accommodations throughout their school careers.

Some hardworking young men and women applying for accommodations to a range of testing agencies—including the SAT, LSAT, MCAT, and U.S. Medical Licensing Exam (USMLE)—and medical certification boards still face a barrage of requests for more and more information, including difficult-to-obtain records, such as demanding the elementary-school records of a thirty-five-year-old applicant. Far too often the litany of requests results in delay after delay, and often expensive retesting for a lifelong disability.

The testing agencies have seemed to be functioning as a law unto themselves. Some students have received accommodations; however, less than 2 percent of college students were applying for accommodations, in contrast to the known 20 percent who are dyslexic. The assumption is that many dyslexic students become discouraged and do not apply for accommodations. Many who did apply found that the agencies seemed to flout the ADA and denied their accommodations. A dyslexic physician was rejected for accommodations by the American Board of Pediatrics with a justification that seems contradictory to the intent and letter of the ADAAA:

> The ADA Amendments Act, like its predecessor law, the Americans with Disabilities Act, covers individuals who are "substantially limited" in a major life activity as the result of a disability. It appears that your history does not reflect the magnitude of symptoms or

global impairment required to support a diagnosis of substantial impairment in academic functioning . . . the ABP will be unable to provide test accommodation for your General Pediatrics Certifying Examination.

Despite the enactment of the ADAAA specifically calling for the determination of who is "substantially limited" to be broadly interpreted, the testing agencies appeared not to be listening. These testing entities, wielding extraordinary power, appeared not to be answerable to any authority, to have no accountability, and rather seemed to function without any external oversight or restraints. There was and continues to be concern, too, over the potential bias of the external reviewers who were recruited and paid by the testing agencies and thus were not independent. Still another concern: If an applicant, following a denial of accommodations, requests a second review, the new reviewer may be biased by previous reviews provided by the testing agency. Perhaps most pernicious, in the denial letters, testing agencies have noted that accommodations are not intended for bright students coming from "elite" schools—students who the testing companies claim just want to increase their scores. The testing agencies appear to adhere to the myth that applicants who request accommodations are trying to game the system, a belief not supported by the evidence.

These agencies' actions and statements reflect an ignorance not only of science but also of the law as expressed in the Final Rule. The agencies often misquote facts and seem to rely instead on personal biases rather than on factual evidence. Agency reviewers, who have never examined or even seen the applicant, second-guess the recommendations of the professional who evaluated that young man or woman—an individual who may have more experience and knowledge and who certainly has a better sense of the applicant gained from spending many hours with him or her. As evidenced by the denial letter to the applicant for accommodations quoted above, the agencies often appear to be misquoting and misapplying the law.

Most upsetting of all, the agencies seemed to be looking for almost any reason or excuse to deny an applicant the requested accommodation. Applicants were frequently placed in a no-win situation in the interpretation of their scores. If their scores were in the average range, they were often told that their scores were not low enough for them to be considered disabled; if their scores were very low, they were accused of malingering. The result: Society was losing valuable human resources and individuals' lives were upended.

Denied accommodation, facing endless delays, or failing unaccommodated tests, applicants would have to drastically lower their ambitions.

A brilliant young woman, Lisa, who had been identified as dyslexic in primary school and in whom a recent evaluation indicated superior intelligence and dyslexia, had acquired an excellent academic record at a rigorous college with accommodations and was encouraged by her admiring instructors to apply to law school. She applied for accommodations and was rejected, even after an appeal. Frustrated and unsure of how long she would have to keep applying for accommodations and of the outcome, Lisa gave up her dream of becoming an attorney. Her mother aptly commented, "If my daughter decides she no longer wants to attend law school, that's her prerogative; however, the decision should not be made for her by the Law School Admissions Council."

If the Final Rule had been published earlier, it would likely have provided Lisa with the backing of the Department of Justice. In fact, in the Final Rule the DOJ specifically takes on many of the arguments that testing agencies such as LSAC and USMLE have used in the past. For example, the Final Rule specifies that the DOJ does not agree with the assertions made by testing and educational entities that evidence of testing and grades is objective and therefore should be weighted more heavily, while evidence of self-mitigating measures, informal accommodations, or recently provided accommodations or modifications is inherently subjective and should be afforded less consideration. Congress's discussion of the relevance of testing outcomes and grades clearly indicates that it did not consider them definitive evidence of the existence or nonexistence of a disability. While tests and grades typically are numerical measures of performance, the capacity to quantify them does not make them inherently more valuable with respect to proving or disproving disability. To the contrary, Congress's incorporation of rules of construction emphasizing broad coverage of disabilities to the maximum extent permitted, its direction that such determinations should neither contemplate ameliorative mitigating measures nor demand extensive analysis, and its recognition of learned and adaptive modifications all support its openness for individuals with impairments to put forward a wide range of evidence to demonstrate their disabilities.

The Final Rule emphasizes that Congress made clear its intention that the ADA's protections should encompass people whose impairment requires an assessment that focuses on *how* they engage in major life activities rather than the ultimate *outcome* of those activities. In addition, the Final Rule clarifies that mitigating measures can include "learned behavioral or adaptive neurological modifications," psychotherapy, behavioral therapy, or physical therapy, and "reasonable modifications" or auxiliary aids and services. The phrase "learned behavioral or adaptive neurological modifications" is intended to include strategies developed by an individual to lessen the impact of an impairment. The phrase "reasonable modifications" is intended to include informal or undocumented accommodations and modifications as well as those provided through a formal process.

The Final Rule recognizes that students may have developed their own ways to support their disability in order to perform major activities required during daily life and that such measures cannot be used to find that the person is not substantially limited. Furthermore, the Final Rule notes that self-mitigating measures or undocumented modifications or accommodations for students who have impairments that substantially limit learning, reading, writing, speaking, or concentrating may include such measures as arranging to have multiple reminders for task completion; seeking help from others to provide reminders or to assist with the organization of tasks; selecting courses strategically (such as selecting courses that require papers instead of exams); devoting a far larger portion of the day, weekends, and holidays to study than students without disabilities; teaching oneself strategies to facilitate reading connected text or mnemonics to remember facts (including strategies such as highlighting and margin noting); being given extra time to complete tests; receiving modified homework assignments; or taking exams in a different format or in a less stressful or anxiety-provoking setting. Each of these mitigating measures, whether formal or informal, documented or undocumented, can improve the academic function of a student having to deal with a substantial limitation in a major life activity such as concentrating, reading, speaking, learning, or writing.

The Final Rule also discusses the effect of the ADA Amendments Act on academic requirements in postsecondary education. Specifically, it notes that nothing in the act changed the existing ADA requirement that covered entities provide reasonable modifications in policies, practices, and procedures unless the entity can demonstrate that making such modifications, including academic requirements in postsecondary education, would fundamentally alter the nature of goods, services, facilities, privileges, advantages, or accommodations involved. The Final Rule emphasizes that Congress noted that the reference to academic requirements in postsecondary education was included "solely to provide assurances that the bill does not alter current law with regard to the obligations of academic institutions under the ADA."

THE DEPARTMENT OF JUSTICE TAKES ON TESTING AGENCIES

In several extremely important cases, the DOJ has demonstrated its power to correct violations. There is the case of the dyslexic Yale medical student Fred Romberg. Currently a practicing anesthesiologist, Fred overcame serious obstacles that would have broken another person: having to support his family from age fourteen after his father returned from the Vietnam War with severe injuries, and being on his own since age sixteen. He worked as a mechanic at a community college garage, where a professor recognized that he was extremely smart and encouraged him to obtain a GED. He was referred to a learning specialist and, following a full evaluation, was diagnosed with dyslexia. Fred is highly intelligent and presents with an unexpected difficulty in reading quickly and automatically. As a result, he requires additional time to read. He can think and reason very quickly, and with some extra time to read, his comprehension is at the highest levels. With accommodations, primarily extra time, Fred rose from a GED to an associate of arts degree, to a BS in electrical engineering at Virginia Polytechnic Institute, to an MS in electrical engineering at Cal Tech with a cumulative GPA of 3.7. He earned a pilot's license, obtained both commercial and flight instructor licenses, and built his own plane, which he enjoys flying.

With accommodations for his dyslexia on his MCAT (Medical College Admission Test), Fred was admitted to Yale Medical School, where he continued to receive accommodations, primarily extra time on tests, and performed well. Despite his prior history of accommodations and the recommendations of his professors and his evaluators, the National Board of Medical Examiners (NBME), the agency responsible for certifying that medical students have satisfactorily completed their initial medical training, via the gatekeeper U.S. Medical Licensing Exam (USMLE), initially refused Fred accommodations. As a result, he failed the USMLE twice. Students must pass the USMLE in order to graduate from medical school. If unable to pass, Fred could not graduate and go on to a residency or, for that matter, a career and life as a physician.

For more than three years the NBME refused to allow Fred accommodations on the USMLE. There was no question of his intelligence (he scored in the superior range), nor of his dyslexia (his history and symptoms as well as very poor reading fluency established that). The NBME simply based its refusal on the fact that Fred had grown up in a very poor household and therefore, they claimed, his reading problems were a result of his disadvantaged upbringing and not dyslexia, somehow perversely asserting that if you grow up poor, you cannot also be dyslexic. It seemed to overlook the fact that he excelled in math and science and had been able, with accommodations, to successfully complete a graduate degree at Cal Tech.

Fred poignantly describes the anguish associated with watching his classmates graduate and move on and the embarrassment associated with having to explain to classmates, faculty, and friends why he was not graduating with his class. Not knowing if he would ever graduate or become a doctor was a persistent worry.

Finally, after thirty-eight months of applying, being turned down, appealing, being turned down, reappealing and at the same time studying nonstop day and night and failing the USMLE without accommodations, Fred filed a complaint with the U.S. Department of Justice's Civil Rights Division. The attorneys there agreed to take his case. The department investigated and in February 2011 announced that a settlement of Fred's case had been reached and he would be allowed to take the test for the third time, this time *with* accommodations. With the accommodation of extra time, he handily passed this same examination without difficulty, showing that it had never been a matter of lack of knowledge but rather lack of time that caused him not to be able to show his ability and as a result to fail test after test. Given the extra time, he passed and was able to go on to a residency and his career. The settlement achieved by the DOJ changed the course of Fred's life. Not only is his future coming to life as he completes his residency, Fred is now also happily married to Elaine, the love of his life, a strong, constant source of support.

Literally as I was writing about Fred my cell phone rang: It was Fred himself, excited to share some good news with me that just happened that day, February 24, 2017. He had finished his last case in the operating room (he was a resident in anesthesiology), turned in his hospital badge, and picked up his Certificate of Completion of Anesthesiology Training. Fred had attained his goal. He is now a practicing anesthesiologist! I must admit a tear or two rolled down my cheeks hearing this. Fred finished by telling me, "You know, you are the very first person I called. You have made such a positive difference in my life."

Receiving accommodation not only made a world of difference to Fred personally but loudly announced that the Department of Justice meant to enforce the law of the land and will defend disabled men and women who request accommodations and whose rights are violated by the testing agencies. In a press release Thomas E. Perez, assistant attorney general for the Civil Rights Division, noted:

In the past, demands for unnecessary or redundant documentation, burdensome and expensive repeated professional evaluations, or irrelevant evaluative testing unrelated to the ability to demonstrate one's knowledge or skills on an examination prevented individuals with appropriately documented disabilities from pursuing their chosen professions. . . . Under the terms of the settlement agreement, NBME is [now] committed to providing reasonable testing accommodations to persons with disabilities who seek to take the USMLE, in accordance with the requirements of the ADA. In addition, it will grant Frederick XX, a Yale Medical School student, the accommodations of double the standard testing time and a separate testing area to take the USMLE.

The requirements under which the NBME will consider requests for accommodations can be found at ada.gov/nbme.htm.

→ It can request documentation only about (1) the existence of a physical or mental impairment; (2) whether the applicant's impairment substantially limits one or more major life activities within the meaning of the ADA; and (3) whether and how the impairment limits the applicant's ability to take the USMLE under standard conditions.

→ It must carefully consider the recommendations of qualified professionals who have personally observed the applicant in a clinical setting and recommended accommodations based on their clinical judgment that the individual is substantially limited in one or more major life activities within the meaning of the ADA and needs the requested test accommodations in order to demonstrate his or her ability and achievement level. Such recommendations are to be based on generally accepted diagnostic criteria and supported by reasonable documentation.

→ It must also consider all evidence indicating whether an individual's ability to read is substantially limited within the meaning of the ADA, including the extent to which it is restricted as to the conditions, manner, or duration as compared to the reading ability of most people.

What kind of physicians do dyslexic men and women make? What do their patients think of them? Figure 51 offers recent insight from a patient of Dr. Romberg's.

Futures are improving, not only for potential physicians but also for other qualified dyslexic professional hopefuls, including attorneys. Despite what attorneys and judges know about the poor predictability of the LSAT, it remains a gatekeeper for prospective attorneys. By many accounts, the Law School Admission Council (LSAC), the organization responsible for the LSAT, has been particularly rigid about accepting requests for accommodations from dyslexic students. A 2011 Government Accountability Office (GAO) report concluded that while there have been significant changes in the law governing accommodations, testing companies have not changed their policies and practice. A major recommendation in the report states that "it is imperative for Justice to establish a credible enforcement presence to detect, correct and prevent violations." Responding to the GAO's rec-

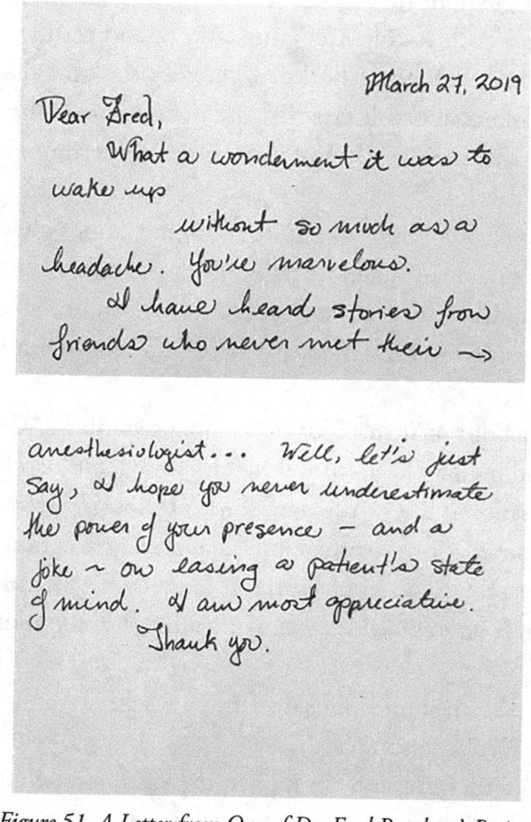

March 27, 2019

Dear Fred,
        What a wonderment it was to wake up
                without so much as a headache. You're marvelous.
                I have heard stories from friends who never met their →

anesthesiologist... Well, let's just say, I hope you never underestimate the power of your presence — and a joke — on easing a patient's state of mind. I am most appreciative.
                Thank you.

Figure 51. A Letter from One of Dr. Fred Romberg's Patients

ommendations, the DOJ opened systematic investigations into the actions of specific testing agencies, focusing on those whose actions have resulted in an especially large number of complaints from disabled applicants. The most serious offender was the LSAC, the first agency the DOJ investigated. As a result, the LSAC has entered into a consent decree with the California Department of Fair Employment and Housing and the United States regarding testing accommodations for individuals with disabilities nationwide. The consent decree states, "LSAC shall ensure that documentation requests are reasonable and limited to the need for the testing accommodation requested."

For those candidates whose documentation establishes that they previously were approved to receive testing accommodations on any standardized examination offered in the United States related to applications for postsecondary admission, and with respect to the testing accommodations for which they were previously approved, the consent decree goes on to say,

> LSAC shall require no more documentation than proof of the approval for such testing accommodations, and certification by the candidate through a checkmark box on the candidate form that the candidate is still experiencing the functional limitations caused by the disability(ies) for which testing accommodations were approved.
>
> Acceptable proof of prior testing accommodations shall consist of a letter or similar documentation from the other test sponsor confirming that testing accommodations were approved and specifically identifying what those approved testing accommodations were.
>
> Upon receipt of such proof in accordance with LSAC's established deadlines, without further inquiry or request for additional documentation, LSAC shall grant those previously approved testing accommodations, or the equivalent testing accommodation offered on the LSAT, with respect to requests for extended time up to double time as well as certain other testing accommodations.

Sadly, there are some indications that the Law School Admissions Council is not fulfilling its obligation.

These changes (despite some bumps in the road) strongly suggest that a new day, a more positive day, is on the horizon for dyslexic students requiring accommodations on high-stakes postsecondary exams. There are many to thank for this progress, and at the very top I would place the U.S. Department of Justice, Civil Rights Division. I must admit that whenever I have had occasion to interact with the individuals employed there or seen the product of their efforts and dedication to justice, I am so proud, as corny as this may sound, to be an American.

THE ELIMINATION OF FLAGGING

The practice of flagging the scores of tests taken with accommodations has been distressing to many students with disabilities. While the purpose of the accommodation is to place those with disabilities on a par with others taking the test, the asterisk placed next to the score telegraphs to all who see it that the test-taker has a disability. The evidence is strong that once an admissions committee sees the asterisk, the tendency is to place that applicant's file on the bottom of the pile. The asterisk identifies the applicant as having an exception and his scores as ones the testing organization cannot validate.

A civil rights law group, Disability Rights Advocates, filed a lawsuit on behalf of a physically disabled man, Mark Breimhorst, who charged that the Educational Testing Service's (ETS) policy of flagging accommodated test scores with the notation "Scores Obtained Under Special Conditions" violated state (California) and federal antidiscrimination laws, "stigmatizing disabled students with a kind of scarlet letter." The judge hearing the case ruled that the tests should "equally measure the skills of disabled and non-disabled test-takers" and then there would be no need to flag these scores. In a settlement reached between the civil rights group and ETS, the testing organization agreed to stop flagging the test scores of individuals with physical and learning disabilities who take these tests with accommodations. Initially the new policy applied to the Graduate Record Exam (GRE), the Graduate Management Admission Test (GMAT), the Test of English as a Foreign Language (TOEFL), and Praxis, a test for teachers. A blue ribbon panel was formed to reexamine flagging and make recommendations for the major standardized test, the SAT I. On July 15, 2002, the panel released its findings: "The majority position of the Panel was to discontinue the practice of flagging the SAT I based on scientific, psychometric, and social evidence." I served on the panel and joined in the majority opinion, which noted the "compelling" evidence against flagging.

WHAT DO STANDARDIZED HIGH-STAKES TESTS PREDICT?

Under scrutiny, these ubiquitous and influential tests do not seem to hold up very well; many of our assumptions concerning the tests and their predictive value turn out to be questionable. A study of over 55,000 students, published in 2020, reported that the predictive power of a student's high school grade point average is five times greater than ACT scores at predicting college graduation, regardless of the high school attended. The imbalance between the power of these tests and their flawed nature is particularly harmful to those who are dyslexic. These tests measure the rapid retrieval of rote facts, which taps directly into the dyslexic's phonologic weakness, while they are mostly unable to measure what may be his strongest assets, namely his ability to reason, to think abstractly, to see the big picture, and to think out of the box. These are the sorts of creative abilities that tend to reflect the dyslexic's sea of strengths, the kinds of cognitive skills in which the dyslexic might score in the ninety-ninth percentile if they were measurable by a simple test.

At the next level, the Law School Admission Tests are not proving themselves to be especially strong predictors of law school performance. One judge proposed that law schools "relax or even eliminate reliance on the LSAT." As reported in *The Wall Street Journal,* the judge wrote that the test "does not predict success in the legal profession at all. . . . One must wonder why the law school concerns itself at all with an applicant's LSAT score."

Law school admissions policies are changing. Students who apply to an ever-growing number of law schools now have the option of taking the GRE instead of the LSAT. Northwestern University Pritzker School of Law will allow all JD applicants to submit either the GRE or the LSAT score for admission. The University of Arizona College of Law and Harvard Law will also accept either the GRE or the LSAT. On October 17, 2017, the *New York Law Journal* reported that "Columbia Law School, Embracing GRE for Fall 2018, Is Latest School to Join Wave." Without doubt, many other law schools will soon follow. Yes, it is once again appropriate to quote Bob Dylan: "The times they are a-changin'."

The United States Medical Licensing Examination has been criticized by those such as Dr. Stephen Smith, professor emeritus of family medicine and formerly associate dean of student affairs at Brown University School of Medicine, and the late Dr. Graeme Hammond, professor emeritus of surgery at Yale University School of Medicine, both of whom served on the National Board of Medical Examiners and participated in the test-making process. According to Dr. Hammond, "I shudder when I think about how those questions were constructed. . . . I believe that the time has come to get rid of these examinations which test the ability to take multiple-choice tests and little more."

A TIME FOR HOPE

Science, morality, and the law are converging to provide the rationale, the societal good, and the legal grounds to support the provision of accommodations to students who are dyslexic. More and more, learned groups are questioning the role of these tests. For example, in 2001, Richard C. Atkinson, then president of the University of California, proposed "an end to the use of SATs as a requirement for admission to the state university system . . . one of the largest and most prestigious." Atkinson indicated that "he would like to move away from numerical measurements of student aptitude and encourage a more 'holistic' approach to evaluating candidates." Moving forward, as noted on page 379, as of 2018, there were over one thousand four-year colleges and universities that did not rely on the SAT or ACT to admit bachelor-degree students.

As for higher levels of education, a comprehensive study published in the *Journal of the American Medical Association* (*JAMA*) suggests that a nontraditional approach to medical school admissions produces physicians whose postgraduate training and career experiences are indistinguishable from those selected through more usual procedures. In this study carried out at the University of California at Davis, the researchers were interested in the consequences of an admissions process in which special consideration was given to a range of factors considered important in choosing future physicians. The goal was to select the "best" applicants but not to be bound by the traditional criteria of grade-point averages and medical school admissions test scores. Students who scored poorly on standardized tests but who demonstrated other, more difficult-to-measure qualities were eligible for consideration by the admissions committee. In this study, which extended from 1968 through 1987, approximately 350 students were admitted by special consideration criteria; 67 percent did not meet the minimum standardized test criterion set by the school for admission, while a smaller number did not meet the minimum undergraduate GPA. The admissions committee was able to disregard the results of standardized tests and select students who went on to successful completion of medical school and residencies. By the time they were residents and in practice, the special group was indistinguishable from their higher-scoring classmates. As a *JAMA* editorial said:

> UC Davis applicants who qualified for the special admissions programs because their Medical College Admissions Tests (MCAT)

scores and GPAs were *below* the "minimum" established for regular admission had excellent outcomes that were comparable to
those admitted in the usual way. The students with higher MCAT
scores and GPAs were not significantly more likely to graduate,
complete residency successfully, become licensed, attain board certification, or enter the full range of career opportunities (including academic medicine) than were the students admitted under
the special admission program. If the highest MCAT scores and
GPAs are not predictive of these outcomes, they are not meaningful admissions factors.

There is accumulating evidence—in undergraduate admissions, in graduate school, and now in law and medical school—that reliance on standardized test scores as reliable predictors of future performance is not fulfilling
its promise. Such scores may not select the best students, and in fact may
be keeping out those students who have the ability not only to graduate
successfully but to contribute in unique ways to their profession. The failure of numerical scores to differentiate between who will and who will not
become a good doctor substantially increases the work of the admissions
committee members; they will clearly need to rely on other, hopefully more
valid criteria.

The importance of the more intangible qualities—of getting beyond the
facts—is of more than theoretical interest. These are the kinds of qualities
that impact our daily lives, that can make the difference between a good
scientist, physician, lawyer, artist, or writer and a great one. It's really all
about the difference between relying on an accumulation of facts compared
to thinking on a higher plane, and this is why we must not cede the selection
of the next generation's leaders to how well they can score on mechanical
standardized tests. We must take in all the information about a person,
including the kinds of data that are not subject to easy reduction. Creativity
is too large and too far-ranging to be fit into the narrow confines of a bubble
response to a multiple-choice question.

34

A PERSON LIKE THAT . . .

Not long ago, at a dinner party, a professor at my table was speaking about dyslexia. "Now dyslexics want to go to law school," he said. "Can you imagine: a person like that as your lawyer?" I smiled at him as I replied, "I would consider it fortunate to have David Boies as my personal attorney. Yes, I would be very lucky to have a person like that." My misguided dinner companion went on. "Well, he couldn't really be dyslexic. He's . . . he's too accomplished—he's just too good to be dyslexic."

Of course this stereotype of dyslexia is false. The very same children who struggle with dyslexia in school often go on to excel in life. I want to introduce you to a diverse group of dyslexics who are successful by anyone's standards in a range of occupations. As children these men and women struggled to read, could not spell, were laughed at when they read aloud in class, rarely finished a book, failed foreign-language courses, and fared miserably on standardized tests. Many dyslexic children and young adults will recognize themselves in these portraits. If so, they will also know that if they are "a person like that," they should be encouraged to dream big, follow their dreams, and know that those dreams can and will come true.

These portraits of highly successful dyslexic men and women provide insight into their early life, challenges, and failures, together with their resilience and their ultimate path to the very top of their fields.

ATTORNEY

DAVID BOIES

David Boies is brilliant and a genuine thinker. His acumen and skill as an attorney have led to honor after honor, and praise such as "Brilliant and tireless, he may be the greatest trial lawyer alive today." You read about his herculean work habits, including forgoing sleep in favor of preparing for a case. "Would you rather sleep or win?" he asks. His stunning win representing the U.S. Department of Justice in its antitrust case against Microsoft was characterized by David Margolick in *Vanity Fair* as "The Man Who Ate Microsoft."

Boies reads slowly but managed to graduate second in his class at Yale Law School. Along with Ted Olson, his co-lead counsel, he represented the plaintiffs in *Perry v. Brown* and was able to win rulings establishing the right of gay and lesbian couples to wed in the state of California. Insights into how this life-changing decision came about are found in Boies and Olson's book, *Redeeming the Dream: The Case for Marriage Equality.* Boies is also the author of *Courting Justice,* providing his unique perspective on his many high-profile cases.

Boies's accomplishments strengthen and add credence to the Sea of Strengths model of dyslexia. He himself stresses the importance of dyslexic children knowing that there is

somebody who has dyslexia who can do this. It gives them a goal. It gives them a certainty that they can do it. And as with almost any characteristic, you have to identify it and identify that it is not disabling, and you have to see people who have that characteristic succeed.

It is a disability, but it does not have to be disabling. And I think that as people see that more and more, society begins to recognize that people with dyslexia are not slow and they're not dumb. They just process information differently. And then as we come to know them as their lawyer and their doctor and their patient and their client we begin to understand the basic reality, which is that we are all the same in every respect that matters. As more and more people have been identified as people with dyslexia and you realize that they are normal in every way that matters and that they can be just as successful and just as contributing to society as everybody else, those prejudices begin to disappear.

Boies stresses that his work in law relies much more on thinking than on reading, and that is something, as a dyslexic attorney, he excels in. He urges people to understand that while *the intake (reading) is impacted by dyslexia, the dyslexic individual's processing of information is a strength,* one enabling him as he matures not only to dream of accomplishing great things but to keep in mind that he *can and will accomplish his dream.*

Boies is particularly concerned for dyslexics during their adolescence:

> In that vulnerable time you have doubts about yourself and your place in the world; it is the time when the input issue has the greatest impact. As you get older, processing takes over in importance.
>
> But when you're in grade school and high school, you're reading, you're learning, you're memorizing. You're taking in all these facts, all these formulas, all these words, vocabulary. And you're not at a point where the processing of that information dominates, as it does when you get out and into the real world. So at the very time that you are the most vulnerable psychologically, you are most inhibited by the dyslexia.

When asked about the seemingly growing impact of having students take standardized tests, Boies expressed deep concern for the harm these can do, especially to students who are dyslexic and to society as a whole.

Reflecting on the frequent sad state of affairs in schools where parents of dyslexic boys and girls are continually being admonished to wait until it "clicks in," Boies says:

> No, the thing is, it doesn't click in. If you've got the patience and your parents and your school have the patience, you will come to a point where your abilities are more important than your disabilities. Dyslexia is not something that indicates a lack of intelligence. If you can just get to the point where what matters is intelligence or what I like to call processing, as opposed to input, like reading, you'll be fine.

Boies is very much aligned with the need for and benefit of specialized schools for dyslexic children, agreeing that in public schools teachers and others too often make incorrect assumptions, believing that if you give dyslexic students the right program, that will do it all. But it won't.

For dyslexic boys and girls, the need is for people who understand in their social studies and their math and their history classes just what it's all about. And teachers can't take a course and say, "Well, this is how you teach it and regurgitate it."

He was impressed when he visited the Lab School to receive an award:

One of the great things about the Lab School in Washington, DC, is that you have a whole group of people that are together and nobody feels inferior. Nobody feels dumb. The school is a true community; nobody is ashamed. There is not the tendency present [among students] at many nonspecialized schools to look for kids who are perceived as different and find a child to shame, to target, essentially saying, "Who can I make fun of because I feel so insecure?"

I grew up in a rural town in the 1940s in northwestern Illinois. I didn't learn to read until I was in third grade. Even if you didn't read, you could pretty well memorize and pretend you were reading. So I didn't grow up with any sense of failure or any sense of not being able to do something that was really important—which gave me a chance to really develop on my own and develop my own set of skills and to adapt.

I think the single biggest misunderstanding is to associate reading with intelligence, to use reading in the early years as a good indicator of somebody's intelligence. I think that a related misunderstanding is a belief that how somebody is able to cope in school when they are in grade school and high school is a good indicator of how they will cope in life. I think that there are some ways in which it is, but I think that, particularly for somebody with dyslexia, it is a very poor indicator because . . . in life, what matters is your judgment, your intelligence, that you're willing to continue to try and learn, how you relate to people, your character and integrity, your work ethic.

How fast you read has very little bearing on how well you do when you get out into the real world. And yet how fast you read, particularly if you're painfully slow, as I was, can have a huge impact on how well you do in school, particularly in grade school and high school.

I think the most important requirement in dealing with dyslexia is *patience*. I think it's very hard for an adolescent to be patient in anything. And it's very hard for parents to be patient about their children because you care so much about them. And yet *patience is, I think, the single most important ingredient of dealing with dyslexia.*

I was in my early thirties when I found out I was dyslexic. To be honest, I didn't think a lot about it. By that time I'd graduated from law school and was already a partner in a law firm.

I asked if law doesn't require a great deal of reading.

I get asked this question a lot by people who are dyslexic and want to go to law school. *First, I'm a very good listener; I learned to listen.* When you don't read a lot, you listen. And when you really learn to listen, you listen to people's words. You listen to their inflection. You can tell what they think is important.

In class, I could listen, I could figure things out just by being there. I'm also a verbal person. I talk. I can interact. And so I can engage the professor. When I don't understand something, I can ask. For me, class and class participation were very important. I really learned an enormous amount there.

When I went to law school, you learned by what is called the case method, which is you read these long cases, and you try to figure out what the law is from that. That's valuable for understanding the legal process, how decisions are made, but if you're smart, you can figure that out without reading book after book in class after class.

I found that I understood the legal method. I understood legal reasoning. I probably read 10 percent as many pages as the average person, maybe even less. But I got a lot out of class, and I got a lot out of my little . . . I called them hornbooks, nutshell books, stuff that sort of told you what the basic principles are. And that allowed me to learn the principles of law, plus legal reasoning, and to apply that in a way that was much more creative than what other people were able to do.

It's not that different [in preparing cases]. There will sometimes be a case where you need to read, and my reading comprehension is really great because I go so slowly that I really learn it. It means I read fewer things. In any area, there are only a handful of cases that are really important. One of the things I've done for forty years is if there's an important principle, I'll say to people, "Give me the three cases that are best for us and three cases that are worst for us." I'll read those cases, and I'll usually understand. As a trial lawyer, what's important is how you present your case. If you're going to go to trial, all you have to do is make your best argument. And you can do that often without knowing every case that's out there. In addition, when you get to a more senior position, what people are looking for is your judgment, your analytical ability. They're not asking you to tell them on the fly the number of years you can have a trust in a particular jurisdiction. That's something somebody can look up. I've said so many times, life is not a timed test. You get time that you need to figure out what you need to know.

In listening to David Boies, I could appreciate his sea of strengths, his big-picture thinking, not being a slave to every little detail or to thinking linearly. For him, it's seeing the whole and figuring out how the parts go together. It's not the intake but the processing of the information, synthesizing the parts and bringing it all together.

I have spoken to many very worried adults who are dyslexic. They have gone through college, law school, or medical school, and now they are so afraid of the workplace. I asked what advice David would give these worried people.

You know, if you made it through school, the workplace is going to be a picnic, because the time that you are most disadvantaged by dyslexia is in the school environment. It is an environment with standardized timed tests. It's an environment that puts a premium on how much information you can absorb and how fast you can absorb it. You get into the workplace, and it puts a premium on how you deal with people and how you can use the information that you have—you're not at a disadvantage in that area. In some cases, I think you may be at an advantage.

We spoke about our grandchildren, and I asked Boies if a future grandchild happened to be dyslexic, what would he wish for that child?

I would just wish patience, just patience and love, support and recognition that this was a passage. This was something you had to go through and *you do emerge at the end, not because you leave your dyslexia behind but because you leave behind the disadvantages of your dyslexia.*

REBECCA ARAGON

Rebecca Aragon sums up her life by embedding within it a very powerful message: "Life has not been easy, that is true, but if it were I wouldn't be here where I am today." Aragon was born to a single mother, a Mexican immigrant, living in a facility for abused and unwed women in Los Angeles, where only about 10 percent of the newborns remained with their biological mothers and the others were given up for adoption. Rebecca's mom, going against her social worker's advice, chose to keep and raise her daughter.

> It was a poor upbringing, but a very enriched one, because I had to learn survival skills to get through elementary school, then through high school. I had reading problems, but I didn't know what it was back then. I thought it was because English was not my first language. Spanish was. I didn't learn to speak English until I was seven years old. I thought it was because it was a poor neighborhood and I didn't have an opportunity to go to very good schools. I thought these were all socioeconomic problems.

This belief that socioeconomic problems were to blame for her struggles is unfortunately widespread and can be harmful. Acceptance of this view takes away the need to try to figure out what is the root cause of a child's reading struggles—to screen for and diagnose the child's reading problem, which can then lead to the provision of evidence-based, effective interventions.

Rebecca Aragon is an attorney and dyslexic. She is a shareholder and an attorney at Littler Mendelson PC, where her practice focuses on leading complex litigation involved in employment law in federal and state courts. For the past several years she has been recognized as one of the top Southern California lawyers and has also been named one of the best lawyers in America in employment law.

> Discrimination, harassment, and other types of employment-related cases require a substantial face-to-face interaction with employees

and opposing counsel. These types of matters require as many people skills as reading skills to resolve issues for clients, which dovetails perfectly with my abilities. I may not have the ability to read words quickly, but I can read people in a nanosecond, which is far more important as a trial lawyer and negotiator.

How did this woman traverse the space from poverty, poor English, and dyslexia to highly lauded attorney?

I pressed forward, even though I was not able to read fast or accurately. I learned how to work hard. I don't know what it was in me. I just never, ever, ever gave up. It was that struggle, that persistence, that helped me push forward, and I learned how to reach out to others for help.

I got into Yale, was a slow reader at Yale, wanted to drop out. I didn't know what was wrong with me. I didn't know I had dyslexia. I got through Yale, went to law school at Georgetown. The same thing happened again. It was very difficult for me. I studied all the time, had no social life but made it through.

I took the California bar exam before the ADA was passed. There was no such thing as accommodations back in those dark ages, so I knew I wouldn't pass the first time. I could not give up. My challenge was to find a position that did not require immediate bar passage, so I had to find a clerkship with a federal judge, which was, ironically, by far the hardest position to achieve after you graduate from law school.

I had to take the bar exam four times. It took me longer to pass the California bar than to go through law school, and that's because I could not read fast enough. It's a timed examination. With persistence and a fighter attitude, I finally passed it.

As a young attorney, I quickly learned I had to develop skills other than reading quickly. What was I good at? What was I not good at? One of the things I learned I was good at was oral advocacy. I was very good with people. I was very good in negotiations. I was very good in depositions, and these all required your ability to think on your feet. I was not good when I had to analyze one of those fine-print legal documents. I was not good at sitting in front of a computer or reading cases and statutes. I could do it, but it took a lot out of me. It meant I was working a full 24/7.

I took my daughter to get a professional diagnosis [when she had reading problems], and I remember the doctor saying, "Rebecca, does anyone else in your family have dyslexia?" I said, "I think I might have." I was tested and I was right. It's been good, because I've been able to identify those things that could help me move along—I'm very proud of that—as well as my daughter, who is dyslexic.

After being a practicing lawyer for thirty years and having worked my way up to being a top partner at an Am Law 100 law firm,* I can now say to my partners and to young associate lawyers, "I'm dyslexic." I bring in the business, I do the work, clients are giving me class actions, so I now have the courage to say that I'm dyslexic. But as a young lawyer I certainly didn't, because I knew that my ability to instill confidence in other partners and in clients was limited. There still is a bias and there still is a prejudice out there. So when I go out on a pitch to a big ABC corporation and they see my track record . . . I have to be very judicious when I say I'm dyslexic.

Aragon credits her success to her leadership ability, a special constellation of skills that creates a team with a range of abilities that complement and extend her own.

I have to surround myself with people much smarter than I am. I don't have an ego. I learned how to delegate. I learned leadership skills. I'm not afraid to ask for help. In fact, I always have my hand out. *Can you help me? Can you help me?* I'm also one to give praise to people and share credit. When we do get a great result, it isn't me who did it; it was the team. So it's all about leadership skills. It's taken a long while, but it's made me who I am and I'm so grateful.

One of the things I think being dyslexic does also, it develops your sense of empathy for others' plights. You learn to look at people and situations in other ways. It's like, *Okay, someone needs help, go ask Rebecca. She won't say no.*

---

* A list of the one hundred largest law firms by gross revenue.

Aragon remembers her failures as a young adult and what they taught her.

> I remember when I was at Yale I needed a summer job, so I man-
> aged to become a receptionist at a law firm. I was fired because I
> had to get the names of callers—spell them right—and then call
> them back. It would take me ten tries before I got the number
> right, because I couldn't get the person's name right. It was humili-
> ating, but I learned not to get that kind of job again. I also observed
> the lawyers—trial lawyers—and I said to myself, *I can do that.*

BUSINESS: FINANCIAL SERVICES

CHARLES SCHWAB

As we learned earlier, Charles Schwab, who revolutionized the financial services business, is dyslexic.

> As a child, schoolwork never came easily for me. Reading and
> writing were particularly difficult. I was pretty good at math. . . .
> Though I didn't know why at the time, I had to work three times
> as hard as other kids to accomplish the same thing.
>
> I flunked English at Stanford not once but twice. . . . I couldn't
> write a composition. I couldn't finish a book. I never read a novel.
> I struggled in French. I had taken Latin in high school, so I could
> memorize the things that were very logical and consistent, it was
> almost a math kind of thing. I didn't know how bad I really was at
> language until I got to Stanford, and then the quantities of reading
> you had to deal with—they were just outrageous for me.

"I couldn't read aloud that well but I could talk and solve problems
pretty well," Schwab said to me, capturing both his difficulties and his sea
of strengths.

> I had this outgoing personality. I was shy but I was confident,
> and so that kept me going. I really just conned my way through it
> in some ways. I had good leadership skills, I had good ways with
> people.
>
> I also had strengths in thinking. I was great at conceptualiz-
> ing. I was naturally good in science and math. And maybe the

creative part of my brain was being used for imagining and for problem-solving. From the earliest part of my life I can remember thinking that I could solve almost anything. I felt that if I just thought about it, I could solve it. But I never pushed it, I never thought about it, because I was so poor at English.

Even though I couldn't read quickly, I could imagine things much faster than some other people who were stuck thinking sequentially. That helped me in solving complex business problems. I could visualize how things would look at the end of the tunnel.

To get around my excruciatingly slow reading I always used, wherever I could, the abbreviated kinds of books, the ones that would give me a five-page synopsis. I learned about Classic Comics back in elementary school. And so as a student I invented my own "accommodations," like relying on these comic books to "read" such assignments as *A Tale of Two Cities* and *Ivanhoe.* I had very poor SATs and I never would have been admitted to Stanford on the basis of those. Fortunately, I was very good at golf.

Schwab knows firsthand about the impact of dyslexia on a child when he must read aloud in front of his classmates, who laugh, and "the kid just dies a thousand deaths." Schwab's greatest concern is the child's self-esteem and the impact of dyslexia on children and families. He has never forgotten the overwhelming helplessness he and his wife, Helen, felt when one of their own children first experienced what was eventually diagnosed as dyslexia. As so often happens, the diagnosis in his child led to the diagnosis of dyslexia in himself. Schwab has been deeply touched by his personal experience with dyslexia and is dedicated to supporting efforts to bring more knowledge and understanding to the condition. I have been personally touched by his extraordinary efforts and generosity. In 2008, he and his wife established the Charles and Helen Schwab Professor in Dyslexia and Learning Development at Yale, which Bennett was appointed to and carries with pride.

"Some of the kids feel as if they're really stupid," says Schwab. "If there is anything I'd want them to know, it is that they're not stupid. Once they understand that they can feel better about themselves."

Schwab credits dyslexia for the kind of out-of-the-box, visionary thinking that led to his own business success. To learn more deeply about Schwab's amazing business insight and the story of his founding of and journey with the Charles Schwab Corporation, including his dyslexia, a must-read is his new book, *Invested: Changing Forever the Way Americans Invest,* published by Currency in 2019. He demonstrates his extraordinary ability to get to the heart of an issue when he describes the rationale for providing accommodations to students who are dyslexic. "What you're trying to do is to convey knowledge to these kids," he says. "It's not about trying to trick them out of success."

Schwab's thinking about the importance of ensuring that dyslexic students receive the time they require on standardized tests is aligned with the opinions of the many accomplished dyslexic men and women who stress this, together with the importance of diagnosis, especially early diagnosis. For example, the Academy Award–winning producer (*A Beautiful Mind*) and dyslexic, Brian Grazer, says,

> The single most important thing is to change the educational system so that people get the diagnosis as early as possible, and they get the help that they need. I think the second thing, which is, I think, less important, but still important, is to break down the hierarchy of the standardized test. The first step is extra time. And extra time doesn't really help a great deal if you're not dyslexic. What it does is it equates the performance with the IQ for somebody who's dyslexic. That's what the test is supposed to do. The test is supposed to be testing your ability. By giving people with dyslexia the extra time, it makes the test more effective.

Most of all, Schwab wants to encourage dyslexic men and women to pursue their dreams.

> People who have a passion and have shown they can do the work should be encouraged to go on to the next level. With their passion, intellect, and compassion, they may be among the best lawyers and doctors.
>
> The important thing is to allow a person to focus on something he has a passion for. The more you focus on something, the more successful you are at it. I'm still a very slow reader of most things. I spend a lot of time on one subject, business, which is pretty boring to some people, but not to me. *The Wall Street Journal* is one thing I know like the back of my hand, because I immerse myself

in it every day, and I know the language of business pretty well now. But when I go on vacation, I'll be packing along a copy of *Shogun,* which is pretty tattered, and for the tenth time I'll try to get through part of it. I'm still working on it, and it may take me another ten years to get through it all.

Charles Schwab is one of an ever-growing list of entrepreneurs and businessmen and -women who are dyslexic, including Sir Richard Branson, founder of Virgin Enterprises; Nancy G. Brinker, founder of the renowned breast cancer advocacy group Susan G. Komen; Gary Cohn, investment banker and former chief economic advisor to President Donald Trump; John Chambers, executive chairman and former CEO of Cisco Systems; Thomas J. Donohue, CEO, U.S. Chamber of Commerce; Ari Gejdenson, founder and president of Mindful Restaurant Group; William Hewlett, cofounder of Hewlett-Packard; Ingvar Kamprad, founder of IKEA; and Ted Turner, founder of Turner Broadcasting.

ECONOMIST

DIANE SWONK

As a student in elementary school in Livonia, Michigan, Diane Swonk struggled to carry out simple arithmetic such as adding a column of numbers or memorizing the times table. Now she is a highly regarded and sought-after economist who is looked up to for her expertise in developing complex economic models based on higher-level mathematics. She is chief economist and managing director for Grant Thornton, a global audit, tax advisory firm, and has earned a reputation as one of the most reliable economic forecasters in finance. Her expertise and judgment are highly valued, as indicated by her multiple top advisory roles for the Federal Reserve Board, the Congressional Budget Office's panel of economic advisors, and the Council on Foreign Relations, among others.

Diane's dyslexia was discovered when she was in fourth grade. As her parents watched her, there was little doubt that Diane was also extremely gifted. Her parents were intent on Diane not revealing her dyslexia for fear she would be placed in a remedial rather than a gifted program in her school. They worried about how their dyslexic daughter would be perceived: As a student requiring remedial instruction? As a student benefiting from placement in a gifted program? Are they mutually exclusive? This experience marks the start of an early education journey, one experienced by many dyslexic boys and girls, of contrasting, perhaps even conflicting approaches. Would her strengths be the focus, so that she was placed in a gifted program, or would the emphasis be on her weakness, so that she was placed in a remedial class? Ideally, as in the sea of strengths approach, educators would like to access her strengths and remediate her weaknesses, as is most often practiced in specialized schools for dyslexia. In Diane's case, she did not have a choice, and she and her parents elected to keep her dyslexia hidden. The cost to Diane was feelings of loneliness and isolation.

Diane reads very slowly, can't spell, has a terrible sense of direction so that she frequently gets lost, and, as noted above, was challenged by carrying out simple addition. Her school experience reveals the mistreatment dyslexics often experience as a result of the widespread ignorance of their sea of strengths profile encompassing both strengths and weaknesses. Diane's combination of strengths as revealed by an intelligence test and weakness as a slow reader and poor speller caused her sixth-grade teacher to interpret this as her not trying hard enough and to seek to punish her for what he perceived as misbehavior.

Characteristic of the experience of many dyslexics was Diane's admission application to the University of Michigan, which presented a seemingly contradictory portrait of low standardized test scores together with perfect grades (the result of hard work and perseverance). Although she had two legacy parents (alumni of the school), she was not accepted at first. Characteristically, she did not give up. Instead, she states, "I fought my way in," successfully petitioning the admissions office for a second chance. As an undergraduate she came close to failing freshman English and failed her foreign-language course. Used to facing adversity, Diane persevered and was elated when she entered the world of higher-level mathematics and economics. Simple arithmetic like memorizing the times table involves storing and retrieving numbers phonologically, something quite difficult for dyslexics, but using higher-level mathematics depends on understanding and facility with complex mathematical concepts. Her brain came into its comfort zone and she felt, at last, that this was where she belonged.

Life as a dyslexic is replete with contrasts between today and yesterday. Diane's life is no exception. "Today I'm standing at the University of Michigan," she says. "I'm about to teach two courses, and [yesterday, when I was a student,] I had to take remedial writing in my first course at the University of Michigan, and it was humiliating."

Swonk was invited to sit on the advisory board of the business school at the University of Chicago, where she received her MBA.

> Every time I have a meeting at the University of Chicago and they show me how good the test scores are of the incoming students, I say, *And that means what?* I mean, I didn't have those test scores and I graduated in the top of the class. Why does that matter?

Having spent over three decades successfully navigating the economics space, she is now at the very top of her field. Yet she still struggles to find her way from the office elevator every day, and to read quickly, and she hopes that spell-check can help her distinguish between *from* and *form*. Although she is frequently asked to give talks, she is unable to read a speech. She jots down notes along with an outline but does not read from these as she speaks.

Swonk grew up ashamed of her dyslexia. Then, in 2002, *Forbes* magazine included her in an article on successful dyslexics in the business world. She views this "outing" as life-changing, allowing her to embrace her dyslexia as part of who she is. She believes, "You learn because you fail a lot."

There are always going to be unexpected failures, which often result, particularly if you are dyslexic, in the development of unparalleled resilience. Diane places high value on endurance, perseverance, and flexibility. Her motto is, *No one succeeds alone.* Not surprisingly, her advice centers on people—understanding their abilities and their challenges and creating a highly effective team that works together. That's what leadership is all about.

> As a dyslexic, I succeed by engaging in ongoing collaborative work.
> I depend on a diverse team whose talents complement one another.
> We find unique answers in our diverse decision making.

Here, I think, as a dyslexic, Diane Swonk may have intuitively hit upon the workplace model of the sea of strengths paradigm. In addition to having a diverse team, she emphasizes the importance of being fearless and not being afraid of taking risks, failing, and picking yourself up and going on.

> As a dyslexic I've learned that the best way to avoid risk is to identify it, and have my team think out of the box—what could go wrong, what could go right? Risk can be on both ends, right? Something really positive could happen, and on the other side something really negative.

Most firms don't want to think about the negative. They want to just kind of cruise on a trend. I think about [fellow dyslexic] Chuck Schwab, who said to me, "You didn't think sequentially." I don't think linearly. I think in reaction and functions. I always tell people, *If you want strategic thinking, come to me. If you want linear thinking, don't come to me.*

To Diane Swonk, the key element in her work is her team and its diversity.

We have a team that sometimes is a little quirky, but we're very diverse and we're all very different, and to me, diversification of a portfolio is the best way to get rid of risk. That's what I'm doing every time I sit down with my people and try to bring as many different views as I can to the table.

Swonk emphasizes, "You learn by doing it. . . . You learn by sheer endurance. You learn from your mistakes."

Swonk joins her big-picture thinking as a dyslexic with her determination to use her deep knowledge of economics to bring about positive change.

The issue in education that I'm very involved in is human capital. Specifically, I've given congressional testimony on income inequalities and economic mobility. The ignorance and lack of action on dyslexia is a major reason why there isn't economic mobility. And understanding the role dyslexia plays in this is a game-changer. Recognizing how prevalent dyslexia is and how often it goes undetected begins to reframe the debate about the causes of economic inequality and serves to shape our nation's potential going forward. Recognizing and understanding dyslexia may be a solution to income inequalities and a solution to the lost human capital.

To Swonk, growing up with dyslexia is a story of endurance and recognizing and finding a comfort zone. And, yes, dyslexia is lifelong—it does not go away. What does change as a life's journey unfolds is that your strengths more and more come to the fore. Weaknesses may remain, but they are far overshadowed by the remarkable strengths that shine brightly and represent who you are.

ENTERTAINMENT

ARI EMANUEL

Growing up, Ari Emanuel experienced the full range of symptoms associated with being dyslexic. In elementary school, shown flash cards, he says, "I knew the word, but I couldn't say the word, so for whatever reason my brain was identifying it but the verbal wasn't working—the word just wouldn't come out of my mouth." Ari knew that there was something amiss; he just didn't know its name, what it was.

> I didn't really know what was going on. I just knew that I couldn't read as fast as other kids. I tried to put myself in the class where I thought that the teacher would overlook me so I wouldn't have to read out loud. And then as I went on in school, people made fun of me. It was completely depressing.

Ari's mother was his great champion and tried in as many ways as she could to support him.

> I didn't know what the problem was; I was just plain stupid. And she would say, "Look, you're such a great athlete." Then I would say, "Mommy, I'm a great athlete, but I still can't read."
>
> My mother would keep on reassuring me and finding ways of helping me. She would get me reading tutors. She never stopped trying to help me. She would stay up very late at night to practice reading with me, and she made me repeat my words over and over again. I hated it at the time. Today, when I think about it, I have to give my mom credit. I am grateful to her.

Other highly successful dyslexics talk about their bouts with low self-esteem and the extraordinary role their champion played. Film and television producer Brian Grazer, who is dyslexic, agrees that one of the most damaging consequences of dyslexia may be its destruction of self-esteem.

> Most or all of my best movies that I've produced, or television shows, all sort of revolve around *self-worth*. Whether it's *Friday Night Lights*—they lose the game, but they become men. They become complete people. Basically, self-worth, self-esteem, and

identity, to me, are probably the foundational and most important thing of all. Dyslexia can break you down, and you can't allow it to break you down. It is so important when you do have one person who can give you the confidence not to allow anything to break you down.

For Emanuel, this person was his mother. For Grazer, it was his grandmother.

I had this little Jewish grandmother, Sonia. And she would always tell me that I was going to be special, that I would become somebody really meaningful. And I just didn't get it. She'd say, "You're going to use curiosity. You ask good questions." And she reinforced that constantly. There was a point when I was always looking at my report cards, and I was always getting D's or straight F's. And I'm thinking, *There is literally no evidence at all that I'm going to be special.* My grandmother would say, "Oh, no, no. You are. You absolutely are. Continue to ask these questions. Define yourself that way." Literally, it was my grandmother and my response to not being able to read that led me to actually refine my curiosity and ultimately use it to gain lots and lots and lots of information. I actually wrote a book titled *A Curious Mind.*

More recently, Grazer has written a book, *Face-to-Face: The Art of Human Connection* (Simon & Schuster, 2019), where he shares a second dyslexic strength—making eye contact with another person and the incredible benefits that come with such direct communication.

For Emanuel, it was not only being dyslexic but being placed in special education when he was in high school that was one of his biggest traumas. While he couldn't read well, he excelled in mathematics and was great at sports, especially wrestling. Still, this did not stop his classmates from making fun of him and calling him all sorts of painful names: "Stupid." "Moron." "Idiot." A trait still with him today is not to take abusive behavior quietly. He describes an incident in the school lunchroom:

I remember all of the kids in special ed would go to lunch together. I can remember the football team would come in there and they would start making fun of me. I just remember taking the tray and throwing it at that bully of a football player. And I would be in the principal's office at the end of every day. I just remember taking the tray and I think I broke his nose with the tray.

For Emanuel, his immediate fear and worry was, would he graduate from high school? He worried "if I was going to make it. Am I ever going to be successful? The toughest thing was being in a special education program in a really high-end [very competitive] school." It is clear in speaking with Emanuel that his hero is his relentless advocate, unabashed supporter, and without exception champion, his mom. She never gave up, and neither did her son. As he recalls, "She put up with all my teachers telling her that I'd never graduate from high school and forget about college. My mom stuck by me, and I didn't want to and would not disappoint her." And he did not disappoint her. Emanuel went on to graduate from Macalester College, where he majored in economics with a double major in computer science. How the future outlook has changed for Ari Emanuel, going from a doubtful future, with teachers predicting he would not graduate from high school, to being labeled by *The New York Times* as "the prominent power player in Hollywood."

Figure 52. Ari Emanuel

He started his career as a talent agent working in television. In 1995, he and three partners founded the Endeavor Agency. In 2001, Patrick Whitesell came on board as co-CEO, and the two have since been described as "rewriting the Hollywood script." In 2009, the Endeavor Agency merged with the long-established William Morris Agency, becoming WME. In 2013, in a bold and what many consider to be an industry-changing disruptive move, WME joined with a very different organization, IMG, the International Management Group, one focused on a very different space, that of representing sports talent and fashion models and selling media rights for major university sporting events. Emanuel believes his dyslexia played a major role in his company's ascendancy.

> Thank God I'm dyslexic. You know, when you're dyslexic, you've got to work really hard. You're okay with failure, because you feel it every day. You see things differently because you have to. You know what failing is? Not doing it. Not that I did it and failed. Failing for me is not doing it. For me, I had these ideas, I want to do it. Let's do it. And guess what? What happens if it doesn't work? You just have to keep on trying until it works.
>
> I spend a lot of time reading with my dyslexic sons as well. Each has something special—for example, unbelievable fortitude. You have to give them all the basics and support that will make a difference to them, tell them about how you dealt with things, and that will then lead them to have the knowledge to make good decisions and succeed.
>
> You're in high school, and somebody says to you, *You're stupid.* There's nothing more embarrassing or hurtful. So now in business I am not afraid. I know I have the ability to think creatively or look around the corner and see things differently. I think it gives you the ability, one, not to be afraid of risk and failing, and two, it makes you try stuff, because you're always trying to do stuff better and differently.

Emanuel, too, stresses the importance of surrounding yourself with a team of people whose skills you can call upon to complement your own needs. Referring once again to his dyslexic son Noah, this totally engaged dad says, "He has an incredible sense of people. The best thing, something dyslexia gives you and I see that Noah has, is empathy."

Emanuel reflects on the importance of failure, and especially a person's response to failure: "Well, there are many lessons, but one lesson for sure—you learn more from failing."

Such is life with dyslexia that when Emanuel was honored by the Lab School of Washington, DC, he learned that he was expected to deliver an acceptance speech. His longtime nemesis, reading aloud before an audience, was now before him as he reflected, "I've always had a horrible time reading aloud, and it takes me a long time to read." As is typical of Emanuel, he overcame his fears. He read his speech and truly enjoyed the entire event.

He appreciates the positive side of dyslexia, which he attributes to helping him to understand that he doesn't know everything, which in turn has led to his "greatest asset": being able to tap into the strengths of those around him and appreciating and utilizing these people to their best ability. His dyslexia has played a major role in building great teams. He further credits his dyslexia with making him more empathetic to those who are struggling. Recall that empathy is a strength in our Sea of Strengths model. (See Figure 12, page 56.)

And empathetic Ari Emanuel is. His empathy, together with his big-picture thinking, has driven him to his latest project, one that will greatly benefit the futures of disadvantaged dyslexic children living in Los Angeles. Emanuel has committed himself and taken major steps to start a specialized public charter school to serve disadvantaged boys and girls who are dyslexic. This school will be a game-changer for dyslexic children whose parents could never afford the tuition of a specialized school for dyslexia. With Louisiana Key Academy as a model of a successful and highly effective specialized public charter school for dyslexia, together with Dr. Laura Cassidy and Dr. Bennett Shaywitz, I am thrilled to be part of Ari's creative and much-needed newest project.

PHYSICIAN: CARDIAC SURGEON

DELOS M. COSGROVE

Dr. Delos (Toby) Cosgrove, who was CEO and president of the Cleveland Clinic from 2004 to 2018, is dyslexic. He is renowned in the field of heart valve repair, where he has worked tirelessly to make cardiac procedures safer and more effective. Known for his innovation and creativity, he has thirty patents for cardiac procedures that have helped to solve many problems encountered in valvular surgery.

Cosgrove's dyslexia caused him to struggle to learn to read throughout his early schooling. In college, his foreign-language class was "almost the end of me as I worked my butt off to obtain three D minuses and a D in remedial

French." Reading, writing a paper, or giving a talk remained "big problems." For young Toby, college meant nonstop work. "All I did was study, even on weekends. While everyone else was partying or going to the movies or sports events, I packed my suitcase and headed off home to Albany, where I studied all weekend. And because I had to work so hard, I earned the nickname Turkey Toby."

Reflecting his reading difficulties, standardized testing, including the SAT and the MCAT, was a disaster for him. As a result of his poor college grades and low MCAT scores, it seemed doubtful that Cosgrove would ever fulfill his dream of becoming a doctor. In fact, he was accepted at only one of the thirteen medical schools to which he applied. The University of Virginia Medical School was eager to admit students from his college, Williams, and so his admission was expedited. As a medical student, Cosgrove received "the first A of my life," in surgery. He was already a physician and completing his surgical residency when he was drafted into the air force. Between 1968 and 1969 he served as chief of the United States Air Force Casualty Staging Flight at Da Nang, Vietnam, and was awarded the Bronze Star. Following his military service, he was accepted into a residency in surgery at Massachusetts General Hospital in Boston, the same residency program that Graeme Hammond (see page 512) had completed several years earlier.

Just about all of the many dyslexic physicians I have come to know agree with Dr. Cosgrove's conclusion that "the hard part with dyslexia is getting through the academics. It's not about practicing medicine." Since the first edition of this book, I have come to know Dr. Cosgrove on a personal level, and my admiration has grown still more.

Figure 53. Dr. Toby Cosgrove and Dr. Shaywitz

In addition to his prominence as a superb surgeon, Cosgrove is an extraordinary, fearless innovator, with accomplishments including creating the very first Global Center for Health Innovation; developing the Cleveland Clinic's Outcomes Book, which measures, details, and makes public outcomes of a range of services; partnering with Case Western Reserve University to start the Cleveland Clinic Lerner College of Medicine, the very first tuition-free medical school in the nation; and, of great benefit to patients, and conforming to the Cosgrove Cleveland Clinic motto, "Patients First," establishing the rare policy of promising patients who call the health system the opportunity to see a physician that very day. The Cleveland Clinic is ranked number two overall nationally by *U.S. News & World Report,* and its heart program is ranked number one. Cosgrove is always seeking new and better ways to improve patients' lives and their experience at the clinic. He is an engaged leader who truly cares deeply about people. Cosgrove has received many honors, including election to the prestigious National Academy of Medicine.

Yet Dr. Cosgrove continued to experience reminders of his dyslexia while practicing medicine. These reminders did not come when he was in the midst of his professional surgical work but emerged during the side times:

> Four or five times each day, I would completely stop a patient's heart from beating for one to two hours, perform some type of surgical intervention, and attempt to restart the heart. In each case, there is a moment of tension while I wait to see if the heart will start beating. As tense as this is, I know I am in control and can deal with any contingency that may arise.
>
> After the surgery is completed, it's a different matter. I pick up the telephone to call the patient's family waiting in the lounge and the nurses break into hysterics as they hear me trying, but unable, to pronounce the patient's family name. I just can't pronounce those silly words.

Of course, family names are the equivalent of nonsense words, and chances are that a dyslexic has not previously encountered many of the names. As a result, he has not memorized their pronunciation and is unable to pronounce the names correctly.

In September 1996, Cosgrove was invited to give the Williams College convocation address to inaugurate the senior year for the class of '97. He chose to speak on "Failure" because "this is a topic with which I am very familiar." He addressed the students and their families:

> My career at Williams was anything but distinguished. Diligent study produced a collection of C's. . . . No honor societies or distinguished offices graced my curriculum vitae. Twelve of thirteen medical schools rejected me. During my residency at Massachusetts General Hospital, I was informed that I was the least talented individual in my residency group and advised not to go into cardiothoracic surgery.

Cosgrove sees failure as the starting point for a process of learning and discovery. He states that he's failed so often he has lost his fear of it. Cosgrove credits his dyslexia with "teaching me to work like crazy." And so even after being rejected by twelve of thirteen medical schools and being told that he was the worst resident in his class at Massachusetts General Hospital and strongly advised not to go into cardiothoracic surgery, he did not give up, nor did he lose sight of his dream. His extraordinary accomplishments, both as a groundbreaking cardiothoracic surgeon and as an unparalleled leader and patient advocate at the Cleveland Clinic, speak to his wisdom of continuing forward and realizing his dream. Dr. Cosgrove's story is a powerful reminder that perhaps we need to rethink how we judge people and determine who has the potential to be a future leader and make a major contribution to society.

GRAEME HAMMOND

The late Graeme Hammond spent his professional career as one of an elite group of surgeons actively engaged in cutting-edge research while maintaining a busy operating schedule. His difficulties in reading were obvious in grade school. His parents took him to the ultimate diagnostician of his time, Dr. Sam Orton, who confirmed that the young boy was dyslexic. He began reading lessons, but they were provided haphazardly and were not terribly effective. It would be years before he was able to read his first book.

"I applied to medical school because I didn't need any foreign languages for medical school," he said, "and I knew I could handle the science." As he progressed, learning medicine in depth, medical school got "better and better every year. Finally, in my last year, I was one of the top ten students of my class."

Following graduation, he was accepted for a surgical residency at Massachusetts General Hospital and then for a faculty position at Yale. Hammond, as judged by his mentors—senior surgeons at both Harvard and Yale—had what it takes to be a top-tier surgeon. However, while he could exercise superior judgment, operate with a sure hand, and in every way more than meet the day-to-day demands of surgery, passing the National Board of Medical Examiners Licensing Examinations was another matter. To him, this seemed more about test-taking skills than about knowledge, but these tests almost curtailed his career:

> When I was a fourth-year medical student I absolutely could not pass the national boards. I still have not passed them. And so I came through an alternate route, the state boards, which consisted of questions to be answered in essay form, not multiple-choice. I have no problem with these, and so I took the New York State boards and sailed through them. And when I was a resident at Mass General, the Massachusetts boards accepted the New York State boards. They never knew I hadn't passed my national boards. Subsequently I was on the National Board of Medical Examiners and have a diploma hanging on my wall from them thanking me for being on their board for so long. But ironically, I never passed their boards.

As much as dyslexia had frustrated him, he believed that dyslexia was responsible for the qualities that ultimately led to his success.

> For me it's been a blessing in disguise. This difficulty reading gave me a tremendous amount of perseverance. My whole life is about striving, because I've had so many roadblocks put in my way. I see a lot of surgeons who do everything well, yet they never, ever question. They've never been forced to try another way to find an answer to a problem. In my own case, I had developed the ability to look for other ways of doing things.

Hammond and Cosgrove follow in the footsteps of many renowned physicians, past and current, who are dyslexic, including Harvey Cushing, who founded modern neurosurgery and was awarded a Pulitzer Prize for his biography of Sir William Osler. The numerous contemporary physicians who are dyslexic include Beryl Benacerraf, professor of radiology at Harvard Medical School, an international authority on prenatal ultrasonography; Tyler Lucas, chief of orthopedic surgery at Metropolitan Hospital in New York City; Karen Santucci, professor of pediatrics at Yale School of Medicine and chief, Pediatric Emergency Department; and Stuart Yudofsky, distinguished professor emeritus and former chairman, Menninger Department of Psychiatry and Behavioral Sciences, Baylor College of Medicine.

SCIENTIST

GEORGE CHURCH

Ask anyone knowledgeable about groundbreaking science whom they most admire for his or her discoveries, insights, and brilliance, and George Church, often referred to as "biotech's genetic prophet," is consistently near or at the top of that list. He is the Robert Winthrop Professor of Genetics at Harvard Medical School and a professor of health sciences and technology at Harvard and MIT, as well as a founding member of Wyss Institute for Biologically Inspired Engineering at Harvard.

When, early on, Church struggled to decipher printed words on the page, he turned to picture books. He had no difficulty interpreting the stories the pictures represented. "If an assignment required a non–picture book, then I would find a picture book that I thought equivalent to it." The picture books were part of a Time-Life series that Church's mother was using to help him compensate for his inability to read words by making sure he had "really interesting" books. Church avidly took to these books, memorizing the twenty volumes so that he could tell you what picture was on each page and what it was about. From Church's perspective, even more than reading difficulty, "I had a worse problem—I was shy."

His caring mom, aware that her son was very bright, tried to unearth the cause of his reading difficulties and periodically had him tested, which eventually led to a diagnosis of dyslexia. Church's early school experiences were a series of ups and downs. As he describes it, "I also managed to repeat ninth grade and flunk out of Duke in 1976." One reason for his leaving the Duke graduate program was that he was so deeply consumed by his groundbreaking research using X-ray crystallography to study the three-dimensional structure of transfer RNA that he often spent hundreds of hours in the lab each week, resulting in his neglecting his classes. As a consequence, he was "withdrawn" from the degree program in January 1976 and presciently told that "whatever problems . . . contributed to your lack of success . . . at Duke will not keep you from a successful pursuit of a productive career." And right they were, with several groundbreaking publications emerging from his graduate work at Duke.

Harvard, wisely seeing past his forced withdrawal from Duke, then accepted Church, and the rest is history. One does not often hear about a scientist anywhere in the world whose mere mention brings out "amazing," "extraordinary," and "genius." Perhaps we should now also knowingly add "dyslexic."

The experience of being asked to leave Duke and then, at age twenty-five, being accepted to a wonderful lab under the direction of Nobel laureate Walter Gilbert at Harvard was like being given a second chance with a breath of invigorating fresh air. As Church describes it, "I had to restart graduate school, and I guess I felt I had a chance to reinvent myself and try much harder to conquer my shyness by being more sociable, and I spent more time reading." As often happens when a dyslexic is interested in a specific topic, he will be drawn to reading more and more about it, and his reading will gradually improve. Church remains a slow reader but one who has improved over time.

> When I read, I might read more things than people in my specialty, but I won't read them the same way they read them. And I don't read them real fast, and I jump around and I look. I spend a lot of time on the figures, still. While they might spend a lot more time, you know, with the words and not the figures. I read incredibly slowly. Because I wasn't trying to read it the way others were trying to read it, I actually got decent comprehension out of it. And then, that made me feel, *Well, maybe I'm not completely broken, I'm just different, you know.*

Church is pleased that bright, successful dyslexics like himself are often cited. At the same time he is concerned that those dyslexics who are not identified during the school years and therefore receive no intervention and no accommodations are far too often ignored. "I think the amount of waste must be enormous," he says.

Is this universally acclaimed scientist still reminded of his dyslexia?

> All the time, all the time. I think that I am uncharacteristically less
> involved in grant paper writing then most professors are. I realize
> that I have those things where I can contribute. But it is very
> strange, because what typically identifies a person as an investigator
> is that they frequently micromanage the entire operation that goes
> on in the lab. But that's not my main contribution.

Here we have a special moment that should not be overlooked. In dyslexia, you may not be perfect, with a number of things you do not do well—and yet even with these imperfections, if you are bright, dedicated, and work very hard, you can succeed at a very high level and make a real difference, a major contribution to society.

Another dyslexic scientist who made major contributions to society is the aforementioned molecular biologist Carol Greider, who was awarded the Nobel Prize in Physiology or Medicine in 2009 for her discovery of telomerase, which plays an important role in age-related diseases. As you read this, keep in mind her earlier experiences: "I had a lot of trouble in school and was put into remedial classes. I thought I was stupid." This pattern for dyslexic scientists seems to repeat itself. Jacques Dubochet, who won the Nobel Prize in Chemistry in 2017 for his work on cryoelectron microscopy, did poorly in school and describes himself as the "first official dyslexic" in the town in Switzerland where he lived. That same year, another dyslexic, Richard H. Thaler, was awarded the Nobel Memorial Prize in Economic Sciences. In awarding the prize, the Nobel jury said, "He's made economics more human," identifying a common denominator of dyslexia.

WRITER

JOHN IRVING

When John Irving was a student at Phillips Exeter Academy, a New England prep school, he demonstrated glimmers of talent only to those few who were able to appreciate it. He was a C– student in English, with an SAT verbal score of 475. Later, of course, he became the best-selling author of *The World According to Garp* and *The Cider House Rules* and an Academy Award–winning screenwriter. As Irving tells it:

I simply accepted the conventional wisdom of the day—I was a struggling student; therefore, I was stupid.

I needed five years to pass the three-year foreign-language requirement. . . . I passed Latin I with a D, and flunked Latin II; then I switched to Spanish, which I barely survived. . . .

It wasn't until my younger son, Brendan, was diagnosed as slightly dyslexic that I realized how I had been given the shaft. His teachers said that Brendan comprehended everything he read but that he didn't comprehend a text as quickly as his peers; they said that he could express himself as well as, or actually better than, his peers but that it took him longer to organize his thoughts on paper. This sounded familiar to me. As a child, Brendan read with his finger following the sentence—as I read, as I *still* read. Unless I've written it, I read whatever "it" is very slowly—and with my finger.

I wasn't diagnosed as dyslexic at Exeter; I was seen as just plain stupid. I failed a spelling test and was put in a remedial spelling class. Because I couldn't learn how to spell—I still can't spell!—I was advised to see the school *psychiatrist*! This advice made no sense to me then—it makes no sense to me now—but if you were a poor student at Exeter, you would develop such a lasting sense of your own inferiority that you'd probably be in need of a psychiatrist one day.

I wish I'd known, when I was a student at Exeter, that there was a word for what made being a student so hard for me; I wish I could have said to my friends that I was dyslexic. Instead I kept quiet, or—to my closest friends—I made bad jokes about how stupid I was. Brendan knows he's not stupid; he knows he's the same kind of student I was.

My only anger was at people who presumed that I was lazy, because I was working harder than anybody else. I had at least one teacher each year who thought that there was nothing the matter with me except that I was lazy.

It was not academics but athletics—wrestling—that was Irving's salvation. Whatever satisfaction he was not receiving in the classroom, he was achieving on the wrestling mat.

My wrestling coach, Ted Seabrooke, kept me in school; he gave me enough confidence in myself, through wrestling, that I was able to take a daily beating in my classes and keep coming back for more. An ironic blessing of my having to repeat my senior year at Exeter was that I finally got to be captain of the wrestling team—my sole distinction, my only honor, in five years at the academy.

How has dyslexia influenced his writing?

It's become an advantage. In writing a novel, it doesn't hurt anybody to have to go slowly. . . . One reason I have confidence in writing the kind of novels I write is that I have confidence in my stamina to go over something again and again no matter how difficult it is—whether it is for the fourth or fifth or eighth time. It's an ability to push myself and not to be lazy. This is something I would ascribe to the difficulties I had to overcome at an early age.

How does Irving, a poor reader, read the voluminous number of pages he produces?

I read slowly and I reread a lot. I read a paragraph, and if I'm not 100 percent concentrating on what I'm reading or if I'm the slightest bit tired, I have a problem. And from that, too, I have learned that if you have to concentrate twice as hard as somebody else, you get twice as tired, twice as soon. And so when I am a little bit tired, the difference is enormous.

I also had to practice. I am actually a very good reader of my own work. But I think that comes from my practice of reading my work aloud and understanding the cadence with which it's written. I read my own work aloud as I write it or I say it aloud as I write it. And in fact, when I am writing in my office at home, people walking by always think I'm on the phone.

I also use my finger, as I did when I was a student. Especially when I'm tired or when the print is a little too small, I just keep my finger on it. And when I do public readings, I never take my finger off where I am.

One could not ask for a more classic description of how a very bright person who is dyslexic reads or of the usefulness of reading aloud.

Time impacted Irving's studies and also his performance on standardized tests:

> The colleges that admitted me out of Exeter all cited my writing sample and a couple of recommendations from teachers who said decent things about my capacity to write. I certainly would have had nowhere to go if I had only been judged on my standardized test scores.

Life was better for Irving as his education progressed. Graduate school was the best, for it was more specialized.

> When I got to graduate school at Iowa in creative writing I was taking one literature course. I could concentrate on one thing for three or four or five hours instead of doing something for an hour and putting it aside and doing something that was completely different for an hour.

Irving's problems with rote memory do not reflect a general memory problem. He has a "wonderful" memory when it comes to remembering the most complex information—as long as it has a meaningful connection. While he cannot remember telephone numbers (which are random numbers), he can recall the most arcane bits and pieces of his novel because they are part of a network of information that forms a meaningful whole. "My interpretation of myself when I was in my late teens was that I had a lousy memory," he says. "I've since discovered that I have a very good memory, because I would say that to write a long, complicated, plot-driven novel, you need to have a wonderful memory."

Irving has ridden the roller coaster of life that you experience as a dyslexic and has been deeply affected by the experience. As a parent, he is also extremely sensitive to the needs of children who are dyslexic. His advice to parents:

> I think the burden on parents and on teachers, when they see that the task of reading is a challenge, is to find ways to reward the child constantly. And you have to not lower your expectations for that child. You have to keep up your belief in the child and maintain your vision of his future.
>
> I had difficulty in school. But I had around me my parents and

my coach to give me support and to recognize that I needed to have more and more confidence in myself because I wasn't getting it from school. And so the harder you have to work at anything, the more important it is that you have somebody telling you you're doing a good job. You need somebody who is giving you a lot of encouragement, a lot of support; someone who knows how hard you're working.

Actually, having to go through something that's a struggle, even at a young age, can be a very positive experience for you—if you're being supported and constantly told that you're doing a good job. I am a believer in telling kids that they're great kids, that you know how hard they're working—you can't overdo that. And, finally, you have to help the child imagine that he can be a success and to sustain that vision of himself. My very first wrestling coach used to say that before you could beat someone better than you, you had to be able to imagine that you could be a winner. You had to believe that you could do it. It sounds like coach-speak on the one hand, but on the other hand, it's very true.

STEPHEN J. CANNELL

Stephen Cannell, Emmy Award–winning television writer of such hits as *The Rockford Files* and *The A-Team,* producer, best-selling author, and actor, was dyslexic. For him, too, school was a nightmare. Especially traumatic to Cannell were the terrible misjudgments of his abilities, particularly his potential as a writer. "I was not very good in writing because one of the things they stressed at private schools was spelling. If I couldn't spell, that meant to them that I couldn't write." Just as an inability to spell some words impacted John Irving, misspellings also constricted Cannell's writing. He recalled that "even the words I thought I could spell I couldn't spell. By the time I got to college I would only use the words I could spell, so I wouldn't use all the words in my vocabulary. That limited my ability to write a good essay."

It was a college instructor, Ralph Salisbury, who finally saw the talent that had remained hidden in Stephen throughout his previous twelve or so years of schooling. A grateful Cannell described what this teacher meant to him: "Ralph was the first instructor I ever had in my entire academic career who said to me, 'It isn't important to me whether you can't spell. I'm interested in your ideas. I'm a creative-writing instructor, not a spelling teacher.' "

Cannell's urgent message—people need to understand what Salisbury clearly understood—was that writing and spelling are not related, and also that writing and dyslexia are not related. He posed a rhetorical question: "How is it to be dyslexic and write? Well, it's easy. People say, you must have climbed this horrible mountain to be able to write with this condition. Writing has nothing to do with dyslexia. Spelling has something to do with it." He was often asked how someone who is dyslexic can choose writing as his career. "We [dyslexics] are very good with abstract thought. And that's the key in writing," he said. You could feel the excitement as he described how his outline, what he referred to as "the cerebral part," allowed "the more visceral part" of his writing to come out:

> Everything I write I see; it's really a written picture. It's like I'm in the scene with the characters. I see it like it was a movie. And I often will describe it the way my imagination is picking it up. For dialogue, I'm everyone in the scene, and I say my lines of dialogue. And so my outline gives structure, it frees me to be all these people, to have a party with my characters.

For Cannell it was the image that drove the writing. He could use words to paint the picture that he saw in his imagination. Words were the servants of the visual images; they were the tools but not the creators of the images that he wrote about.

His dyslexia did not prevent Cannell from being a prolific and highly regarded writer, scripting more than 450 episodes of the nearly forty television series he created or cocreated. In addition he published at least eighteen novels. I was surprised and extremely touched that in his novel *At First Sight,* Cannell chose to write these words:

> There was important research on dyslexia being conducted at Yale by Drs. Bennett and Sally Shaywitz using fMRIs to determine what part of the brain was activated when reading. Chandler had been excited with Sally and Bennett's work and had invested foundation money to support their research.

Stephen Cannell was a friend, and it saddens me to report that he died in September 2010.

Other dyslexic writers include Max Brooks, Octavia Butler, John Cheever,

Fanny Flagg, Richard Ford, John Grisham, and the children's book authors
Dav Pilkney, Avi, and Patricia Polacco.

After reading these stories, no one should ever doubt "a person like that" can
be a winner in any field in which he or she has an interest and talent. These
are stories of triumph over innate difficulties as well as those created by misunderstandings, misjudgments, and stereotyped views of what constitutes
real ability or talent. For these dyslexics and thousands of others, the obstacle
to success is often not their inherent physiologic weakness but the misguided
perceptions of others—those who believe that spelling has something to do
with creative writing, that slow reading is incompatible with sharp thinking,
that scores on standardized tests predict performance in real life for dyslexics.
Each of these remarkably talented individuals almost didn't realize his or
her life's goals. Each was made to feel stupid or incapable. So-called experts
erected barriers and gave bad advice. And yet each ultimately triumphed,
viewing dyslexia as a gift more than a burden. As Sir Richard Branson often
says, "If I hadn't been dyslexic, I would never have done it."

Dyslexics think differently. They are intuitive and excel at problem-solving,
seeing the big picture, and simplifying. They feast on visualizing, abstract
thinking, and thinking out of the box. They are poor rote reciters but inspired
visionaries. Adult dyslexics are tough—having struggled, they are used to
adversity; hard work and perseverance come naturally. Having experienced
failure, they are fearless, undaunted by setbacks. Repetition and practice are
a way of life. Each was rescued by a special person—a parent, a grandparent,
a teacher—who saw the raw talent and nurtured it in the midst of all the
naysayers. Hope was sustained by a taste of success in sports or some other
activity. Yes, the symptoms of dyslexia persist, but they needn't interfere
with success.

Success is waiting for your child, and now you know what to do to help
her achieve it. You don't have to rely on chance. You know how to identify a
problem early and how to get the right help to ensure that it is your child's
strengths and not the misperceptions of others that ultimately define her.
You know what is possible and how to nourish it. Children blossom with
reward and praise, flourish because of high expectations. Above all, you
must maintain your belief in your child, provide unconditional support for
her, and hold true to a bright vision of her future. The rewards will be great.

Today all dyslexic children should know that they can succeed. Dyslexia
can be overcome.

A final word: I often compare dyslexia to an iceberg that is 90 percent hidden underwater, as shown in Figure 54. We see many of the successful individuals you have just read about—Emanuel, Boies, Swonk, Cosgrove, and Schwab—visible, standing atop the iceberg. What we must not forget about are the 90 percent of dyslexics who are underwater and unseen, literally millions of dyslexic boys and girls who one might say are drowning, struggling, not knowing that they have dyslexia and that they are smart. It is this hidden 90 percent that I worry about so much. It is these invisible dyslexic boys and girls who are currently ignored by their schools, losing self-esteem, often becoming anxious, fearing the future and far too often dropping out of school. It does not have to be this way. We know better and must do better. We have the knowledge. We must act. We must ensure that schools screen for and identify dyslexia and provide evidence-based effective interventions that will give these currently hidden children a chance at the American dream. Our dyslexic children and our country demand nothing less.

LET'S NOT FORGET ABOUT THE UNSEEN 90 PERCENT

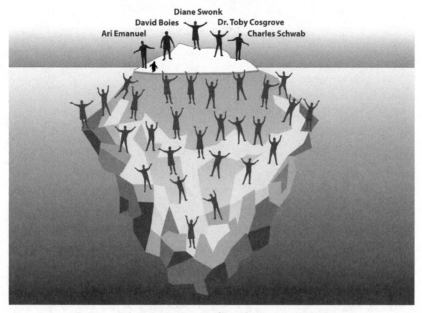

Figure 54. Let's Not Forget About the Unseen 90 Percent

APPENDIX

APPENDIX A: STATE-OF-THE-ART DEFINITION OF DYSLEXIA

1. IN FEDERAL LAW: FIRST STEP ACT OF 2018. 115TH CONGRESS. SIGNED INTO LAW BY PRESIDENT TRUMP ON DECEMBER 21, 2018, AND BECAME PUBLIC LAW NO. 115-391.

In this law, dyslexia is defined:

(1) DYSLEXIA.—The term "dyslexia" means an unexpected difficulty in reading for an individual who has the intelligence to be a much better reader, most commonly caused by a difficulty in the phonological processing (the appreciation of the individual sounds of spoken language), which affects the ability of an individual to speak, read, and spell.

(2) DYSLEXIA SCREENING PROGRAM.—The term "dyslexia screening program" means a screening program for dyslexia that is—

(A) evidence-based (as defined in section 8101(21) of the Elementary and Secondary Education Act of 1965 (20 U.S.C. 7801(21) with proven psychometrics for validity;

(B) efficient and low-cost; and

(C) readily available.

2. IN SENATE RESOLUTION (S. RES.) 680.

IN THE SENATE OF THE UNITED STATES OCTOBER 11, 2018 Mr. CASSIDY (for himself, Mr. MURPHY, Mrs. CAPITO, Ms. WARREN, and Mr. VAN HOLLEN) submitted the following resolution; which was considered and agreed to RESOLUTION Calling on Congress, schools, and State and local educational agencies to recognize the significant educational implications of dyslexia that must be addressed, and designating October 2018 as "National Dyslexia Awareness Month." Whereas dyslexia is— (1) defined as an unexpected difficulty in reading for an individual who has the intelligence to be a much better reader; and (2) most commonly caused by a difficulty in phonological processing (the appreciation of the individual sounds of spoken language), which affects the ability of an individual to speak, read, and spell, and often, the ability to learn a second language; Whereas dyslexia is the most common learning disability and affects 80 to 90 percent of all individuals with a learning disability; Whereas dyslexia is persistent and highly prevalent, affecting as many as 1 out of every 5 individuals; Whereas dyslexia is a paradox, in that an individual with dyslexia may have both— (1) weaknesses in decoding that result in difficulties in accurate or fluent word recognition; and (2) strengths in higher-level cognitive functions, such as reasoning, critical thinking, concept formation, and problem solving; Whereas great progress has been made in understanding dyslexia on a scientific level, including the epidemiology and cognitive and neurobiological bases of dyslexia; Whereas the achievement gap between typical readers and dyslexic readers occurs as early as first grade; and Whereas early screening for, and early diagnosis of, dyslexia are critical for ensuring that individuals with dyslexia receive focused, evidence-based intervention that leads to fluent reading, promotion of self-awareness and self-empowerment, and the provision of necessary accommodations that ensure success in school and in life: Now, therefore, be it Resolved, That the Senate— (1) calls on Congress, schools, and State and local educational agencies to recognize that dyslexia has significant educational implications that must be addressed.

APPENDIX B: CHILDREN'S BOOKS

CHILDREN'S BOOKS THAT JOYFULLY PLAY
WITH THE SOUNDS OF LANGUAGE

Ada, Alma Flor, and F. Isabel Campoy. *Pío Peep! Traditional Spanish Nursery Rhymes*. New York: HarperCollins, 2003.

Arnold, Tedd. *Parts*. New York: Dial, 1997.

Baker, Keith. *Hickory Dickory Dock*. Orlando: Harcourt, 2007.

Bemelmans, Ludwig. *Madeline*. New York: Penguin, 2012.

Bingham, Kelly. *Z Is for Moose*. New York: Greenwillow, 2012.

Buell, Janet. *Sail Away, Little Boat*. Minneapolis: Carolrhoda, 2006.

Christelow, Eileen. *Five Little Monkeys Jumping on the Bed*. New York: Clarion, 1989.

Craig, Lindsey. *Dancing Feet!* New York: Knopf, 2010.

Crews, Nina. *The Neighborhood Mother Goose*. New York: Greenwillow, 2004.

Crow, Kristyn. *Cool Daddy Rat*. New York: Putnam, 2008.

Degan, Bruce. *Jamberry*. New York: Harper & Row, 1983.

dePaola, Tomie. *Tomie dePaola's Mother Goose*. New York: Putnam, 1985.

Fischer, Scott M. *Jump!* New York: Simon & Schuster, 2010.

Fleming, Candace. *Muncha! Muncha! Muncha!* New York: Atheneum, 2002.

Fleming, Denise. *In the Tall, Tall Grass*. New York: Henry Holt, 1995.

Fox, Mem. *Where Is the Green Sheep?* Orlando: Harcourt, 2004.

Galdone, P. *Henny Penny*. Boston: Houghton Mifflin, 1984.

Gravett, Emily. *Orange Pear Apple Bear*. New York: Simon & Schuster, 2007.

Krauss, Ruth. *Bears*. New York: Scholastic, 1993.

————. *I Can Fly*. New York: Golden, 1999.

LeSieg, Theo. *The Eye Book*. New York, Random House, 1968.

————. *Please Try to Remember the First of Octember!* New York: Random House, 1988.

Lewis, K., and D. Kirk. *Chugga-Chugga Choo-Choo*. New York, Hyperion, 1999.

Litwin, Eric, and James Dean. *Pete the Cat and His Four Groovy Buttons*. New York: Harper, 2012.

Martin, Bill, Jr. *Brown Bear, Brown Bear, What Do You See?* New York: Henry Holt, 1992.

Martin, Bill, Jr. and John Archambault. *Chicka Chicka Boom Boom*. New York: Simon & Schuster, 1989.

Miranda, Anne. *To Market to Market*. San Diego: Harcourt Brace, 1997.

Ochs, C. P. *Moose on the Loose.* Minneapolis: Carolrhoda, 1991.

Park, Linda Sue. *Bee-bim Bop!* New York: Clarion, 2005.

Rinker, Sherri Duskey. *Goodnight, Goodnight, Construction Site.* San Francisco, Chronicle, 2011.

Rosen, Michael. *We're Going on a Bear Hunt.* New York: Margaret K. McElderry, 2009.

Seuss, Dr. *Hop on Pop.* New York: Beginner Books, 1963.

———. *One Fish Two Fish Red Fish Blue Fish.* New York: Random House, 1988.

———. *There's a Wocket in My Pocket.* New York: Random House, 1996.

Shaw, Nancy. *Sheep in a Jeep.* Boston: Houghton Mifflin, 1986.

Sierra, Judy. *Wild About Books.* New York: Knopf, 2004.

Silverstein, Shel. *A Giraffe and a Half.* New York: Harper & Row, 1964.

Thomas, Jan. *Rhyming Dust Bunnies.* New York, Atheneum, 2009.

Van Allsburg, C. *The Z Was Zapped.* Boston: Houghton Mifflin, 1987.

Wheeler, Lisa. *Jazz Baby.* Orlando: Harcourt, 2007.

Wood, Audrey. *Silly Sally.* San Diego: Harcourt Brace Jovanovich, 1992.

JUST RIGHT FOR A CHILD ON THE CUSP OF READING

Cox, Phil Roxbee. *Ted and Friends.* London: Usborne, 2002.

Cuyler, Margery. *Tick Tock Clock: A Phonics Reader.* New York: Harper, 2012.

Degan, Bruce. *I Said, "Bed!"* New York: Holiday House, 2014.

Eastman, P. D. *Go, Dog, Go!* New York: Beginner, 1961.

Grant, Judyann. *Chicken Said, "Cluck!"* New York: HarperCollins, 2008.

Hutchins, Pat. *Rosie's Walk.* New York: Macmillan, 1968.

Kertell, Lynn Maslen. *Bob Books: Sight Words Kindergarten.* New York: Scholastic, 2010.

Klassen, Jon. *I Want My Hat Back.* Somerville, MA: Candlewick, 2011.

Klein, Adria F. *Max Goes to School.* Minneapolis: Picture Window, 2006.

Martin, Bill, Jr. *Brown Bear, Brown Bear, What Do You See?* New York: Henry Holt, 1992.

Pizzoli, Greg. *The Watermelon Seed.* New York: Disney Hyperion, 2013.

Seuss, Dr. *The Foot Book.* New York: Random House, 1968.

Tafuri, Nancy. *The Ball Bounced.* New York: Greenwillow, 1989.

Willems, Mo. *There Is a Bird on Your Head.* New York: Disney Hyperion, 2007.

Wood, Julie M. *Learn to Read with Tug the Pup and Friends!* New York, HarperCollins, 2014.

PICTURE BOOKS YOU CAN READ TO A YOUNG CHILD

Allard, Harry. *Miss Nelson Is Missing.* Boston: Houghton Mifflin, 1977.

Bridwell, Norman. *Clifford the Big Red Dog.* New York: Scholastic, 1985.

Brown, Margaret Wise. *Goodnight Moon.* New York: Harper & Row, 1975.

Brown, Peter. *The Curious Garden.* New York: Little, Brown, 2009.

Burton, Virginia Lee. *The Little House.* Boston: Houghton Mifflin, 1942.

Cronin, Doreen. *Click, Clack, Moo: Cows That Type.* New York: Simon & Schuster, 2000.

Freeman, Don. *Corduroy.* New York: Viking, 1968.

Gravett, Emily. *Monkey and Me.* New York: Simon & Schuster, 2008.

Henkes, Kevin. *Chrysanthemum.* New York: Greenwillow, 1991.

Jeffers, Oliver. *The Incredible Book Eating Boy.* New York: Philomel, 2007.

Keats, Ezra Jack. *The Snowy Day.* New York: Viking, 1962.

Levy, Ganit and Adir, *What Should Danny Do?* Los Angeles: Elon Books, 2017.

———. *What Should Danny Do? School Day.* Los Angeles: Elon Books, 2018.

Morales, Yuyi. *Niño Wrestles the World.* New York: Roaring Brook, 2013.

Polacco, Patricia. *Thunder Cake.* New York: Philomel, 1990.

Rathmann, Peggy. *Officer Buckle and Gloria.* New York: Putnam, 1995.

Reynolds, Peter. *The Dot.* Cambridge, MA: Candlewick, 2003.

Rosenthal, Amy Krouse. *Little Pea.* San Francisco: Chronicle, 2005.

Say, Allen. *Grandfather's Journey.* Boston: Houghton Mifflin, 1993.

Scieszka, Jon. *The True Story of the 3 Little Pigs.* New York: Viking Kestrel, 1989.

Seeger, Laura Vaccaro. *Lemons Are Not Red.* Brookfield, CT: Roaring Brook, 2004.

Sendak, Maurice. *Where the Wild Things Are.* New York: Harper & Row, 1963.

Stead, Philip. *A Sick Day for Amos McGee.* New York: Roaring Brook, 2010.

Van Allsburg, Chris. *Jumanji.* Boston: Houghton Mifflin, 1981.

Viorst, Judith. *Alexander and the Terrible, Horrible, No Good, Very Bad Day.* New York: Atheneum, 1972.

Williams, Vera B. *A Chair for My Mother.* New York: Greenwillow, 1982.

Wilson, Karma. *Whopper Cake.* New York: Margaret K. McElderry, 2007.

Yaccarino, Dan. *The Fantastic Undersea Life of Jacques Cousteau.* New York: Knopf, 2009.

EASY READERS

Adler, David A. *Bones and the Dinosaur Mystery.* New York: Viking, 2005.

Arnold, Ted. *Hi! Fly Guy.* New York: Scholastic, 2005.

Bell, Cece. *Rabbit & Robot: The Sleepover.* Somerville, MA: Candlewick, 2012.

Lin, Grace. *Ling & Ting: Not Exactly the Same.* New York: Little, Brown, 2010.

Lobel, Arnold. *Frog and Toad Are Friends.* New York: Harper & Row, 1970.

Manushkin, Fran. *Katie Finds a Job.* Mankato, MN: Picture Window, 2011.

Mills, Claudia. *Gus and Grandpa Ride the Train.* New York: Farrar, Straus and Giroux, 1998.

Rylant, Cynthia. *Puppy Mudge Finds a Friend.* New York: Simon & Schuster, 2004.

Schwartz, Alvin. *In a Dark, Dark Room, and Other Scary Stories.* New York: Harper & Row, 1984.

Seuss, Dr. *Mr. Brown Can Moo! Can You?* New York: Random House, 1970.

Sharmat, Marjorie Weinman. *Nate the Great.* New York: Coward, McCann & Geoghegan, 1972.

Willems, Mo. *My Friend Is Sad.* New York: Hyperion, 2007.

Ziefert, Harriet. *The Little Red Hen.* New York: Viking, 1995.

POETRY

Florian, Douglas. *Laugh-eteria: Poems and Drawings.* San Diego: Harcourt Brace, 1999.

———. *Mammalabilia.* San Diego: Harcourt Brace, 2000.

Katz, Bobbi. *Pocket Poems.* New York: Dutton, 2004.

Levine, Gail Carson. *Forgive Me, I Meant to Do It: False Apology Poems.* New York: Harper, 2012.

Prelutsky, Jack. *It's Raining Pigs and Noodles: Poems.* New York: Greenwillow, 2000.

———. *A Pizza the Size of the Sun: Poems.* New York: Greenwillow, 1996.

Prelutsky, Jack, and Arnold Lobel. *Random House Book of Poetry for Young Children.* New York: Random House, 1983.

Rex, Adam. *Frankenstein Takes the Cake.* Orlando: Harcourt, 2008.

Shange, Ntozake. *Ellington Was Not a Street.* New York: Simon & Schuster, 2004.

Sidman, Joyce. *Red Sings from Treetops: A Year in Colors.* Boston: Houghton Mifflin, 2009.

Silverstein, Shel. *Where the Sidewalk Ends: The Poems & Drawings of Shel Silverstein.* New York: Harper & Row, 1974.

Singer, Marilyn. *Mirror Mirror: A Book of Reversible Verse.* New York: Dutton, 2010.

Wilson, Karma. *Outside the Box.* New York: Margaret K. McElderry, 2014.

PICTURE BOOKS WITH RHYMES/PATTERN REPETITION

Aylesworth, Jim. *The Gingerbread Man.* New York: Scholastic, 1998.

Brett, Jan. *The Mitten: A Ukrainian Folktale.* New York: Putnam, 1989.

Cooney, Barbara. *Miss Rumphius.* New York: Viking, 1982.

Foley, Greg. *Thank You Bear.* New York: Viking, 2007.

Himmelman, John. *Chickens to the Rescue.* New York: Holt, 2006.

Kimmel, Eric A. *Anansi and the Moss-Covered Rock.* New York: Holiday House, 1988.

London, Jonathan. *Froggy Gets Dressed.* New York: Viking, 1992.

MacDonald, Margaret Read. *The Squeaky Door.* New York: HarperCollins, 2006.

Root, Phyllis. *Looking for a Moose.* Cambridge, MA: Candlewick, 2006.

Scanlon, Elizabeth Garton. *All the World.* New York: Beach Lane, 2009.

Shannon, David. *A Bad Case of the Stripes.* New York: Blue Sky, 1998.

Sloat, Teri. *Sody Sallyratus.* New York: Dutton, 1997.

Stein, David Ezra. *Interrupting Chicken.* Somerville, MA: Candlewick, 2010.

Swanson, Susan Marie. *The House in the Night.* Boston: Houghton Mifflin, 2008.

Taback, Simms. *Joseph Had a Little Overcoat.* New York: Viking, 1999.

Tucker, Kathy. *The Seven Chinese Sisters.* Morton Grove, IL: A. Whitman, 2003.

Willey, Margaret. *Clever Beatrice: An Upper Peninsula Conte.* New York: Atheneum, 2001.

Winter, Jeanette. *The House That Jack Built.* New York: Dial, 2000.

Wood, Audrey. *The Napping House.* San Diego: Harcourt Brace Jovanovich, 1984.

TRANSITIONAL CHAPTER BOOKS

Adler, David. *Cam Jansen and the Catnapping Mystery.* New York: Viking, 1998.

Cleary, Beverly. *Ramona's World.* New York: Morrow, 1999.

DiCamillo, Kate. *Mercy Watson to the Rescue.* Cambridge, MA: Candlewick, 2005.

Gannett, Ruth Stiles. *My Father's Dragon.* New York: Random House, 1948.

Jenkins, Emily. *Toys Go Out: Being the Adventures of a Knowledgeable Stingray, a Toughy Little Buffalo, and Someone Called Plastic.* New York: Schwartz & Wade, 2006.

Joyner, Andrew. *Boris on the Move.* New York: Branches, 2013.

Kline, Suzy, and Frank Remkiewicz. *Horrible Harry in Room 2B.* New York: Viking Kestrel, 1988.

Pinkwater, Daniel Manus. *The Hoboken Chicken Emergency.* Englewood Cliffs, NJ: Prentice-Hall, 1977.

Roy, Ron. *The Absent Author.* New York: Random House, 1997.

Scieszka, Jon. *Knights of the Kitchen Table.* New York: Viking, 1991.

Smith, Alexander McCall. *The Great Cake Mystery: Precious Ramotswe's Very First Case.* New York: Anchor, 2012.

Smith, Robert Kimmel. *Chocolate Fever.* New York: Putnam, 1972.

Warner, Sally. *Ellray Jakes Is Not a Chicken!* New York: Viking, 2011.

Wight, Eric. *Frankie Pickle and the Closet of Doom.* New York: Simon & Schuster, 2009.

Consider this selection a starting point for further exploration. Your local library has a treasury of good books, and you might also want to consult the lists of recommended works in such guides as:

GUIDES/ BOOKLISTS

A to Zoo: A Subject Access to Children's Picture Books, by Carolyn and John Lima. Westport, CT: Libraries Unlimited, 2006.

Books to Build On: A Grade-by-Grade Resource Guide for Parents and Teachers, edited by John Holdren and E. D. Hirsch, Jr. New York: Dell, 1996.

Boys and Literacy: Practical Strategies for Librarians, Teachers, and Parents, by Elizabeth Knowles and Martha Smith. Westport, CT: Libraries Unlimited, 2005.

How to Get Your Child to Love Reading: Inspiration and Ideas from a Teacher Who Really Knows, by Esmé Raji Codell. Chapel Hill: Algonquin, 2003.

The New York Times Parent's Guide to the Best Books for Children, by Eden Ross Lipson. New York: Three Rivers, 2000.

100 Best Books for Children, by Anita Silvey. Boston: Houghton Mifflin, 2004.

APPENDIX C: A WORD ABOUT COMMON CORE

Common Core State Standards (CCSS) in English and mathematics were proposed in 2009 by the National Governors Association and the Council of Chief State School Officers, and these have been adopted either wholly or partially by most states. For reading, there is a focus on ensuring that "all students must be able to comprehend texts of steadily increasing complexity as they progress through school." Reading comprehension, particularly of expository or informational text, is the focus of CCSS in reading, and the goal is to, "by the end of grade 12, read and comprehend literature [informational texts, history/social studies texts, science/technical texts] at the high end of the grades 11–CCR text complexity band independently and proficiently." Text complexity is determined both qualitatively and quantitatively. Qualitative measures include features such as structure, language, and knowledge demands. Quantitative measures of text complexity relate to factors such as word frequency, sentence length, word familiarity, words per sentence, average grade level of words, and characters per word that are more efficiently measured by computer programs. An example of a quantitative measure is the Lexile score. CCSS text complexity grade bands are available for grades two through twelve and are linked to Lexile levels, which allow teachers to select particular texts at complexity levels recommended for students at any grade between two and twelve. Standards recommend that quantitative measures such as Lexile be used to determine the level of complexity of a particular text and that results be confirmed or overruled by a qualitative analysis of the text in question.

Note that the Standards' text complexity requirements do not begin until grade two. The Standards creators were aware that students in kindergarten and grade one are just beginning to acquire the reading code and there is great variability in mastering beginning reading. Furthermore, K–1 texts are not amenable to quantitative measure. CCSS expect second-graders to read and comprehend expository, informational literature in the grades 2–3 text complexity band proficiently by the end of the year.

> *CAUTION:* At the same time, CCSS instructional guidance does not include a focus on teaching decoding skills in early grades. In fact, such instruction is relegated to an appendix, with little emphasis on the importance of such instruction at the beginning of reading instruction.

APPENDIX D: THE WISC-V

The WISC-V test typically comprises 10 primary or individual subtests assessing various aspects of intelligence. These subtests group into five major categories or Index Scores:

→ Verbal Comprehension
→ Visual Spatial
→ Fluid Reasoning
→ Working Memory
→ Processing Speed

You may also be given a score referred to as the Full Scale IQ, a measure obtained by an algorithm based on virtually all subtests. If there is fairly good consistency among the five index scores, it is reasonable to consider the Full Scale IQ an appropriate measure of overall intellectual ability.

> *CAUTION:* If, as often happens, there is great variability among these index scores, then it is *not* appropriate to use the Full Scale IQ score as a valid measure of intelligence. In such cases I find it is often helpful to focus on the Verbal Comprehension Index (VCI), which reflects language and verbal reasoning proficiency and knowledge and is often associated with academic achievement. Others may use the Fluid Reasoning or, especially if the student excels in perceptual tasks, may find it useful to turn to the Visual Spatial Index as a measure of the child's ability.

APPENDIX E: PROGRAMS FOR
ADULT DYSLEXICS FROM THE XPRIZE COMPETITION

Learning Upgrade, led by Vinod Lobo, helps students learn English through songs, video, games, and rewards and is available for both Android and iOS devices at no cost. People ForWords, led by Southern Methodist University's Simmons School of Education and Human Development, uses a mobile game based on an archaeological adventure storyline to help dyslexic adults improve their English reading skills and is available for Android devices at no cost. AmritaCREATE, led by Dr. Prema Nedungadi, uses engaging, culturally appropriate e-content linked to life skills and is available for Android and iOS devices at no cost. Cell-Ed, led by Dr. Jessica Rothenberg-Aalami, offers on-demand essential skills, microlessons, and personalized coaching for any mobile device, without Internet. AutoCognita, led by Frank Ho, applies a constructivist learning approach to engage dyslexic adults to acquire basic literacy through a comprehensive curriculum. These programs allow adult struggling readers to have an app that they can use autonomously to learn to read.

ACKNOWLEDGMENTS

How do you best acknowledge the myriad of people whose support, work, and wise counsel have made this book what it is? I am incredibly grateful to the very many men and women who have contacted me, thanking me for the first edition of *Overcoming Dyslexia* while also sharing with me their or their children's, or someone they care deeply about's, life as a dyslexic. The knowledge and wisdom (I hope) that this has imparted to me is immeasurable and deeply valued and appreciated. Each of these interactions and all of them together have given me insight and more: hope that there is a much-deserved better future ahead for all those who are dyslexic.

For me, this most recent journey has taken five years. All the while I have worked, pursuing important unanswered questions concerning dyslexia and carrying out research at our Yale Center for Dyslexia & Creativity (YCDC) to address each. Concurrently, following my own advice given in congressional testimony, I have worked hard—through lectures, testimonies to Congress and educating those in Congress, published papers, and our YCDC website and newsletters—to try to ensure that scientific knowledge of dyslexia is translated into action. Here I must acknowledge the thoughtful and proactive legislators who listened and took action. First and foremost is Senator Bill Cassidy (R-LA), who, as you will have read in this second edition, is the father of a wonderful, bright dyslexic daughter and the spouse of Dr. Laura Cassidy, a devoted mother and the founder and leader of an extraordinary, highly effective, life-changing public charter school, LKA, in Baton Rouge, LA. To share but one example, Senator Cassidy championed the groundbreaking First Step Act, which brought to life the very first evidence-based definition of dyslexia in federal legislation while also defining the criteria for a valid dyslexia screening instrument. Senator Chris Murphy (D-CT), my extraordinary Representative Rosa DeLauro (D-CT), together with Representatives Dr. Larry Buschon (R-IN), Joe Courtney (D-CT), and Bruce Westerman (R-AR), Dutch Ruppetsberger (D-MD) and his chief of staff, Tara Oursler, and former California representative Pete Stark, provided valuable support, along with others I may have inadvertently left out. State officials have played a significant role, too. I must single out New York state assemblyman Robert Carroll and California governor Gavin Newsom, both dyslexic themselves and caring greatly about all those who are, and who have worked exceptionally hard to make scientifically based, positive policy changes that support dyslexic students.

To say that I have been blessed does not do justice to the role my family and particularly my spouse of fifty-seven years and work partner of forty-seven years, Bennett, have played in my life and certainly in my pursuit of the science and heart of dyslexia.

This pursuit—some refer to it as an obsession in the most positive sense of the word—has been carried out jointly with Bennett, a child neurologist, who holds the Charles and Helen Schwab Chair in Dyslexia and Learning Development at Yale. We are truly one in every sense of the word: the psychological support, the intellectual rapport and exchange of knowledge, and, most of all, sharing a love of science and an even greater passion to bring newly minted scientific knowledge to the dyslexic children and adults who can most benefit from it. We were so excited when we saw our longitudinal data validating the unexpected nature of dyslexia—that is, a person can be both dyslexic and highly intelligent. Seeing this, we have worked to ensure that all of society is aware of this relationship and are grateful that dyslexia's critical unexpected nature is now part of the federal definition of dyslexia. Similarly, data from the Connecticut Longitudinal Study showing clearly that dyslexia impacts students as early as first grade, where a large and persistent achievement gap between typical readers and dyslexic readers is already present, drove us to action. Deeply concerned about this large early gap, we developed the evidence-based Shaywitz DyslexiaScreen™ to screen and identify as early as kindergarten and first grade those young students who are at-risk for dyslexia. My gratitude goes to Tina Eichstadt and her colleagues at Pearson Publishers for making this critical instrument a practical reality. I also want to acknowledge and thank our incredible hardworking team at our Yale Center for Dyslexia & Creativity—my irreplaceable longtime assistant, Carmel Lepore; Melanie Loeber; Sally Sherman; John Holahan; Reissa Michaels; Mary Losee; Christine Colonis Babinski; Rebecca Sandmann; and Nancy Pelegano—who are dedicated to making sure our clinical studies are carried out with care and a sense of urgency. I am, indeed, fortunate to have the support of Dr. Cliff Bogue, our Chair of Pediatrics, and the very able Konstantza Popova and John Palmieri in making sure our center runs so well.

Much of the exciting new information in the second edition of *Overcoming Dyslexia* has come from the Connecticut Longitudinal Study. This unique study of entering kindergarteners has now continuously been followed by me and our team for over three decades. I want to thank and praise the dedicated group of colleagues who worked so hard to help decipher and make sense of the study's data. In particular, we are extremely fortunate to work with and benefit from the expertise of our collaborators and colleagues Emilio Ferrer and Eduardo Estrada, Jim Hammitt and Daniel Herrera-Araujo, John Gore and Baxter Rogers, Catherine Lebel, John Gabrieli, Irwin Kirsch, Don Rock, Kentaro Yamamoto, Larry Weiss, Don Saklofske, Reid Lyon, and Dr. Fred Romberg. I must also note a very productive collaboration with Eli Lilly and Company, and particularly the role of Linda Wietecha. I must acknowledge scholar Cecil Reynolds who initially introduced me to and then taught me so much about education, especially RTI.

Yale has been my academic home and provided comfort as well as challenges, but overall incredible support and stimulation. Yale president Peter Salovey is an extraordinary leader who provides leadership and direction in the best sense of the word: leadership that cares deeply about Yale's students. This commitment and openness touches all Yale students and has led to Yale's embarking on new, previously uncharted pathways based on scientific evidence, most notably in providing a partial foreign-language waiver that has been a blessing for dyslexic students. At Yale I must single out and share that there is not enough praise to do justice to Linda Lorimer, whom you will have read more about here in this book, for her dedication to and for what she has made possible for dyslexic students at our university and how she has both inspired and provided me with exceptional support to carry out new ideas. I also want to extend my deep appreciation to Kate Stith, the Lafayette S. Foster Professor of Law and the former acting dean of Yale Law School, for her wisdom and guidance. She has an unusual depth of feeling that aligns so well with her brilliant mind, much to the benefit of all members of the law school and Yale community. At the same time, I must note the incredible support I have received from Sheree Carter-Galvan, former Yale attorney and now senior vice president and general counsel at the American Museum of Natural History. One can have good ideas, but you also need faculty and staff to make these a positive reality. Here at Yale it has been my pleasure, and indeed an honor, to work with Dean Mark Schenker, who chairs the Committee on Honors and Academic Standing, on ensuring that every step in the process of reviewing requests for a partial language waiver by dyslexic students is carried out in a scrupulous manner, equitably and with care, fair to all concerned. My hope is that by reading about how meaningful and successful this accommodation process has been at Yale, other universities and, yes, high schools as well will give thought to incorporating this much-needed accommodation for dyslexic students into their schools.

It has been my honor and good fortune to interact with Dean Schenker and his committee and learn from each the true meaning of collegiality. In a similar vein, interacting with our Yale admissions heads, initially Jeff Brenzel and more recently Jeremiah Quinlan, has taught me much about the care and dedication that goes into the review process for each applicant. Since schools are not permitted to ask about conditions such as dyslexia, I worked with then dean Brenzel to help review applicants and educate the committee members on how some of the students' attributes appearing on an application might misleadingly be misinterpreted as negative signs when they most likely are symptoms of dyslexia in an otherwise very bright and promising applicant. I plan to and look forward to sharing this information with ongoing committees. This experience has greatly influenced me and encouraged me to incorporate sections on high school and beyond in this edition.

Not only the university but sections of our Yale Medical School have been exceptionally welcoming and eager to learn more about their students, residents, and faculty who are dyslexic. Outstanding here is the Chair of Surgery, Dr. Nita Ahuja, with whom I've had highly stimulating discussions leading to a session with departmental members on the science and clinical impact of dyslexia. I have learned, as I've discussed in this book, so much about how slow readers can be excellent, fast thinkers from my many interactions with the brilliant, highly regarded dyslexic surgeons Dr. Toby Cosgrove and the late Dr. Graeme Hammond.

A major question facing educators and parents concerns school—specifically, where can I send my dyslexic son or daughter to ensure understanding and addressing of his or her educational, social, and emotional needs? My frequent visits to and interactions with educators at a range of schools have been not only incredibly informative but also inspirational. I have learned so much from my direct interactions with students, parents, and educators at Park Century School, where I now sit on the board as a trustee. I am fortunate to be able to interact with, learn from, and exchange ideas with school head Judith Fuller and fellow trustees (and parents) Hilary Garland, Bess Armstrong, Jennie Burkart, Dr. Isla Garraway, Leslie Goodman, Paul Jennings, Trish Torrey, Sally Weil, and others. Visiting and interacting with school head Jay Russell, associate head Jon Rosenshine, and admissions head Maureen Sweeney only make my admiration for the Windward School and its approach to its dyslexic students grow more and more. I have been fortunate to work with, advise, and serve on the board of a very new and special school model, that of a specialized public charter school for dyslexic students, Louisiana Key Academy (LKA). Started, as I have noted, by Dr. Laura Cassidy, this school is making a huge difference, especially for disadvantaged dyslexic boys and girls, who are thriving in this environment. Dr. Cassidy, board chair at LKA, along with Kayla Reggio, director of the associated Dyslexia Resource Center, are intensely involved not only in the school but in outreach about dyslexia to the public and parochial schools in Baton Rouge and elsewhere. I also want to acknowledge Dacia Toll, co-CEO and president of Achievement First, a nonspecialized charter school network, for her desire to learn more about dyslexia and support a special project I headed designed to help teach dyslexic children to read. I have also learned much and benefited from visiting and interacting with schools in the Los Angeles Unified School District (LAUSD). In particular, I want to express my admiration for the efforts of Mary Ann Sullivan and Vy Nguyen at the LAUSD Intensive Diagnostic Educational Center, or IDEC as it is commonly referred to, in working hard to initiate forward-looking intervention programs serving a group of their dyslexic students.

Both visiting and speaking at Microsoft headquarters and having unusually productive exchanges with Mike Tholfsen and, previously, Tim Hopper not only provided me with excellent information on advances in technology helpful to dyslexics but, as importantly, gave me insight into Microsoft's serious concern for and dedication to advancing the interests of all those who are dyslexic.

Nate Hundt is a person whom I greatly admire and who is very special to me. I have known Nate (and his wonderful parents, Elizabeth Katz and Reed Hundt) since he was diagnosed as dyslexic as a young boy, and I got to know and admire him even more as he advanced through Yale College and then Yale School of Management. Nate is not only dyslexic but also brilliant and caring. He is now principal products manager at Workday, which provides administrative software for large organizations such as Yale. Nate invited Bennett and me to speak to senior administrators at the company, which was a powerful and enlightening experience. We chose to speak on "Dyslexia, Diversity, and Team Building—the Case for Cognitive Diversity," which was very well received and taught us much as well. Following our formal talk, we were treated to a more informal and highly stimulating conversation with Nate, Sarah Ospina, and Greg Pryor. Sarah in particular spoke passionately and persuasively about the role of empathy in dyslexia and in team building. Our response to hearing from Sarah, Greg, and Nate has been to add empathy to the strengths represented in our Sea of Strengths model of dyslexia.

Bennett and I were privileged to be invited to serve as advisors to the XPRIZE, which provided a rare inside look at how this acclaimed high-level challenge process works. Even more important, this experience introduced us to an unusually talented and dedicated leader, Shlomy Kattan, executive director at the XPRIZE Foundation. In an event organized and led by Shlomy, we had an incredibly positive experience interacting with and learning about the goal and progress of software developed for teaching adults to read using handheld devices, as noted in our adult chapter.

I can write so definitively about brilliant, caring, and highly successful and resilient dyslexics because I have met and gotten to know each at a very personal level. These inspirational men and women, including David Boies, Dr. Toby Cosgrove, Ari Emanuel, and Diane Swonk, have become close friends whom I cherish. I have also been extremely fortunate to meet and get to know the highly accomplished dyslexic attorney Rebecca Aragon, the acclaimed dyslexic scientist Frank Church, and the Academy Award–winning producer Brian Grazer. Each has taught me something new about dyslexia and resilience.

I must note how proud I am to have Ari Emanuel as a friend. Ari is creative and brilliant and has an incredibly large heart, which has led him to work to establish a public charter school in Los Angeles for disadvantaged boys and girls who are dyslexic. I am honored to be part of this incredible endeavor, where Bennett and I are working with his very able assistant, Romola Ratnam.

So many others have made a difference in their support, including Valerie Jarrett and her mom, the distinguished educator Barbara Bowman, a founder and now president of the Erikson Institute, a major leader in early childhood education. I very much enjoyed getting to know Barbara and exchanging views when we both served on the Committee on the Prevention of Reading Difficulties in Young Children. I have greatly benefited from Valerie's generosity and commitment to helping children from the time Bennett and I discussed dyslexia with her in her office in the West Wing of the Obama White House, and later when she graciously agreed to be the introductory speaker for our 2015 Yale Conference "Dyslexia: Slow Readers, Fast Thinkers—It Takes a Dyslexic Brain." I want to acknowledge the passionate and meaningful support of Sam Fox, Deborah Stark, and Pleasant Rowland. They care deeply about dyslexia and have made a real difference in our work.

I also want to acknowledge other supporters who care and are knowledgeable about dyslexia: Eileen Marzola, Donald Takacs, Sheryl Knapp, and parent Don Vail. A warm and grateful shout-out and thank-you to my longtime dyslexia experts Helaine Schupack and Gail Hirsch. A special thank-you to Barbara Wilson, founder and developer of the Wilson Reading System, which has made and is making such a difference to students who are dyslexic.

My ongoing and frequent interactions with the Lucas family—dad Dr. Tyler Lucas, who is dyslexic; brilliant daughter Skye, who is both dyslexic and a Penn student; son Hayden, who is also very bright and dyslexic; and mom, Geralyn, who is the epitome of an effective, loving, persevering champion for her dyslexic children—have been inspirational. I am so happy and fortunate to know you and to learn from you over the years what it means to grow up very bright and dyslexic and what it means to be a dyslexic child's family.

My desire to inform parents, educators, dyslexics themselves, and society as a whole about the true nature of dyslexia has been strongly advanced by *New York Times* reporter Katie Hafner, who is a superb communicator, conveying facts and feelings accurately and at the highest level. I have also been fortunate to work with CBS television producer Amiel Weisfogel on an up-to-date, unusual cover story on dyslexia on *CBS Sunday Morning* with Jane Pauley. Bennett and I were honored to be selected as the recipients of the Liberty Science Center 2019 Genius Award "in recognition of your inspiring accomplishments and your pioneering work in advancing our understanding of dyslexia." To receive this award was wonderful and filled us with gratitude.

Equally inspiring and rewarding was getting to know the leaders of the Liberty Science Center, beginning with its head, Paul Hoffman, and associates Elise Nolan and Christine Arnold-Schroeder, who care deeply about dyslexia and are committed to educating society about the latest scientific knowledge.

That there is a book at all and one that has connected to so many people is the singular responsibility of my patient and highly skilled editor, Jonathan Segal. How do I thank you, Jonathan, first for being the one to recognize the potential of the first edition of this book and for gently but often firmly leading me through its creation and completion? Jonathan Segal is a master editor who knows every facet of editing, most of which were and some of which may remain unknown to me. In all, the process of writing, editing, and rewriting each edition has been a long one, and I have been incredibly fortunate to have this brilliant, caring editor at my side. I owe so much for his insights, his skill, and his ability to take me along on his editing journey with this edition. Knowing him more personally now, I can share that Jonathan's skill as an editor is equaled by his compassion as a human being. Not surprisingly, I have also benefited from the attention and support of Jonathan's very able assistants, Sam Aber and more recently Erin Sellers, and superb production editor Victoria Pearson.

The hub and indeed the heart of the work Bennett and I carry out is the Yale Center for Dyslexia & Creativity. Notable accomplishments by us and in the field are reported on our website, yale.dyslexia.edu. We have been incredibly fortunate to have the dedicated support of Kathy Crockett, Lynn Waymer, Amber Bobin, Wendy Pinto, and Barbara Pearce. More recently, Michael Mershon and his Hatcher group have ably taken on the responsibility of managing our communications, including our website.

Charles Schwab has been highly supportive in meaningful ways, such as endowing the Charles and Helen Schwab Professor in Dyslexia and Learning Development, which has meant so much to Bennett and me. Schwab stands out as an inspiring example of what a dyslexic can accomplish. He also cares deeply about ensuring that dyslexia is better understood so that dyslexic students are acknowledged and supported.

I am indebted to Karen Pritzker for so many meaningful contributions that it is difficult to know where to begin. First, she has endowed the Audrey G. Ratner Chair in Learning Development and Dyslexia, which I proudly hold. This has provided me with an opportunity to get to know Audrey (who is Karen's mother) on a personal level and learn from her in so many ways. She was drawn to reading and how best to teach reading early on, and it is such a pleasure to discuss this together. Audrey is also, not surprisingly, very wise, and I have gained so much from our conversations and exchanges.

It is really difficult for me to do justice to what Karen Pritzker's contributions have meant to me, scientifically, academically, and personally. Karen is an extraordinary person in every sense of the word: She is brilliant, brimming with original ideas and solutions; modest; a deeply caring mother; and she has a huge heart. I have been fortunate to be able to plan with her, bounce ideas off her, and benefit from her insights. Following a conversation with her, I typically say to myself, "Now, why didn't I think of that?" She is a leader in thinking but never overpowering or boastful. No, she gently tries to guide you to see the light. I feel privileged to know not only Karen and her late, wonderful husband, Michael Vlock, but also her sweet, smart, and loving children as well. Having known them and cared about them since they were younger, I view them as part of my loving extended family.

Earlier, when speaking about my spouse, Bennett, I mentioned that I have been blessed. Here I want to share that I indeed am blessed to have the family that I have, my husband and our three sons and their families. Having such a caring family has made all the difference to me and given me all the support, strength, and reassurance needed to carry out stressful and yet rewarding activities such as writing a book like *Overcoming Dyslexia*. They are responsible for my remaining true to my three goals in writing: to share cutting-edge scientific knowledge, to translate scientific knowledge into action, and to ensure that compassion and empathy are integral to the information I share and advice I give in this book.

Family has always been my primary focus. I had three children, including a set of twins, in under three years and chose to stay at home with my kids when they were young. I loved being with them and am so grateful to see them now as caring, successful adults and fathers. Together we are a tightly connected, incredibly loving and caring family who put one another first. I note all this because this is what is responsible for any success I have had. I have a loving safety net that is always there to support and protect me. Whatever I need, they are there for me; I am fortunate because they are so bright and caring. I don't have to ask—they are there for me, especially as physicians themselves. They are my comfort, my home editors with whom I can discuss topics and who help me decide what direction a particular discussion might take. They are my comforters, my reminders, my encouragers and are always there for me. David is consistently there, watching out for his parents and brothers and, along with his wife, Diana, makes sure his three remarkable daughters always know that they are the center of their parents' lives. Yet David, an extraordinary, knowledgeable, and caring physician-scientist, still makes time both to write incredibly thoughtful pieces synthesizing and clarifying complex issues at the cutting edge of health and technology and to create intriguing and highly stimulating, not-to-be-missed *Tech Tonics* podcasts with Lisa Suennen, all the while maintaining a very active domestic and international schedule. I am particularly honored and proud that my son Jonathan, a remarkable, incredibly knowledgeable, and deeply caring psychiatrist, has joined me to coauthor this book and share his deep expertise in areas such as anxiety and attention-deficit disorder, which are so common and so disruptive to dyslexic children and adults. Jonathan is the psychiatrist everyone wishes they had and could speak with. Jonathan is not only a superb psychiatrist, he is a wonderful dad who is always there for his sweet, smart young daughter. Adam, Jonathan's twin, is a brilliant, dedicated physician-scientist, relentlessly exploring every possible pathway to finding cures for rare diseases affecting children. Given his unusual ability to address and truly consider each issue from multiple perspectives, both scientific and humanistic, Adam is a much-admired leader in his field. As busy as he and his wife, Marina, are, they always prioritize making time for their adorable daughter and wonderful, curious son. Adam is an incredibly caring individual, always looking after me. He is the consulting physician for everyone in our family whom we all rely on for the most thoughtful, empathetic, and truly helpful advice. What is so amazing to me and brings me so much joy is that as incredibly busy as our children are, there is always a call or an email asking how I'm doing, asking how they might help, and suggesting that I am working too hard and should take time for myself. Being surrounded by a loving and incredibly caring family is my magic charm that cheers and energizes me so that I can go forward and realize my dream.

Speaking of family, I must acknowledge and pay tribute here to the unfailing devotion of my parents, Dora and Meyer Epstein, who first gave me love, strength, and compassion, and my sister, Irene, who continues to do so. She and her wonderful late husband, Marvin, nurtured me when I lost my mother and have always been there for me. Fortunately, the feeling and love between us is strong and mutual.

NOTES

CHAPTER 2: THE HISTORICAL ROOTS OF DYSLEXIA

14 "He has always been": W. Pringle Morgan, "A Case of Congenital Word Blindness," *British Medical Journal* (1896): 1378.

15 What is perhaps the earliest: A good discussion of the earliest reports of acquired reading problems can be found in Arthur L. Benton and Robert J. Joynt, "Early Descriptions of Aphasia," *Archives of Neurology* 3 (August 1960): 109/205–126/222.

16 "I can see [the words] but": Sir William Broadbent, "Cerebral Mechanisms of Speech and Thought," *Transactions of the Royal Chirigal Society* 55 (1872): 145–94.

16 "a complete text-blindness": Dr. Adolf Kussmaul, *Die Storungen Der Spache* (Leipzig: Verlag Von F.C.W. Vogel, 1877).

16 "there may be very great": in R. Berlin, "*Eine besondre Art der Wortblindheit (Dyslexia)*" (Weisbaden: Verlag von J. F. Bergmann, 1887), as quoted by James Hinshelwood, M.D., in "A Case of Dyslexia: A Peculiar Form of Word-Blindness," *The Lancet* 2 (November 21, 1896): 1451–54. Here, too, can be found Berlin's description of his patient, Herr B.'s, difficulty in reading: "reading of printed and written."

17 "he could not read": J. Hinshelwood, M.D., "Word-Blindness and Visual Memory," *The Lancet* (December 21, 1895): 1564–70.

19 "he could not be taught to read": J. Hinshelwood, M.D., "Congenital Word-Blindness," *The Lancet* (1900): 1506–8. All quotations describing both Case 1 and 2 are from this early article by Hinshelwood.

20   "after four years of laborious effort": J. Hinshelwood, "Congenital Word-Blindness, with Reports of Two Cases," *Ophthalmic Review* 21 (1902): 91–99. The following quotations are also from this 1902 paper.

21   Both Hinshelwood's and Morgan's reports: Binet was first commissioned by the French government to develop a test in 1903; he did not publish his methods until 1905: A. M. Binet and T. Simon, "Methodes nouvelles pour le diagnostic du niveau intellectuel des anormaux," *L'Année Psychologique* 11: 191–224.

21   "were frequently overlooked": Quoted in discussion following J. H. Fisher, "Congenital Word-Blindness (Inability to Learn to Read)," *Transactions of the Ophthalmological Society of the United Kingdom* 30 (1910): 216–25. In addition to his clinical position at the ophthalmic hospital, Dr. Treacher Collins had been a professor at the Royal College of Surgeons.

21   "The father . . . who is": In J. Lange, *Agnosia and Apraxia,* as quoted in J. W. Brown (ed.), *Agnosia and Apraxia: Selected Papers of Liepman, Lange, and Potzl* (Hillsdale, NJ, Erlbaum, 1988), p. 85.

22   "It is a matter of": Hinshelwood, "Congenital Word-Blindness, with Reports of Two Cases."

22   "In these days": J. Hinshelwood, "A Case of Congenital Word-Blindness," *The Ophthalmoscope* 2 (1904): 399–405.

23   "On my advice": Hinshelwood, "Congenital Word-Blindness, with Reports of Two Cases."

23   "I advised them to": Hinshelwood, "A Case of Congenital Word-Blindness." *The Ophthalmoscope,* 1904, Vol. II, pp. 399–405.

23   "The detection of congenital": E. Nettleship, "Cases of Congenital Word-Blindness (Inability to Learn to Read)," *Ophthamic Review* 20 (1901): 61–67.

24   By 1909, E. Bosworth McCready: McCready reviews the number of reported cases of congenital word-blindness in "Congenital Word-Blindness as a Cause of Backwardness in School Children. Report of a Case Associated with Stuttering," *Pennsylvania Medical Journal* 13 (1910): 278–84. The cases from the United States are described by E. Jackson in "Developmental Alexia (Congenital Word-Blindness)," *American Journal of the Medical Sciences* 131 (1906): 843–49.

24   "While the majority of cases": McCready, "Congenital Word-Blindness as a Cause of Backwardness."

24   "could not read but little": E. Bosworth McCready, "Defects in the Zone of Language (Word-Deafness and Word-Blindness) and Their Influence in Education and Behavior," *American Journal of Psychiatry* 6 (old series 83) (1926–1927).

CHAPTER 3: THE BIG PICTURE: WHO IS AFFECTED
AND WHAT HAPPENS OVER TIME

27   "Classification is the art of carving": R. E. Kendell, *The Role of Diagnosis in Psychiatry* (Oxford: Blackwell Scientific, 1975), as quoted by R. K. Blashfield, *The Classification of Psychopathology: Neo Kraepelinian and Quantitative Approaches* (New York: Plenum, 1984), p. 65.

28   According to the most recent: U.S. Department of Education, "The condition of education—Participation in education—Elementary/secondary—Children and youth with disabilities—last updated May (2019)." Available at https://nces.ed.gov/programs/coe/indicator_cgg.asp, accessed 9/12/19.

29 Since dyslexia is estimated to comprise: J. Lerner, "Educational Interventions in Learning Disabilities," *Journal of the American Academy of Child and Adolescent Psychiatry* 28 (1989): 326–31. https://nces.ed.gov/programs/coe/indicator_cgg.asp, accessed 1/13/2018.

29 Data from our Connecticut Longitudinal Study: E. Ferrer, B. Shaywitz, et al., "Uncoupling of Reading and IQ Over Time: Empirical Evidence for a Definition of Dyslexia," *Psychological Science* (2010): 93–101; S. Shaywitz, J. Fletcher, and B. Shaywitz, "Issues in the Definition and Classification of Attention Disorder," *Topics in Language Disorders* 14 (1994): 1–25.

30 The apparent large-scale under-identification of dyslexic children: E. Ferrer, B. A. Shaywitz, et al., "Achievement Gap in Reading Is Present as Early as First Grade and Persists through Adolescence," *Journal of Pediatrics* 167 (2015): 1121–25.

31 In 1996 I published an article: S. Shaywitz, "Dyslexia," *Scientific American* 275 (1996): 98–104.

33 In contrast, we found no significant difference: S. E. Shaywtiz et al., "Prevalence of Reading Disability in Boys and Girls: Results of the Connecticut Longitudinal Study," *Journal of the American Medical Association* 264 (1990): 998–1002.

34 *dyslexic readers do not catch up*: Ferrer et al., "Achievement Gap in Reading Is Present as Early as First Grade."

CHAPTER 4: WHY SOME SMART PEOPLE CAN'T READ

36 Gregory, a medical student: For Gregory's story, see S. Shaywitz, "Dyslexia," *Scientific American* 275 (1996): 98–104.

39 The problem is a linguistic one: Evidence that the difficulty in dyslexia is within the language system is based on findings described by J. Torgesen, R. Wagner, and C. Rashotte, "Longitudinal Studies of Phonological Processing and Reading," *Journal of Educational Psychology* 27 (1994): 276–86; and J. Fletcher et al., "Cognitive Profiles of Reading Disability: Comparisons of Discrepancy and Low Achievement Definitions," *Journal of Educational Psychology* 84 (1994): 6–23.

40 Think of the language system: Explanations of the modular system can be found in D. Shankweiler and S. Crane, "Language Mechanisms and Reading Disorder: A Modular Approach," *Cognition* 24 (1986): 139–68; J. A. Fodor, *The Modularity of Mind* (Cambridge, MA: MIT Press), 1983.

CHAPTER 5: EVERYONE SPEAKS, BUT NOT EVERYONE READS

45 As linguists Noah Chomsky and Steven Pinker: Steven Pinker, *The Language Instinct* (New York: Morrow, 1994).

45 John DeFrancis, professor: John DeFrancis, *Visible Speech* (Honolulu: University of Hawaii Press, 1989).

46 In 1989, linguist William Abler: W. Abler, "On the Particulate Principle of Self-Defining Systems," *Journal of Social and Biological Structures* 12 (1989): 1–13.

47 A speaker can generate phonemes: A. Liberman, "How Theories in Speech Affect Research in Reading and Writing," in B. Blachman (ed.), *Foundations of Reading Acquisition and Dyslexia: Implications for Dyslexia* (Mahwah, NJ: Erlbaum, 1997).

50 "Writing is not language": Leonard Bloomfield, *Language* (New York: Holt, Rinehart and Winston, 1933), p. 21.

50 Suddenly the child appreciates: *Phonemic awareness* refers to the ability to hear, notice, identify, and manipulate the individual sounds—phonemes—in spoken words. See J. Torgesen and P. Mathes, *A Basic Guide to Understanding, Assessing and Teaching Phonological Awareness* (Austin: Pro-Ed, 2000), p. 2.

54 By age six and a half to seven years: I. Liberman et al., "Explicit Syllable and Phoneme Segmentation in the Young Child," *Journal of Experimental Child Psychology* 18 (1974): 201–12.

54 British researchers Lynette Bradley and Peter Bryant: L. Bradley and P. Bryant, "Categorizing Sounds and Learning to Read—a Causal Connection," *Nature* 301 (1983): 419–21.

54 In the 1990s we and other research groups: J. M. Fletcher et al., "Cognitive Profiles of Reading Disability: Comparisons of Discrepancy and Low Achievement Definitions," *Journal of Educational Psychology* 86 (1994): 6–23. See also K. E. Stanovich and L. S. Siegel, "Phenotypic Performance Profile of Children with Reading Disabilities: a Regression-Based Test of the Phonological-Core Variable-Difference Model," *Journal of Educational Psychology* 86 (1994): 24–53.

54 One type of test in particular: R. Wagner, J. Torgesen, et al., *CTOPP-2: Comprehensive Test of Phonological Processing* (Austin: Pro-Ed, 2013).

55 Even in high school students: S. Shaywitz et al., "Persistence of Dyslexia: The Connecticut Longitudinal Study at Adolescence," *Pediatrics* 104 (1999): 1351–59.

58 The two-step mechanism: W. J. Levelt, A. Roelofs, and A. S. Meyer, "A Theory of Lexical Access in Speech Production," *Behavioral and Brain Sciences* 22 (1999): 1–75. S. Hanly, "Tip-of-the-Tongue and Word Retrieval Deficits in Dyslexia," *Journal of Learning Disabilities* 43 (2010): 15.

CHAPTER 6: READING THE BRAIN

61 The earliest conceptions of the mind: Early conceptualizations of the mind are discussed in Stanley Finger, *Origins of Neuroscience* (New York: Oxford University Press, 1994), p. 8.

62 Galen viewed the brain: Ibid., p. 18.

62 It was just such a circumstance: Localization of expressive language, including the derivation of the term *aphasia,* occurred around 1861–65. See Finger, ibid., p. 379.

64 The apothecary crisply clipped Harvey's frenulum: Ibid., quoted in A. R. Lecours, J-L. Nespoulous, and D. Pioger, "Jacques Lordat or the Birth of Cognitive Neuropsychology," in E. Keller and M. Gopnik (eds.), *Motor and Sensory Processes of Language* (Hillsdale, NJ: Erlbaum, 1987), p. 10.

66 In 1973, with the advent: A. B. Wolbarst, *Looking Within* (Berkeley: University of California Press, 1999), has user-friendly discussions of the various imaging modalities.

67 "These facts seem to indicate": C. S. Roy and C. S. Sherrington, "On the Regulation of the Blood Supply of the Brain," *Journal of Physiology* 1 (1890): 85.

67 "It is, therefore, clear": L. Sokoloff, "Relationships Among Local Functional

Activity, Energy Metabolism, and Blood Flow from the Central Nervous System," *Federation Proceedings* 40 (1981): 2311–16, quote on page 2315.

CHAPTER 7: THE WORKING BRAIN READS

70 Immediately after imaging: It takes more time to complete the processing necessary for the functional MRI. Remember, for dyslexics, the problem is in the brain wiring—a glitch exposed by a functional MRI. The basic anatomic structures are intact.

73 To begin we recruited: B. A. Shaywitz et al., "Sex differences in the Functional Organization of the Brain for Language," *Nature* 373 (1995): 607–9. The finding of a sex difference in the organization of the brain for language has been replicated by other investigators. See J. J. Jaeger et al., "Sex Differences in Brain Regions Activated by Grammatical and Reading Tasks," *Neuroreport* 9 (1998): 2803–7; and K. Kansaku et al., "Sex Differences in Lateralization Revealed in the Posterior Language Areas," *Cerebral Cortex* 10 (2000): 866–72.

73 This represented the first demonstration: Shaywitz et al., "Sex Differences in the Functional Organization of the Brain."

74 First, the neural systems for reading: B. A. Shaywitz et al., "Disruption of Posterior Brain Systems for Reading in Children with Developmental Dyslexia," *Biological Psychiatry* 52 (2002): 101–10.

75 As early as 1891: J. Dejerine, "Sur un cas de cécité verbale avec agraphie, suivi d'autopsie," *C.R. Société du Biologie* 43 (1891): 197–201; see also his "Contribution à l'étude anatomopathologique et clinique des differentes variétés de cécité verbale," *Mémoires de la Société de Biologie* 4 (1892): 61–90. These brain regions were first proposed on the basis of postmortem studies of adults with acquired inability to read (acquired dyslexia).

76 It responds very rapidly: C. Price, C. Moore, and R. S. J. Frackowiak, "The Effect of Varying Stimulus Rate and Duration of Brain Activity During Reading," *Neuroimage* 3 (1996): 40–52.

76 This conceptualization has now been strongly challenged: C. J. Price and J. T. Devlin, "The Interactive Account of Ventral Occipito Temporal Contributions to Reading," *Trends in Cognitive Science* 15, no. 6 (June 2011): 246–53.

79 These data from fMRI studies: C. Price and A. Mechelli, "Reading and Reading Disturbance," *Current Opinion in Neurobiology* 15 (2005): 231–8; F. Richlan, M. Kronbichler, and H. Wimmer, "Functional Abnormalities in the Dyslexic Brain: A Quantitative Meta-analysis of Neuroimaging Studies," *Human Brain Mapping* 30, no. 10 (2009): 3299–308. F. Richlan, M. Kronbichler, and H. Wimmer, "Meta-Analyzing Brain Dysfunctions in Dyslexic Children and Adults, *Neuroimage* 56 (2011): 1735–42; R. Peterson and B. Pennington, "Developmental Dyslexia," *Lancet* (2012); S. Shaywitz and B. Shaywitz, "Dyslexia (Specific Reading Disability)," *Biological Psychiatry* 57 (2005): 1301–9.

79 Similar findings have been reported: M. Kronbichler, F. Hutzler, et al., "Evidence for a Dysfunction of Left Posterior Reading Areas in German Dyslexic Readers," *Neuropsychologia* 44, no. 10 (2006): 1822–32; S. Brambati, C. Termine, et al., "Neuropsychological Deficits and Neural Dysfunction in Familial Dyslexia," *Brain Research* 1113, no. 1 (2006): 174–85.

79 Some studies in Chinese: W. Siok, Z. Niu, et al., "A Structural-Functional Basis for Dyslexia in the Cortex of Chinese Readers," *Proceedings of the National Academy of Science-USA* 105 (2008): 5561–66; W. Siok, C. Perfetti, et al., "Biological Abnormality of Impaired Reading Is Constrained by Culture," *Nature* 431 (2004): 71–76; W. Siok, J. Spinks, et al., "Developmental Dyslexia Is Characterized by the Coexistence of Visuospatial and Phonological Disorders in Chinese Children," *Current Biology* 19 (2009): R890–92. Other studies of Chinese dyslexics find other differences in brain activation. For example, one study demonstrated reduced activation for Chinese dyslexics in the right occipital cortex, consonant with a language-specific role of the right visual cortex in Chinese reading and suggesting a deficit in holistic visuo-orthographic analysis in Chinese dyslexi. See L. Liu, W. Wang, et al., "Similar Alterations in Brain Function for Phonological and Semantic Processing to Visual Characters in Chinese Dyslexia," *Neuropsychologia* 50 (2012): 2224–32.

81 When does the neural signature: N. M. Raschle, J. Zuk, and N. Gaab, "Functional Characteristics of Developmental Dyslexia in Left-Hemispheric Posterior Brain Regions Predate Reading Onset," *Proceedings of the National Academy of Sciences USA* 109, no. 6 (2012): 2156–61; Z. M. Saygin, E. S. Norton, et al., "Tracking the Roots of Reading Ability: White Matter Volume and Integrity Correlate with Phonological Awareness in Prereading and Early-Reading Kindergarten Children," *Journal of Neuroscience* 33 (2013): 13251–58.

82 In fact, this synchrony between anatomically distinct regions: M. P. van den Heuvel and H. E. Hulshoff Pol, "Exploring the Brain Network: A Review of Resting-State fMRI Functional Connectivity," *European Neuropsychopharmacology* 20, no. 8 (2010): 519–34.

82 Although there have been some functional connectivity: B. Horwitz, J. Rumsey, and B. Donohue, "Functional Connectivity of the Angular Gyrus in Normal Reading and Dyslexia," *Proceedings of the National Academy of Sciences USA* 95 (1998): 8939–44; M. Koyama et al., "Cortical Signatures of Dyslexia and Remediation: An Intrinsic Functional Connectivity Approach," *PLOS One* 8:e55454 (2013); L. Stanberry et al., "Low-Frequency Signal Changes Reflect Differences in Functional Connectivity Between Good Readers and Dyslexics During Continuous Phoneme Mapping," *Magnetic Resonance Imaging* 24 (2006): 217–29; S. van der Mark et al., "The Left Occipitotemporal System in Reading: Disruption of Focal fMRI Connectivity to Left Inferior Frontal and Inferior Parietal Language Areas in Children with Dyslexia," *Neuroimage* 54 (2011): 2426–36; A. Vogel, F. Miezin, et al., "The Putative Visual Word Form Area Is Functionally Connected to the Dorsal Attention Network," *Cerebral Cortex* 22 (2012): 537–49.

83 Rather than examining connectivity: E. Finn et al., "Disruption of Functional Networks in Dyslexia: A Whole-Brain, Data-Driven Approach to fMRI Connectivity Analysis," *Biologic Psychiatry* 76 (2013): 397–404.

84 New findings: M. Reynolds and D. Besner, "Reading Aloud Is Not Automatic: Processing Capacity Is Required to Generate a Phonological Code from Print," *Journal of Experimental Psychology: Human Perception and Performance* 32 (2006): 303–23; S. Shaywitz and B. A. Shaywitz, "Paying Attention to Reading:

The Neurobiology of Reading and Dyslexia," *Developmental Psychopathology* 20 (2008): 1329–49.

84 This region has long been known to: I. Kovelman, E. S. Norton, et al., "Brain Basis of Phonological Awareness for Spoken Language in Children and Its Disruption in Dyslexia," *Cerebral Cortex* 22, no. 4 (2012): 754–64.

CHAPTER 8: DIAGNOSING DYSLEXIA—WHY, WHAT, AND HOW

93 "Learning disabilities, as described": The Individuals with Disabilities Education Act (IDEA), enacted in 1975 as the Education for All Handicapped Children Act (EAHCA) and then updated as IDEA in 1990 and updated again in 2004, is designed to ensure that grade-school and secondary-school students with a disability are provided a free appropriate public education (FAPE) in the least restrictive environment (LRE) tailored to their individual needs.

94 In September 2014, when I testified: The full testimony can be found at http:// dyslexia.yale.edu/sept18Hearing.php.

102 These data provide the long-sought-after: E. Ferrer, B. Shaywitz, et al., "Uncoupling of Reading and IQ over Time: Empirical Evidence for a Definition of Dyslexia," *Psychological Science* 21 (2010): 93–101.

107 So it should come as no surprise: S. E. Shaywitz et al., "Neural Systems for Compensation and Persistence: Young Adult Outcome of Childhood Reading Disability," *Biological Psychiatry* 54 (2003): 25–33.

107 I can think: Interview with Steven Spielberg by Quinn Bradley, posted online September 29, 2012, https://www.youtube.com/watch?v=4N6RKHOHMJQ, accessed 9/12/19.

CHAPTER 9: RECOGNIZING DYSLEXIA'S
EARLY IMPACT: SPOKEN LANGUAGE

113 In England, researchers asked: P. E. Bryant et al., "Nursery Rhymes, Phonological Skills and Reading," *Journal of Child Language* 16 (1989): 407–28.

114 The British researchers: L. Bradley and P. E. Bryant, "Difficulties in Auditory Organization as a Possible Cause of Reading Backwardness," *Nature* 271 (1978): 746–47.

118 Much to the chagrin: E. L. Meaburn et al., "Quantitative Trait Locus Association Scan of Early Reading Disability and Ability Using Pooled DNA and 100K SNP Microarrays in a Sample of 5760 Children," *Molecular Psychiatry* 13 (2008): 729–40.

118 "If you told a modern geneticist": E. Yong, "What If (Almost) Every Gene Affects (Almost) Everything?," *The Atlantic*, June 16, 2017, https://www.theatlantic.com/science/archive/2017/06/its-like-all-connected-man/530532/, accessed 9/13/19.

118 "the popular misconception": See https://researchtheheadlines.org/2018/09/10/talking-headlines-simon-fisher, accessed 9/10/19.

118 "that common diseases involve": D. Duncan, "Scientist at Work: Eric Shadt, Enlisting Computers to Unravel the True Complexity of Disease," *New York Times*, August 24, 2009.

118 As conceptualized by Dorothy Bishop: Dorothy V. Bishop, *Bishopblog,* June 16, 2013, http://deevybee.blogspot.com/2013/06/overhyped-genetic-findings -case-of.html, accessed 9/10/19.

119 "it would be ineffective": D. V. M. Bishop, "The Interface Between Genetics and Psychology: Lessons from Developmental Dyslexia," *Proceedings of the Royal Society* 282 (2014): 31–39.

CHAPTER 10: RECOGNIZING DYSLEXIA'S IMPACT ON LEARNING TO READ

121 When a child first begins: U. Frith, "Beneath the Surface of Developmental Dyslexia," in K. Patterson, J. Marshall, and M. Coltheart (eds.), *Surface Dyslexia* (London: Erlbaum, 1985), 301–30.

121 We now know that: P. Maisonheimer, P. Drum, and L. Ehri, "Does Environmental Print Identification Lead Children into Word Reading?," *Journal of Reading Behavior* 16 (1984): 363–67.

121 Children can memorize: W. Nagy and P. Herman, "Breadth and Depth of Vocabulary Knowledge: Implications for Acquisition and Instruction," in M. McKeown and M. Curtis (eds.), *The Nature of Vocabulary Acquisition* (Hillsdale, NJ: Erlbaum, 1987), 19–35.

122 If a child knows: L. C. Ehri and L. S. Wilce, "Movement into Reading: Is the First Stage of Printed Word Learning Visual or Phonetic?," *Reading Research Quarterly* 20 (1985): 163–79.

122 Such children might read: For research on the tendency to confuse words that have some similar letters but no phonologic similarities, see M. Stuart, "Factors Influencing Word Recognition in Pre-Reading Children," *British Journal of Psychology* 81 (1990): 135–46; and M. Stuart and M. Coltheart, "Does Reading Develop in a Sequence of Stages?," *Cognition* 30 (1988): 139–81.

124 Reading comprehension develops: T. G. Sticht et al., *Auding and Reading: A Developmental Model* (Alexandria, VA: Human Resources Research Organization, 1974); What Works Clearinghouse, "Foundational Skills to Support Reading for Understanding in Kindergarten Through 3rd Grade," July 2016.

125 Reading researchers Anne Cunningham and Keith Stanovich: A. E. Cunningham and K. E. Stanovich, "What Reading Does for the Mind," *American Educator* 22 (1998): 8–15.

125 Books offer almost three times: Ibid.

126 Benchmarks: These markers are adapted from C. E. Snow, M. S. Burns, and P. Griffin (eds.), *Preventing Reading Difficulties in Young Children* (Washington, DC: National Academy Press, 1998), 80–83.

134 So it is easy: E. Ferrer, B. A. Shaywitz, et al., "Achievement Gap in Reading Is Present as Early as First Grade and Persists Through Adolescence," *Journal of Pediatrics* 167 (2015): 1121–25.

CHAPTER 11: SHOULD MY CHILD BE EVALUATED FOR DYSLEXIA?

140 "Once children fall behind": J. K. Torgesen, "Catch Them Before They Fall: Identification and Assessment to Prevent Reading Failure in Young Children," *American Educator* 22 (1998): 32–39, quote on p. 32.

140 Without a specific diagnosis: The persistence of dyslexia has been demonstrated

in the Connecticut Longitudinal Study sample. Among the many published reports of this study are B. A. Shaywitz et al., "A Matthew Effect for IQ but Not for Reading: Results from a Longitudinal Study," *Reading Research Quarterly* 30 (1995): 894–906; S. E. Shaywitz, J. M. Fletcher, et al., "Persistence of Dyslexia: The Connecticut Longitudinal Study at Adolescence," *Pediatrics* 104 (1999): 1351–68; E. Ferrer, B. A. Shaywitz, et al., "Uncoupling of Reading and IQ over Time: Empirical Evidence for a Definition of Dyslexia," *Psychological Science* 21 (2010): 93–101; E. Ferrer, B. A. Shaywitz, et al., "Achievement Gap in Reading Is Present as Early as First Grade and Persists Through Adolescence," *Journal of Pediatrics* 167 (2015): 1121–25.

151 For the first time, federal law: The law defines dyslexia as "unexpected difficulty in reading for an individual who has the intelligence to be a much better reader."

CHAPTER 12: DIAGNOSING DYSLEXIA IN THE SCHOOL-AGE CHILD

158 Today's evaluation for dyslexia: S. Shaywitz, "Dyslexia," *New England Journal of Medicine* 333 (1998): 307–12.

158 This unexpected nature of dyslexia: E. Ferrer, B. A. Shaywitz, et al., "Uncoupling of Reading and IQ over Time: Empirical Evidence for a Definition of Dyslexia," *Psychological Science* 21, no. 1 (2010): 93–101.

160 This is measured: K. Schrank, N. Mather, and R. Woodcock, *Woodcock-Johnson-IV* (Riverside, 2014); *Wechsler Individual Achievement Test,* 3rd ed. (WIAT-III) (Wechsler, 2009); A. S. Kaufman and N. L. Kaufman, *Kaufman Test of Educational Achievement,* 3rd ed. (KTEA-3) (Boston: Pearson, 2014); V. L. Brown, J. L. Wiederholt, D. D. Hammill, et al., *Test of Reading Comprehension,* 4th ed. (TORC-4) (Austin: Pro-Ed, 2009).

161 The Test of Word Reading Efficiency: J. Torgesen, R. Wagner, and C. Rashotte, *TOWRE-2: Test of Word Reading Efficiency* (Austin: Pro-Ed, 2013).

162 Oral reading fluency: J. Wiederholt and B. Bryant, *Gray Oral Reading Test-5 Examiner's Manual* (Austin: Pro-Ed, 2012).

162 Helpful spelling tests: S. C. Larsen, D. Hammill, and L. Moats, *Test of Written Spelling-5th ed.* (TWS-5) (Austin: Pro-Ed, 2013).

162 Because dyslexia is defined: D. Wechsler, *Wechsler Intelligence Scale for Children,* 5th ed. (Boston: Pearson, 2014).

164 The Comprehensive Test of Phonological Processing: R. Wagner, J. Torgesen, et al., *CTOPP-2: Comprehensive Test of Phonological Processing* (Austin: Pro-Ed, 2013).

164 In the most commonly used test: D. M. Dunn, *Peabody Picture Vocabulary Test,* 5th ed. (Bloomington, MN: NCS Pearson, 2019).

CHAPTER 13: IDENTIFYING THE AT-RISK CHILD

170 Researchers like myself have recognized: D. L. Compton et al., "Selecting At-Risk Readers in First Grade for Early Intervention: Eliminating False Positives and Exploring the Promise of a Two-Stage Gated Screening Process," *Journal of Educational Psychology* 102 (2010): 327–40; M. J. Snowling, "Early Identification and Interventions for Dyslexia: A Contemporary View," *Journal of Research in Special Educational Needs* 13 (2013): 7–14.

170 We found that: The Multigrade Inventory for Teachers was developed for the Connecticut Longitudinal Study to capture the valuable observations of the classroom teacher. The MIT's development and use are described in M. E. Agronin, J. M. Holahan, et al., "The Multi-Grade Inventory for Teachers (MIT): Scale Development, Reliability, and Validity of an Instrument to Assess Children with Attentional Deficits and Learning Disabilities," in S. E. Shaywitz and B. A. Shaywitz (eds.), *Attention Deficit Disorder Comes of Age: Toward the Twenty-First Century* (Austin: Pro-Ed, 1992), pp. 89–116; and J. M. Holahan et al., "Developmental Trends in Teacher Perceptions of Student Cognitive and Behavioral Status as Measured by the Multi-Grade Inventory for Teachers: Evidence from a Longitudinal Study," in D. M. Molfese and V. J. Molfese (eds.), *Developmental Variations in Learning: Applications to Social, Executive Function, Language, and Reading Skills* (Mahwah, NJ: Erlbaum, 2002), 23–55.

170 This screener: The clinical validity of a screener is evaluated based on sensitivity, or its strength in finding true positives (individuals who have dyslexia); specificity, or its strength in eliminating true negatives (individuals who do not have dyslexia); and AUC (Area Under the Receiver Operating Characteristics curve). For a screening measure to be useful it must yield a high percentage of true positives, that is, sensitivity rates of .70 and above, and a high percentage of true negatives, that is, specificity rates of .70 and above. For the Shaywitz DyslexiaScreen™, sensitivity = .74; specificity = .81. AUC estimates range from .50 (chance accuracy) to 1.00 (perfect accuracy). Many of the strongest-performing clinical assessment inventories deliver AUC estimates in the 0.7 to 0.8 range. AUC estimates of .81 (kindergarten) and .89 (first grade) for the Shaywitz DyslexiaScreen™ indicate that this screener has good accuracy in separating children at-risk for dyslexia from those not at-risk.

171 New data from the Connecticut Longitudinal Study: E. Ferrer, B. A. Shaywitz, et al., "Achievement Gap in Reading Is Present as Early as First Grade and Persists Through Adolescence," *Journal of Pediatrics* 167 (2015): 1121–25.

172 Most clinicians and researchers: R. Wagner, J. Torgesen, et al., *Comprehensive Test of Phonological Processing,* 2nd ed. (CTOPP-2) (Austin: Pro-Ed, 2013).

173 *I strongly recommend not delaying kindergarten entry*: Compelling data regarding delayed kindergarten entry are presented by F. Morrison, E. Griffith, and D. Alberts, "Nature-Nurture in the Classroom: Entrance Age, School Readiness, and Learning in Children," *Developmental Psychology* 33 (1997): 254–62; and F. Morrison, L. Smith, and M. Dow-Ehrenberger, "Education and Cognitive Development: A Natural Experiment," *Developmental Psychology* 31 (1995): 78–79. These are discussed by V. W. Berninger in *Guides for Intervention: Reading, Writing* (San Antonio: Psychological Corporation, 1998), pp. 30–31. Another, more recent article supports this conclusion: M. Andrew, "The Scarring Effects of Primary-Grade Retention? A Study of Cumulative Advantage in the Educational Career," *Social Forces* 93, no. 2 (2014): 653–85.

CHAPTER 14: DIAGNOSING BRIGHT YOUNG ADULTS

178 As comedian Richard Pryor: Ascribed to Richard Pryor by Bob McTeer, president and CEO, Federal Reserve Bank of Dallas, in remarks before the Cato

Institute conference "Monetary Policy in the New Economy," Washington, DC, October 19, 2000.

178 In 1996, in celebration: S. Shaywitz, "Dyslexia," *Scientific American* 275 (1996): 98–104.

179 Numerous studies conducted: S. Shaywitz et al., "Persistence of Dyslexia: The Connecticut Longitudinal Study at Adolescence," *Pediatrics* 104 (1999): 1351–59; M. Bruck, "Persistence of Dyslexics' Phonological Awareness Deficits," *Developmental Psychology* 28 (1992): 874–86; R. Felton, C. Naylor, and F. Wood, "Neuropsychological Profile of Adult Dyslexics," *Brain and Language* 39 (1990): 485–97; and R. Davidson and J. Strucker, "Patterns of Word Recognition Error Among Adult Basic Education Students," *Scientific Studies of Reading* (2002): 299–316.

181 Studies have confirmed: E. Paulesu et al., "Dyslexia-Cultural Diversity and Biological Unity," *Science* 291 (2001): 2165–67; F. Richlan, M. Kronbichler, and H. Wimmer, "Meta-Analyzing Brain Dysfunctions in Dyslexic Children and Adults," *Neuroimage* 56 (2011): 1735–42.

187 This was legally affirmed in a 2014 settlement: U.S. Department of Justice, Office of Public Affairs, "Law School Admission Council Agrees to Systemic Reforms and $7.73 Million Payment to Settle Justice Department's Nationwide Disability Discrimination Lawsuit," May 20, 2014, https://www.justice.gov /opa/pr/law-school-admission-council-agrees-systemic-reforms-and-773-million -payment-settle-justice, accessed 9/11/19.

CHAPTER 15: RECOGNIZING DYSLEXIA IN ADULT MIDLIFE WOMEN

192 As noted in the report: T. Wizemen and M. Pardue, "Report of Committee on Understanding the Biology of Sex and Gender Differences: Exploring the Biological Contributions to Human Health: Does Sex Matter?," Board of Health Sciences Policy, Institute of Medicine (Washington, DC: National Academies Press, 2001).

192 The initial study: S. E. Shaywitz, B. A. Shaywitz, et al., "Estrogen Alters Brain Activation Patterns in Postmenopausal Women During Working Memory Tasks," *Journal of the American Medical Association* 281 (1999): 1197–1202.

193 In our next study: S. E. Shaywitz, F. Naftolin, et al., "Estrogen Improves Oral Reading and Short-Term Memory in Midlife Postmenopausal Women," *Menopause* 10 (2003): 420–26.

CHAPTER 16: ALL CHILDREN CAN BE TAUGHT TO READ

199 Strong scientific data: U.S. Department of Education, Institute of Education Sciences, "Evaluation of Response to Intervention Practices for Elementary School Reading," November 2015. See also "RTI Practice Falls Short of Promise, Research Finds," *Education Week,* November 11, 2015.

201 There is now evidence: J. K. Torgesen, "The Prevention of Reading Difficulties," *Journal of School Psychology* 40 (2002): 7–26.

203 For the first time: U.S. Department of Health and Human Services, National Reading Panel, *Teaching Children to Read: An Evidence-Based Assessment of the*

Scientific Research Literature and Its Implications for Reading Instruction (Washington, DC: National Institute of Child Health and Human Development, 2000). The report is available on the panel's website: http://www.nationalreadingpanel .org, and at http://www.nichd.gov/. It is available in its entirety or as an abbreviated (thirty-three-page) executive summary.

205 "solid answers": Gina Kolata, "Guesses and Hype Give Way to Data in Study of Education," *New York Times,* Sept. 2, 2013, https://www.nytimes .com/2013/09/03/science/applying-new-rigor-in-studying-education.html, accessed 9/13/19.

205 During the time I served: "The Impact of Two Professional Development Interventions on Early Reading Instruction and Achievement," What Works Clearinghouse, September 2008, p. xv: "Although there were positive impacts on teachers' knowledge of scientifically based reading instruction and on one of the three instructional practices promoted by the study, neither PD intervention resulted in significantly higher student test scores at the end of the one-year treatment. . . . The added effect of the coaching intervention on teacher practices in the implementation year was not statistically significant. . . . There were no statistically significant impacts on measured teacher or student outcomes in the year following the treatment."

CHAPTER 17: HELPING YOUR CHILD BREAK THE READING CODE

215 Teachers and parents preferring: These programs have been evaluated positively by the What Works Clearinghouse as having evidence to support their claims that they will improve phonemic awareness and phonics in young children.

217 Brain imaging technology: Expertise depends on practice, elegantly demonstrated by T. Elbert et al., "Increased Cortical Representation of the Fingers of the Left Hand in String Players," *Science* 270 (1995): 305–7.

CHAPTER 18: HELPING YOUR CHILD BECOME A READER

229 As children reach higher grades: J. M. Holahan, E. Ferrer, et al., "Growth in Reading Comprehension and Verbal Ability from Grades 1 through 9," *Journal of Psychoeducational Assessment* 36, no. 4 (2016): 307–21.

230 A child *learns to read*: Noted initially by educator Jeanne Chall.

235 As the National Reading Panel noted: But note that one of the best specialized schools for dyslexic children, Windward School (see pages 327–333), does not typically offer one-to-one instruction.

245 More extensive word lists: Fry instant words are the most common words used in English. They were created in 1996 by Dr. Edward Fry, who expanded on the Dolch word list, a list compiled by Edward William Dolch in 1948.

248 "curiosity about the content": Professor Tim Shanahan, http://www.shanahanon literacy.com/blog/further-explanation-of-teaching-students-with-challenging -text, accessed 9/14/19.

CHAPTER 19: HELPING YOUR CHILD BECOME A SKILLED READER

252 The most recent National Assessment: "NAEPFact: Listening to Children Read Aloud: Oral Fluency," https://nces.ed.gov/pubs95/web/95762.asp, accessed

06/09/19. Fifty-five percent of fourth-graders were considered to be fluent; 13 percent met the criteria for the highest rating in their oral reading.

254 More recent studies replicate: T. Rasinski, C. Blachowicz, and K. Lems (eds.), *Fluency Instruction*, 2nd ed. (New York: Guilford, 2012).

254 Results indicate that while both methods: D. V. Dennis, K. L. Solic, and R. L. Allington, "Hijacking Fluency and Instructionally Informative Assessments," in Rasinski, Blachowicz, and Lems, *Fluency Instruction*; C. M. Connor, P. A. Alberto, et al., "Improving Reading Outcomes for Students with or at Risk for Reading Disabilities: A Synthesis of the Contributions from the Institute of Education Sciences Research Centers" (Washington, DC: National Center for Special Education Research, 2014), NCSER 2014-3000.

255 "while reading aloud": M. Kuhn, "Helping Students Become Accurate, Expressive Readers: Fluency Instruction for Small Groups," *Reading Teacher* 58, no. 4 (December 2004/January 2005): 342.

258 As the child's fluency improves: *Shanahan on Literacy*, June 26, 2016; O'Connor, Swanson, and Geraghty, "Improvement in Reading Rate Under Independent and Difficult Text Levels: Influences on Word and Comprehension Skills," *Journal of Educational Psychology* 102 (2010): 1–19.

259 The size of a child's vocabulary: First-grade vocabulary predicted students' reading achievement in their junior year of high school; see A. E. Cunningham and K. E. Stanovich, "Early Reading Acquisition and Its Relation to Reading Experience and Ability 10 Years Later," *Developmental Psychology* 33, no. 6 (1997): 934–45.

259 Language measures: This statement reflects substantial research: Z. O. Weizman and C. E. Snow, "Lexical Input as Related to Children's Vocabulary Acquisition: Effects of Sophisticated Exposure and Support for Meaning," *Developmental Psychology* 37, no. 2 (2001): 265–79; F. J. Zimmerman, J. Gilkerson, et al., "Teaching by Listening: The Importance of Adult-Child Conversations to Language Development," *Pediatrics* 124, no. 1 (2009): 342–49; V. A. Marchman and A. Fernald, "Speed of Word Recognition and Vocabulary Knowledge in Infancy Predict Cognitive and Language Outcomes in Later Childhood," *Developmental Science* 11, no. 3 (2008): F9–16; S. Q. Cabell, L. M. Justice, et al., "Teacher–Child Conversations in Preschool Classrooms: Contributions to Children's Vocabulary Development," *Early Childhood Research Quarterly* 30, Part A (2015): 80–92; M. L. Rowe, S. W. Raudenbush, and S. Goldin-Meadow, "The Pace of Vocabulary Growth Helps Predict Later Vocabulary Skill," *Child Development* 83, no. 2 (2012): 508–25; D. K. Dickinson and M. V. Porche, "Relation Between Language Experiences in Preschool Classrooms and Children's Kindergarten and Fourth-Grade Language and Reading Abilities," *Child Development* 82, no. 3 (2011): 870–86; V. N. Durand, I. M. Loe, et al., "Effects of Early Language, Speech, and Cognition on Later Reading: A Meditation Analysis," *Frontiers in Psychology* 4 (2013): Article ID 586; C. Hulme, H. M. Nash, et al., "The Foundations of Literacy Development in Children at Familial Risk of Dyslexia," *Psychology of Science* 26, no. 12 (2015): 1877–86.

259 In a recent report: J. M. Holahan, E. Ferrer, et al., "Growth in Reading Comprehension and Verbal Ability from Grades 1 through 9," *Journal of Psychoeducational Assessment* 36, no. 4 (2016): 307–21.

260 Isabel Beck, professor emerita: I. L. Beck, M. G. McKeown, and L. Kucan, *Bringing Words to Life*, 2nd ed. (New York: Guilford, 2013).

260 Vocabulary is best learned: Experience has shown that it is best for the teacher to lead the discussion of a new word rather than have children guess at what the word means. Having students guess what the word means can lead them to give the wrong definition or remember the wrong definition rather than the correct one.

262 This latter program: The Academic Word List was released in the year 2000 and is available at https://www.victoria.ac.nz/lals/resources/academicwordlist.

270 "the most critical element": S. Graham, C. A. MacArthur, and J. Fitzgerald (eds.), "Designing an Effective Writing Program," in *Best Practices in Writing Instruction,* 2nd ed. (New York: Guilford, 2013), pp. 3–29. These authors have also contributed to the Educator's Practice Guide "Teaching Elementary School Students to Be Effective Writers," published by the What Works Clearinghouse, Jessup, MD, in 2012.

CHAPTER 20: MICHELE'S PROGRAM: A MODEL THAT WORKS

282 The Shaywitz DyslexiaScreen™: The Shaywitz DyslexiaScreen™ is distributed by Pearson Publishing (www.pearsonclinical.com/shaywitz). Please see pages 170–174 for a full description.

283 Good evidence highlights the difference: J. K. Torgesen et al., "Intensive Remedial Instruction for Children with Severe Reading Disabilities: Immediate and Long-Term Outcomes from Two Instructional Approaches," *Journal of Learning Disabilities* 34 (2001): 33–58.

283 In another study computers were used: B. Wise and R. K. Olson, "Computer-Based Phonological Awareness and Reading Instruction," *Annals of Dyslexia* 45 (1995): 99–122. This study demonstrated the importance of teachers even in computer-based reading instruction.

CHAPTER 21: TEACHING THE DYSLEXIC CHILD TO READ

285 The first two elements: The What Works Clearinghouse found that phonological awareness training plus letter knowledge training had positive or potentially positive effects on print knowledge, phonological processing, and early reading and writing. At the same time, the WWC found it difficult to find commercially available programs incorporating the principles of phonological awareness training plus letter knowledge training that met their inclusion standards. In their words, "Phonological Awareness Training plus Letter Knowledge Training is a practice that does not have a single developer responsible for providing information or materials."

286 For children in preschool or those in kindergarten: These programs have been evaluated positively by the What Works Clearinghouse as having evidence to support their claims that they will improve phonemic awareness and phonics in young children.

287 As I noted earlier: The WWC found that phonological awareness training plus

letter knowledge training had positive or potentially positive effects on print knowledge, phonological processing, and early reading and writing.

288　Certified Wilson instructors are required: The What Works Clearinghouse Intervention Report—Wilson Reading System, p. 3, July 2, 2007.

290　"unappreciated paucity of empirical support": C. R. Reynolds and S. E. Shaywitz, "Response to Intervention: Ready or Not? Or, From Wait-to-Fail to Watch-Them-Fail," *School Psychology Quarterly* 24, no. 2 (2009): 130–45.

291　The article highlighted: "RTI Practice Falls Short of Promise: First Graders Who Were Identified for More Help Fell Further Behind," *Education Week,* November 11, 2015. For the report, see Institute of Education Sciences, National Center for Educational Evaluation and Regional Assistance, "Evaluation of Response to Intervention Practices for Elementary School Reading," November 2015.

291　It is important that: The purpose and contribution of screening instruments compared to progress-monitoring instruments should be differentiated. Screening instruments are developed to capture the major aspects of a condition. For example, in dyslexia it would be relating to academic, phonologic, and linguistic performance based on ongoing teacher observations from working with the student daily and validated by evidence—a study that compared the performance of dyslexic and nondyslexic students on this instrument. In contrast, progress-monitoring instruments track a student's performance, typically on one aspect of his or her reading over time. In this case, the student is compared to himself or herself; there is no comparison between dyslexic and typical readers.

292　I am proud to say: The child's teacher takes five to ten minutes to complete the Shaywitz DyslexiaScreen™, sharing her perceptions gained from working closely with the child. The teacher's response on a tablet with immediate feedback of "at-risk" or "not at-risk" accurately and reliably identifies those children at-risk for dyslexia.

293　Recent research emphasizes the importance of fluent reading: Y. S. G. Kim and R. K. Wagner, "Text (Oral) Reading Fluency as a Construct in Reading Development: An Investigation of Its Mediating Role for Children from Grades 1 to 4," *Scientific Studies of Reading* 19 (2015): 224–42.

293　"While reading aloud to students": M. Kuhn, "Helping Students Become Accurate, Expressive Readers: Fluency Instruction for Small Groups," *Reading Teacher* 58, no. 4 (2004/January 2005): 342.

294　After sufficient practice: D. R. Reutzel, "Hey Teacher, When You Say Fluency, What Do You Mean?," in T. Rasinski, C. Blachowicz, and K. Lems (eds.), *Fluency Instruction,* 2nd ed. (New York: Guilford, 2012).

295　Comparisons showed that children who read aloud: K. J. Topping, "Paired Reading: Impact of a Tutoring Method on Reading Accuracy, Comprehension and Fluency," in Rasinski, Blachowicz, and Lems, *Fluency Instruction.*

298　I earlier detailed Isabel Beck's three-tier framework: I. L. Beck, M. G. McKeown, and L. Kucan, *Bringing Words to Life,* 2nd ed. (New York: Guilford, 2013).

299　To summarize: Ibid.

301　"The two primary goals": The program is available from Cambium/Sopris West, and Hochman teaches the course to teachers during a summer session at Windward School in White Plains, New York.

CHAPTER 26: SUCCEEDING IN POST-SECONDARY EDUCATION

368 Instead, share with your child: Yale Center for Dyslexia & Creativity, *Dyslexia: Profiles of Success,* 3rd ed. (New Haven, 2016).

373 If your child appears stressed: Based on review that appeared by D. M. Davis and J. A. Hayes, "What Are the Benefits of Mindfulness: A Practice Review of Psychotherapy-related Research," *Psychotherapy* 48, no. 2 (2011): 198–208.

CHAPTER 27: SELECTING A COLLEGE

379 In a far larger follow-up study: W. Hiss and V. Franks, "Defining Promise: Optional Standardized Testing Policies in American College and University Admissions," National Association for College Admissions Counseling, 2014.

380 This has the unfortunate potential consequence: Renee Dudley, "Despite Warnings, College Board Redesigned SAT in Way That May Hurt Neediest Students," September 21, 2016, http://www.reuters.com/investigates/special -report/college-sat-redesign/, accessed 11/24/19.

CHAPTER 29: HELPING ADULTS BECOME BETTER READERS

404 Lack of literacy is a major national concern: M. Kutner, E. Greenberg, et al., "The Health Literacy of America's Adults: Results from the 2003 National Assessment of Adult Literacy" (NCES 2006-483). U.S. Department of Education, Washington, DC: National Center for Education Statistics.

405 At least a ninth-grade: As defined by the National Center for Education Statistics, low literacy is typically regarded as having below a fifth-grade reading level.

405 In a recent paper Bennett, I: D. Herrera-Araujo, B. A. Shaywitz, et al., "Evaluating Willingness to Pay as a Measure of the Impact of Dyslexia in Adults," *Journal of Benefit and Cost Analysis* 8, no. 1 (2017): 24–48; Renee Dudley, "Despite Warnings, College Board Redesigned SAT in Way That May Hurt Neediest Students," September 21, 2016, https://www.reuters.com/investigates/special -report/college-sat-redesign, accessed 9/11/19.

407 There are two national, online, Zip Code–based directories: Acknowledgment is made to Heidi Silver-Pacuilla, Ph.D. (team leader, Applied Innovation and Improvement) at the U.S. Department of Education, and to Brenda Dann-Messier, formerly of this section at the DOE, for their insights and information.

409 Studies have found: D. Greenberg, J. C. Wise, et al., "Persisters and Nonpersisters: Identifying the Characteristics of Who Stays and Who Leaves from Adult Literacy Interventions," *Reading & Writing* 26, no. 4 (2013): 494–514.

412 Reflecting the imprint of the CCSS: Taken from Susan Pimentel, "College and Career Readiness Standards for Adult Education," U.S. Department of Education, Office of Vocational and Adult Education, 2013.

412 A GED practice test is available: A note of possible concern for how the revised CCSS GED test will affect dyslexic test-takers: The CCSS is based on students' having very good decoding and fluency skills, leading to a high level of comprehension. Many, including myself, believe that the CCSS-based PARCC and Smart Balance tests given to students in early grades are inappropriate and damaging because these students, in, for example, third grade, have not yet mastered

fluency. The worry is that many of the adults taking the GED did not attend school when teaching basic reading skills was a focus and may be penalized by the new GED. The "College and Career Readiness Standards for Adult Education" states that "the Standards are not meant to specify the full spectrum of support and interventions appropriate for English language learners and students with special needs to meet these standards, nor do they mirror the significant diversity of students' learning needs, abilities, and achievement levels." It is unclear what this means and, most important, how it will be interpreted in terms of the needs of and what is actually available to dyslexic test-takers.

415   "The diagnosis changed every way": Camp Spring Creek blog, April 15, 2014, Interview: Esteemed author Amanda Kyle Williams, http://campspringcreek .org/our-blog/2014/04/15/interview-esteemed-author-amanda-kyle-williams, accessed 9/11/19.

CHAPTER 30: ANXIETY, ADHD, AND DYSLEXIA

419   Anxiety and ADHD so commonly occur: Dyslexia and ADHD occur together more frequently than expected by chance, with 25 to 40 percent of children with one disorder meeting criteria for the other; see G. J. August and B. D. Garfinkel, "Comorbidity of ADHD and Reading Disability Among Clinic-Referred Children," *Journal of Abnormal Child Psychology* 18 (1990): 29–45; M. Semrud-Clickeman, J. Biederman, et. al., "Comorbidity Between ADHD and Learning Disability: A Review and Report in a Clinically Referred Sample," *Journal of the American Academy of Child & Adolescent Psychiatry* 31 (1992): 439–48; E. G. Willcutt and B. F. Pennington, "Psychiatric Comorbidity in Children and Adolescents with Reading Disability," *Journal of Child Psychology and Psychiatry* 41 (2000): 1039–48. Similarly, good evidence indicates that anxiety disorders are more common in children with dyslexia; see J. Carroll, B. Maughan, et al., "Literacy Difficulties and Psychiatric Disorders: Evidence for Comorbidity," *Journal of Child Psychology and Psychiatry* 46 (2005): 524–32.

421   "Marked fear": *Diagnostic and Statistical Manual of Mental Disorders,* 5th ed., 2013.

421   Sam's problems with word retrieval: W. J. Levelt, A. Roelofs, and A. S. Meyer, "A Theory of Lexical Access in Speech Production," *Behavioral and Brain Sciences* 22 (1999): 1–75; S. Hanly, "Tip-of-the-Tongue and Word Retrieval Deficits in Dyslexia," *Journal of Learning Disabilities* 43 (2010): 15.

422   The medications effective: The neurotransmitters are released at the ends of nerve terminals into a place between nerves called the synaptic cleft. The transmitters released then stimulate another nerve to fire. The excess neurotransmitters released are then inactivated by being taken up by the nerve terminal. The neurotransmitters important in anxiety include serotonin and the excitatory catecholamine neurotransmitters, including norepinephrine and dopamine.

422   The core first-line medications: SSRIs inhibit serotonin reuptake in the presynaptic cleft and thus increase the extracellular level of serotonin in the brain.

423   Another class of medications: Benzodiazepines work by enhancing the effect of the inhibitory neurotransmitter gamma-aminobutyric acid (GABA) at the GABA-A receptor. This results in an anxiolytic and calming effect.

423 An equally important treatment: S. N. Compton et al., "Cognitive-Behavioral Psychotherapy for Anxiety and Depressive Disorders in Children and Adolescents: An Evidence-Based Medicine Review," *Journal of the American Academy of Child and Adolescent Psychiatry* 43 (2004): 930–59.

425 As discussed in Carol Dweck's insightful book: C. Dweck, *Mindset: The New Psychology of Success* (New York: Ballantine, 2018).

425 At the same time: For children who have had reading struggles and improve significantly in their reading, parents (and teachers) looking back at times wonder if attentional difficulties were there all along but were pushed to the background because of the dyslexic student's overwhelming reading difficulties.

429 Recent reports suggest that: B. Bental and E. Tirosh, "The Effects of Methylphenidate on Word Decoding Accuracy in Boys with Attention-Deficit/Hyperactivity Disorder," *Journal of Clinical Psychopharmacology* 28 (2008): 89–92; N. Grizenko, M. Bhat, et al., "Efficacy of Methylphenidate in Children with Attention-Deficit Hyperactivity Disorder and Learning Disabilities: A Randomized Crossover Trial," *Journal of Psychiatry and Neuroscience* 31 (2006): 46–51; E. H. Keulers, J. G. Hendriksen, et al., "Methylphenidate Improves Reading Performance in Children with Attention Deficit Hyperactivity Disorder and Comorbid Dyslexia: An Unblinded Clinical Trial," *European Journal of Paediatric Neurology* 11 (2007): 21–28; D. Williamson, D. W. Murray, et al., "Methylphenidate in Children with ADHD With or Without Learning Disability," *Journal of Attention Disorders* 18 (2014): 95–104.

429 Two small clinical trials: C. R. Sumner, S. Gathercole, et al., "Atomoxetine for the Treatment of Attention-Deficit/Hyperactivity Disorder (ADHD) in Children with ADHD and Dyslexia," *Child and Adolescent Psychiatry and Mental Health* 3 (2009): 40; C. G. de Jong, S. Van De Voorde, et al., "Differential Effects of Atomoxetine on Executive Functioning and Lexical Decision in Attention-Deficit/Hyperactivity Disorder and Reading Disorder," *Journal of Child and Adolescent Psychopharmacology* 19 (2009): 699–707.

429 More recently, we examined the effect: S. Shaywitz, B. Shaywitz, et al., "Effect of Atomoxetine Treatment on Reading and Phonological Skills in Children with Dyslexia or Attention-Deficit/Hyperactivity Disorder and Comorbid Dyslexia in a Randomized, Placebo-Controlled Trial," *Journal of Child and Adolescent Psychopharmacology* 27, no. 1 (2017): 19–28.

CHAPTER 31: THE ROLE OF TECHNOLOGY

430 "Many schools still use": "Designing an Effective Writing Program," in *Best Practices in Writing Instruction,* S. Graham, C. A. MacArthur, and J. Fitzgerald (eds.), 2nd ed. (New York: Guilford, 2013), pp. 3–29. These authors have also contributed to the Educator's Practice Guide "Teaching Elementary School Students to Be Effective Writers," published by What Works Clearinghouse in 2012.

431 In studies of eighth- and ninth-graders: J. Montali and L. Lewandowski, "Bimodal Reading: Benefits of a Talking Computer for Average and Less Skilled Readers," *Journal of Learning Disabilities* 29, no. 3 (1996): 271–79.

431 But a negative effect: E. Staels and W. Van den Broeck, "Orthographic Learning and the Role of Text-to-Speech Software in Dutch Disabled Readers," *Journal of Learning Disabilities* 48, no. 1 (2015): 39–50.

CHAPTER 32: ACCOMMODATIONS: BUILDING A BRIDGE TO SUCCESS

438 If you are a dyslexic reader, accommodations: According to the U.S. Department of Justice, Civil Rights Division, Disability Rights Section, "Testing accommodations are changes to the regular testing environment and auxiliary aids and services that allow individuals with disabilities to demonstrate their true aptitude or achievement level on standardized exams or other high-stakes tests. Examples of the wide range of testing accommodations that may be required include: Screen reading technology; Extended time; and Distraction-free rooms," https://www.ada.gov/regs2014/testing_accommodations.html, accessed 9/14/19.

440 "The patterns of deficits": M. Bruck, "Outcomes of Adults with Childhood Histories of Dyslexia," in C. Hulme and R. M. Joshi (eds.), *Reading and Spelling: Development and Disorders* (Mahwah, NJ: Erlbaum, 1998), pp. 179–200, quote on p. 197.

440 Brain imaging studies in adults with dyslexia: F. Richlan, M. Kronbichler, and H. Wimmer, "Meta-analyzing Brain Dysfunctions in Dyslexic Children and Adults," *Neuroimage* 56 (2011): 1735–42.

459 Only students diagnosed as dyslexic: M. K. Runyan, "The Effect of Extra Time on Reading Comprehension Scores for University Students With and Without Learning Disabilities," in S. E. Shaywitz and B. A. Shaywitz (eds.), *Attention Deficit Disorder Comes of Age: Toward the Twenty-First Century* (Austin: Pro-Ed, 1992), pp. 185–95; S. M. Weaver, "The Efficacy of Extended Time on Tests for Postsecondary Students with Learning Disabilities," *Learning Disabilities* 10 (2000): 47–56.

460 The Blue Ribbon Panel: "The Flagging Test Scores of Individuals Who Are Granted the Accommodation of Extended Time: A Report of the Majority Opinion of the Blue Ribbon Panel on Flagging," July 15, 2001, https://dralegal .org/case/breimhorst-v-ets/, accessed 9/11/19.

461 In a recent article: F. Romberg, B. A. Shaywitz, and S. E. Shaywitz, "How Should Medical Schools Respond to Students with Dyslexia?," *AMA Journal of Ethics* 18, no. 10 (2016): 975–84.

CHAPTER 33: THE LAW IS ON YOUR SIDE

466 This legal definition: U.S. Senate Resolution 680 (2018) was sponsored by Senators Cassidy, Murphy, Capito, Warren, and Van Hollen, and was passed by unanimous consent. The full text is in the Appendix.

467 This up-to-date definition: U.S. House of Representatives Committee on Science, Space, and Technology hearing, "The Science of Dyslexia," September 18, 2014, dyslexia.yale.edu/advocacy/national-advocacy/science-of-dyslexia-hearing /accessed 9/14/19.

467 Senate HELP Committee: U.S. Senate Health Education Labor and Pensions (HELP) Committee, "Understanding Dyslexia: The Intersection of Scientific Research & Education," May 10, 2016, dyslexia.yale.edu/ABOUT_Shaywitz_ May2016.html, accessed 9/14/19.

468 One of those fourteen: The historical background and current issues of IDEA are reviewed in Professor Ruth Colker's excellent book *Disabled Education* (New York: New York University Press, 2013).

468 The Americans with Disabilities Act (ADA): 42 USCA § 12132 is the codification of Title II of the ADA, which prohibits disability discrimination by public entities. The implementing regulations are found at 28 CFR § 35.101 et seq. The Department of Justice is the responsible agency. 42 USCA § 12182 is the codification of Title III of the ADA, which prohibits disability discrimination by private entities that operate places of public accommodation. The implementing regulations are found at 28 CFR § 36.101 et seq. Again, the DOJ is the primary agency.

468 Section 504: 29 USCA § 794 is the codification of Section 504 of the Rehabilitation Act of 1973 (as amended). Section 504 prohibits discrimination on the basis of disability by recipients of federal funds. The regulations at 34 CFR part 104 et seq. are the implementing regulations for Section 504. The responsible agency is the Department of Education's Office for Civil Rights (OCR).

469 The first step: Helpful information about the IEP can be found in R. Colker and J. K. Waterstone, *Special Education Advocacy* (Albany, NY: Mathew Bender, 2011), in L. M. Siegel, *Nolo's IEP Guide* (Berkeley: LexisNexis Group, 2011).

470 The law requires: These requirements are from Colker and Waterstone, *Special Education Advocacy,* pp. 159–60.

471 In 2016, the U.S. Department of Justice: See https://www.ada.gov/regs2016/adaaa.html, accessed 9/14/19.

471 Here are its most important elements: Ibid., p. 3.

471 "dyslexia and other learning disabilities": Ibid., p. 76.

472 *"should not demand extensive analysis"*: Ibid., p. 81.

476 "this bill makes the definition of disability more generous": Ibid., pp. 82–83.

481 In a press release: Settlement Agreement Between United States of America and National Board of Medical Examiners, DJ# 202-16-181, https://www.ada.gov/nbme.htm, accessed 9/14/19.

483 A 2011 Government Accountability Office: Higher Education and Disability: Improved Federal Enforcement Needed to Better Protect Students' Rights to Testing Accommodations," https://www.gao.gov/products/GAO-12-40, November 29, 2011, accessed 11/24/19.

484 "LSAC shall ensure": "Law School Admission Council Agrees to Systemic Reforms and $7.73 Million Payment to Settle Justice Department's Nationwide Disability Discrimination Lawsuit," May 20, 2014, https://www.ada.gov/dfeh_v_lsac/lsac_consentdecree.htm, accessed 9/14/19.

484 Sadly, there are some indications: As reported in the *Daily Journal California Lawyer* (July 19, 2018), LSAC is being sued by the California Department of Fair Employment and Housing for violating an agreement to "better handle" accommodation requests for the LSAT, Lyle Moran, "State Seeks Lawyer Fees from LSAT Administrator."

485 A blue ribbon panel: The findings of the panel are reported in "The Flagging of Test Scores of Individuals Who Are Granted the Accommodation of Extended Time: A Report of the Majority Opinion of the Blue-Ribbon Panel on Flagging," available at https://dralegal.org/case/breimhorst-v-ets/, accessed 9/11/19.

485 On July 15, 2002: Ibid.

486 At the next level: Lani Guinier, a professor at Harvard Law School, reports that differences in scores obtained on the LSAT explain less than 14 percent of the differences in student grades during the first year of law school. See L. Guinier, M. Fine, and J. Balin, *Becoming Gentlemen* (Boston: Beacon Press, 1997).

486 "relax or even eliminate": U.S. District Judge Bernard A. Friedman as quoted in "Law School Admission Council Aims to Quash Overreliance on LSAT," Jess Bravin, *Wall Street Journal,* March 29, 2001.

487 "an end to the use": Diana Jean Schemo, "Head of U. of California Seeks to End SAT Use in Admissions," *New York Times,* February 17, 2001.

487 Elaine M. Allensworth and Kallie Clark, "High School GPAs and ACT Scores as Predictors of College Completion: Examining Assumptions About Consistency Across High Schools," *Educational Researcher,* January 27, 2020, DOI: 10.3102/0013189X20902110

487 as of 2018, there were over: "More Than 1000 Accredited Colleges and Universities That Do Not Use ACT/SAT Scores to Admit Substantial Numbers of Students into Bachelor-Degree Programs, FairTest," The National Center for Fair and Open Testing, current as of Fall 2019, https://www.fairtest.org/university/optional, accessed 9/14/19.

487 As for higher levels: R. C. Davidson and E. Lewis, "Affirmative Action and Other Special Consideration Admissions at the University of California, Davis, School of Medicine," *Journal of the American Medical Association* 278 (1997): 1153–58.

487 "UC Davis applicants who qualified": Editorial, Davidson and Lewis, "Medical School Admissions Criteria: The Needs of Patients Matter," *Journal of the American Medical Association* 278 (1997): 1196–99.

CHAPTER 34: A PERSON LIKE THAT . . .

490 David Boies is brilliant: D. Margolick, "The Man Who Ate Microsoft," *Vanity Fair,* March 2000.

495 This belief that socioeconomic problems: The harm that this mindset can lead to is exemplified by the experience of Dr. Fred Romberg, whom the USMLE refused to provide accommodations because he was born and grew up "disadvantaged" and this explained all his reading problems. As you may recall, we tested him and found he was incredibly intelligent and very dyslexic. This opened up a whole new world for him. He was able to receive accommodations and with the additional time passed his USMLE exam and is now a practicing physician.

514 Ask anyone knowledgeable: Matthew Herper, "Going to Church," *Forbes,* August 9, 2015.

515 As a consequence: Peter Miller, "News, The Innovators Project: George Church, The Future Without Limits," *National Geographic,* 2015, https://www.nationalgeographic.com/news/innovators/2014/06/140602-george-church-innovation-biology-science-genetics-de-extinction/, accessed 2/26/15.

516 That same year: M. Lewis, *The Undoing Project: A Friendship That Changed Our Minds* (New York: Norton, 2007), p. 280.

516 "He's made economics more human": Peter Gärdenfors of Lund University, announcing the award of the Nobel Memorial Prize in Economic Sciences to Richard H. Thaler, https://ccl.ht.lu.se/2017/10/hes-made-economics-more-human, accessed 9/14/19.

INDEX

Page numbers in *italics* refer to figures and illustrations.

PERMISSIONS ACKNOWLEDGMENTS

ILLUSTRATION CREDITS

A NOTE ABOUT THE AUTHORS

SALLY E. SHAYWITZ, M.D., is the Audrey G. Ratner Professor in Learning Development at Yale University and cofounder and codirector of the Yale Center for Dyslexia & Creativity. She is the author of more than 350 scientific articles and book chapters, and the creator of the Shaywitz DyslexiaScreen™, a new tool used by teachers to reliably screen young children for dyslexia. Dr. Shaywitz is an elected member of the National Academy of Medicine, is regularly selected as one of the "Best Doctors in America," and was chosen, along with Dr. Bennett Shaywitz, her husband, as a recipient of the Liberty Science Center 2019 Genius Award for "advancing our understanding of dyslexia." She lives in Woodbridge, Connecticut, and Martha's Vineyard, Massachusetts.

JONATHAN SHAYWITZ, M.D., is a Harvard Medical School graduate and UCLA-trained board-certified psychiatrist. He was the director of the Anxiety Disorders Program at Cedars Sinai in Los Angeles and more recently the medical director of behavioral health at Mission Hospital Laguna Beach and Mission Hospital Mission Viejo. He lives in Los Angeles, California.